PRACTICAL
HEAT TRANSFER

LICENSE, DISCLAIMER OF LIABILITY, AND LIMITED WARRANTY

PRACTICAL
HEAT TRANSFER
USING MATLAB® AND COMSOL®

Layla S. Mayboudi, Ph.D.

MERCURY LEARNING AND INFORMATION

Dulles, Virginia
Boston, Massachusetts
New Delhi

Publisher: David Pallai
MERCURY LEARNING AND INFORMATION
22841 Quicksilver Drive
Dulles, VA 20166
info@merclearning.com
www.merclearning.com
800-232-0223

L. S. Mayboudi. *Practical Heat Transfer Using MATLAB® and COMSOL®.*
ISBN: 978-1-68392-633-7

The publisher recognizes and respects all marks used by companies, manufacturers, and developers as a means to distinguish their products. All brand names and product names mentioned in this book are trademarks or service marks of their respective companies. Any omission or misuse (of any kind) of service marks or trademarks, etc. is not an attempt to infringe on the property of others.

Library of Congress Control Number: 2021951258

222324321 Printed on acid-free paper in the United States of America.

Our titles are available for adoption, license, or bulk purchase by institutions, corporations, etc. For additional information, please contact the Customer Service Dept. at 800-232-0223(toll free).

All of our titles are available in digital format at *academiccourseware.com* and other digital vendors. *Companion files are also available for downloading by writing to info@merclearning.com.* The sole obligation of MERCURY LEARNING AND INFORMATION to the purchaser is to replace the book or disc, based on defective materials or faulty workmanship, but not based on the operation or functionality of the product.

To the scientist, mathematician, engineer, physician, linguist, aviator, healer of the broken wings... To whomever wishes to improve methods, mend fences, be a guiding light, brighten lives, listen, write, set the record straight, and free lamenting souls... To the one who does not submit meekly to power... To the still-to-be-known warrior without borders, the one who honors professional oaths, preserves dignity, defends the rights to human empowerment, knowledge, and freedom; who is demeaned and shunned for being the maverick, works ethically and diligently, perseveres, endures, and lives happily forever and ever after... To the trailblazer, the unknown traveler...

CONTENTS

PREFACE

This book guides the reader through the subject of heat transfer, covering the analytical, coding, finite element, and hybrid methods of thermal modeling. Cylindrical pipes are the focus of this work for their widespread use and multitude of engineering applications. This book also gives the reader background information about pipes and their use in various fields as well as differential equations [1]. Examples are given throughout where different pipe geometry configurations are created, and models are built and analyzed.

This book can stand on its own, but it can also be treated as a companion to the previous publications by the author. *Using COMSOL® in Heat Transfer Modeling from Slab to Radial Fin* [2] will be of interest to those looking for the detailed exploration of the heat transfer modeling of extended surfaces (fins). *COMSOL Multiphysics® Geometry Creation and Import* [3] provides in-depth review of the geometry creation techniques in COMSOL Multiphysics. *COMSOL Heat Transfer Models (Multiphysics Modeling)* [4] is appropriate for readers who have the basic knowledge about modeling and would like to develop their skills further; comprehensive case studies covering a variety of subject matter, such as heat transfer in car seats, water boiling in a kettle, and a complex rotini fin, are included. *Mechanical Engineering Exam Prep: Problems and Solutions* [5] provides an opportunity to test the reader's knowledge in the field of heat transfer as well as other mechanical engineering curriculum areas, including over 1,500 innovative problems in these subject areas. The rest of this section provides a framework for this work, a roadmap to enhance the learning experiences along this journey. The purpose is to become familiar with the field of heat transfer modeling through the focused examples with significant applications such as transferring energy in the form of heat (and matter) in the pipes; the information can be expanded to transfer this knowledge to the constricted bodies (pipes with noncylinder cross sectional areas).

The book's primary focus is on the MATLAB® (R2021) and COMSOL Multiphysics (5.6) software packages; however, the learning gained here can be transferred to other FEA tools. Both software packages consist of a core module and numerous, specialized add-on toolboxes (MATLAB) or modules (COMSOL). For MATLAB, the reader needs to have access to the MathWorks® *Partial Differential Equation (PDE) Module*. For COMSOL, the reader needs to have access to the *Heat Transfer* and *CAD Import*

Modules (the *CAD Import Module* functionality can be also obtained via any of the *LiveLink™ Modules* that connect COMSOL Multiphysics with a specific CAD tool).

Chapter 1 examines the numerous applications of pipes, such as those found in nature like blood vessels, as well as man-made ones that address the transfer of water (e.g., irrigation and portable water), waste (e.g., sewer waste), chemicals (e.g., ammonia), and beverages (e.g., milk). Pipe use in thermal management and performance monitoring as it relates to the pipes is covered. Finally, the science of heat transfer is briefly introduced from a historical perspective.

Chapter 2 provides insight into heat transfer modeling. Basic concepts such as the laws of thermodynamics are discussed. The thermal sciences are reviewed, introducing the effects on the molecular scale and their role in identifying the heat transfer regime; the concept of the *Knudsen Number* (Kn) is introduced. Energy and heat are discussed and the related processes, such as isenthalpic and isochoric, are presented. As the next step, the thermal analyses of systems at the component and subcomponent levels are reviewed. This journey starts by introducing the thermophysical properties of materials—density, thermal conductivity, and heat capacity— that are key parameters of these models. The concepts of homogeneous versus nonhomogeneous materials and static versus dynamic systems are also discussed.

Thermal flow and test management are the essential elements of managing heat sources—both hot and cold. This subject is touched upon in this book, because heat transfer modeling is often done with the purpose of better understanding and improving the thermal management of a process. Defining the physics for the modeling tool, including the methods of deriving thermophysical relations by balancing energy, is the first step when modeling thermal systems. The modes of heat transfer, including the conduction, convection, and radiation, are examined along with some of the dimensionless numbers such as the *Biot Number*. Balances of energy to obtain temperature and temperature gradient profiles are presented, with accounting for major heat transfer regimes. The concept of thermal resistance is discussed; it considers the heat flow as an analog to the electric current and rewrites the heat transfer equations for the three heat transfer modes in terms of thermal resistance. The chapter ends with a comprehensive derivation of the energy balance equations expressed in Cartesian, cylindrical, and spherical coordinate systems.

Chapter 3 provides an overview of finite element modeling (FEM) as applied to heat transfer problems. Starting with the basic concepts of

FEM, the chapter proceeds to describe the stages of the process: geometry creation, definition of material properties, selection of appropriate analysis type, definition of boundary, initial, and domain conditions, meshing, and the solution.

Chapter 4 is an introduction to the MATLAB environment that should help readers quickly acquire familiarity with it. The chapter starts by reviewing the MATLAB interface components, such as the *Workspace, Command Window, and EDITOR.* The definitions and manipulation of different variable types (numeric, logical, and character) are described. Understanding how to use logical variables is needed to extract the sought-after information from data arrays. This is an important skill, as it will be needed in later chapters where heat transfer modeling with the MATLAB *PDE Toolbox* is covered. Matrices, the foundation of MATLAB, are covered next. Their efficient use is key to unlocking the full power of this software. This chapter presents commonly used built-in functions, the writing of scripts, and creation of user-defined functions. Finally, data plotting methods are covered.

Chapter 5 discusses how heat transfer models can be created and solved in the MATLAB environment. The discussion encompasses Ordinary Differential Equations (ODE) and progresses toward the Partial Differential Equations (PDE). The focus is on the latter approach (PDE) because it is more versatile and applicable to heat transfer problems. Use of the MATLAB *PDE Modeler* application, which is part of the PDE Toolbox, is covered in detail. The chapter presents an overview of the MATLAB *PDE Modeler* interface, its components, and how to set up a two-dimensional model.

Next, the chapter looks at how the MATLAB *PDE Toolbox* can be used to solve partial differential equations, with the focus on heat transfer modeling. A connection is shown between this tool and commercial FEA tools (such as COMSOL Multiphysics) that is discussed in later chapters. The MATLAB *PDE Toolbox* provides much flexibility to those interested in the low-level control offered by custom code development. Later chapters show how the benefits of low-level control of coding can be combined with the convenience of the graphical user interface and other functionalities of a dedicated FEA tool.

Attempting heat transfer models using the MATLAB script is an important part of this work, extensively employed alone or in combination with the MATLAB *PDE Modeler.* All the steps required to set up a model in the MATLAB environment by creating scripts are discussed. Setting up a model in MATLAB is like setting up models in any modeling environment. Therefore, concepts such as geometry import or creation, material properties, analysis type, boundary and initial conditions (e.g., heat generation), mesh, solver options, and solution postprocessing are reviewed.

Chapter 6 shows how the material from the previous chapters can be applied to five case studies completed in the MATLAB environment. They include axisymmetric single and multiple domains set in both steady-state and transient settings and relate to pipe applications. The case studies are presented in multiple steps that include model setup, results, and validation by analytical or other tools. Heat loss calculations are conducted and are later employed to investigate the sensitivity of the analyses to the materials or geometry configurations (e.g., use of insulation and extended surfaces). Pipe models with variable temporal and spatial boundary conditions are introduced, with one case showing how to implement a moving heat source.

The chapter ends with description of a hybrid thermal model that incorporates the MATLAB *PDE Modeler* and *Toolbox* functions. It is shown that the variables of a model, created in the MATLAB *PDE Modeler, can* be exported to the *PDE Toolbox* script.

Chapter 7 briefly reviews modeling with COMSOL Multiphysics. More comprehensive coverage of this subject can be found in the author's previous works [2,3,4]. The steps required to set up and solve models within this FEA tool, such as creating or importing geometry, adding materials and physics, defining boundary conditions, meshing, and running the solutions, are discussed herein. Also, the *LiveLink for MATLAB Module,* an add-on module provided by COMSOL which links the two tools, is reviewed.

Chapter 8 presents five case studies where the heat transfer in solids and non-isothermal flow inside the pipe are combined and analyzed using COMSOL Multiphysics. The studies have a common theme: the heat dissipation from pipes with extended surfaces of varying geometry. Internally finned, externally finned, and internally-externally finned pipes are investigated, as well as the special fin shaped like a rotini pasta introduced in the author's earlier publication [4]. The solution time and computational requirements are given, as well as the results in the form of contour plots and spatial and transient temperature profiles. A comparison is made between the case studies to show their relative effectiveness in heat dissipation.

Chapter 9 presents several exercises that the reader can use to practice the heat transfer concepts introduced earlier in this work. The exercises are presented in order of increasing complexity, starting from a base condition and gradually adding features, such as spatial and temporal variations in the thermophysical properties and boundary conditions. Most exercises provide a sample solution result so the reader can confirm the correctness of their own work. In another set of suggested follow up work, it is proposed to apply alternative solution techniques to the case studies presented earlier; for example, the reader can attempt to solve the problems in the MATLAB case studies using the COMSOL FEA tool.

Chapter 10 provides examples of how one can apply the Lean Six Sigma concepts to the subjects related to this work. The first step is to be able to decide upon some *critical-to-quality* variables that help assess the system's progress or success. The term *system* can apply to a range of subject areas connected with the modeling, preprocessing, solution, and postprocessing of heat transfer problems. Depending on the situation, the reader may focus on the hardware, human resources, computing facilities, analysis techniques, and tools. Any of the said items are part of the *5M* principles that relate to the Lean Six Sigma implementation, aiming to make any process both effective and efficient. Some may choose to improve the solution time, for instance, and to achieve that, they may either opt in or out of using certain modeling techniques, given the available resources or restrictions. Eventually, the attempt is made to balance the cost, quality, and time when performing certain tasks as individual pieces, but also when assembled with the rest of the tasks. The choice of the FEA tool and good practices involved are also examined in this chapter, with the objective of improving the solution time and performance.

Chapter 11 concludes this work. It redefines the known-and-tested-method concept in industrial or engineering applications and elaborates on if adhering to the *status quos* is something to be proud of, especially when dealing with critical or ethical situations concerning time, product quality, and human resources. It is a reminder that *practice makes perfect*. Even though the process may be grounded in historically-practiced concepts, it should be reviewed occasionally to sustain the systems and revised often based on a control-loop approach to improve the system. Ethical design and responsible approaches are the main emphases of this chapter, especially when decisions are to be made under direct *peer pressure* or apparent *status quos*. Examples of responsible designs in the form of the Leadership in Energy and Environmental Design (LEED) projects are presented herein.

The publication ends with the appendices, summarizing the analytical approaches and applicable governing equations. Appendix A provides analytical approaches and mathematical methods to solve heat and wave equations. Simplifying the problem is the first step to consider when attempting to solve complex physics analytically. In many scenarios, symmetry can be taken advantage of to reduce model dimensions. Even after the problem is simplified or reduced to lower dimensions, the scholar will not be able to address them the same way; therefore, mathematical techniques are required to address these scenarios efficiently. Some analytical approaches are more versatile and effective tools than the rest, including the separation of variables, variation of parameters, Duhamel's theorem, complex combinations, superposition, Laplace transform, and the integral and perturbation methods, which are presented along with examples to facilitate their understanding. Appendix B provides a summary of the governing equations related to dimensionless analysis and application of analytical relations to extended surfaces with general curves.

Before the advent of the electronic computation tools, partial differential equations were attempted using the analytical methods developed by the physicists and mathematicians who introduced techniques such as the separation of variables and the Fourier transform. As technological capabilities increased, prompted in large measure by NASA's work on the space program, the computational needs grew. Before the electronic computers, human "calculators" were employed by NASA in the 1950s to perform the necessary computations, such as those for launch trajectories. Even when the earliest electronic computers were introduced, these "calculators" continued making contributions to the successful implementation of this new technology, mainly in service of the space program. These first electronic computers themselves owed their existence in large part to the work of Alan Turing—a British mathematician, computer scientist, and cryptologist in the twentieth century. From the start of World War II, he developed electromechanical devices to help decode Nazi communications and, after the war, he worked on the theory and design of the electronic computing devices [6,7]. Appendices C and D provide the reader with the list of figures and tables in the book.

Note that whatever endeavor you undertake, never cease asking questions, learning, and trying to understand things. Having an inquisitive mind is the essence of the humanity. The main lesson the author would like the reader to take from this publication is to never stop being curious, learning, and striving against any obstacles.

Acknowledgments

My teachers, mentors, family, spouse, and publisher; I am infinitely grateful to you for your generous support and the positive influence on my life.

Layla S. Mayboudi
January 2022

End Notes

[1] *"Differential Equation, Partial," Encyclopedia of Mathematics*, EMS Press, 2020; http://encyclopediaofmath.org/index.php?title=Differential_equation,_partial&oldid=46668

[2] Layla S. Mayboudi, *Heat Transfer Modelling Using COMSOL: Slab to Radial Fin*, p. 250, Mercury Learning and Information, 2018.

[3] Layla S. Mayboudi, *Geometry Creation and Import with COMSOL Multiphysics®*, p. 250, Mercury Learning and Information, 2019.

[4] Layla S. Mayboudi, *COMSOL Heat Transfer Models (Multiphysics Modeling)*, p. 400, Mercury Learning and Information, 2020.

[5] Layla S. Mayboudi, *Mechanical Engineering Exam Prep: Problems and Solutions*, p. 300, Mercury Learning and Information, 2021.

[6] https://www.britannica.com/biography/Katherine-Johnson-mathematician

[7] https://www.britannica.com/biography/Alan-Turing

CHAPTER 1

PIPES, THEIR APPLICATIONS, AND HEAT TRANSFER

Pipelines have been used to transport a wide variety of substances (liquids, gases, and solids) over short and long distances. Most commonly, the substances are carried from their place of origin to wherever they need to be processed further or as end-products to be used for their intended purpose. For water or any fluid to flow inside a conduit, the pressure difference is the driving force. This can be created by a pump, gravity, or due to a temperature difference. If the walls of the pipe are exposed to heat at different rates, this temperature variation results in the fluid flow, with its direction moving from areas with a higher temperature to those with a lower temperature. The non-isothermal flow inside the ducts can be characterized by different flow regimes. In a laminar regime, the fluid moves smoothly along its flow lines. In a turbulent regime, however, current eddies are formed, and the fluid undergoes considerable mixing, with variations of the flow direction and speed. A flow within a conduit transitions from laminar to turbulent above a certain flow velocity; this limit depends on the fluid properties and the conduit's cross-sectional area. Any sudden disruption in the fluid flow due to the barriers or sharp corners causes the formation of local turbulent flows, with the potential to produce excessive noise in some applications.

The gravity-driven pressure difference is a very common way to move liquid through pipes. That is how water is transported in most homes. The water comes from elevated water tower tanks. Until the nineteenth century, fountains, such as the famous ones in Versailles, France, used to work only with gravity. A source of water higher than the fountain was required to convert the potential energy of the height into the kinetic energy of the water exiting the fountain's nozzle. Many such fountains are still part of the English countryside, perhaps placed there by the famous eighteenth-century British landscape architect Capability Brown.

1.1   Artificial Systems

The main application of pipelines is to transport matter over a certain distance. Possibly the oldest industrial pipeline can be found in the village of Hallstatt in Austria, which has a rich history of salt mining. The 40-km pipeline, originally built in 1595, transports the brine from Hallstatt to Ebensee. The pipes are made of 13,000 hollowed out tree trunks. Until 1994, 30,000 liters of milk from Ameland Island were transported daily to the Netherlands mainland by means of an 8-km pipeline laid at the bottom of the Wadden Sea. Another pipeline in Brazil is used to transport coal, liquified into a slurry, from a Minas-Rio mine to a port in Açu. In Germany, the pubs located throughout the Veltines-Arena stadium are connected by the 5-km pipelines to several large underground distribution tanks, where the beer is kept cool.

Pipes do not just serve utilitarian purposes. They have been used in some modern art creations, showing the infinite creativity of the human mind. Pipes are used in fountains, which can be said to embody the human spirit and its love for purity (water) and life (movement). Water, the source of life, is the most responsive being to forces and energies, seen during its crystallization process, forming ice, melting to bring life to the Earth, and flowing to clear the mundane. The soothing sound of the water flowing inside the underground pipes and finding its way to the exterior environment is affected by the pipes' characteristics—from the material to the length and shape. French designers must have conducted research during the development and construction of the 50 working fountains in the Gardens of Versailles.

If you have witnessed water moving harmoniously to music during a water show, you see that water exits the pipe outlets at different heights with patterns that are affected by the size of the outlet nozzles. In these light shows, water is moved by means of pumps that power the super and mini shooters, delivering water in mist or liquid forms. Modern fountain installations can use large quantities of water and electrical power. The iconic Bellagio Hotel fountains in Las Vegas, NV, reportedly contain about

20 million gallons of water, which are delivered via 12,000 nozzles. Thus, resource management, such as water quantity, pressure, and temperature, is an important element of running such shows [8,9,10].

One interesting application of pipes where their purpose as a conduit of liquid was combined with a structural function was for the construction of the Beesat Bridge on the Southern section of the River Arvand by the Iranian engineers in 1986 during the first Gulf war (Dawn 8 Operation). This river starts at the confluence of the Tigris and Euphrates rivers and empties into the Persian Gulf about 160 km downstream. The southern section of this river forms the border between Iraq and Iran. In 1986, a bridge crossing needed to be quickly constructed across the River Arvand. The river at this point was flowing at 11 km/h; it was 1-km wide, 12-m deep, with a 3- to 5-m tidal depth variation. To address this challenge, the engineers assigned to the task had an innovative idea. They developed the Beesat Bridge structure, made of pipes that were 1.42-m in diameter and 12-m in length, and had a 16-mm wall thickness (Figure 1.1). Approximately 5,000 7-ton steel pipes (ST67, 35,000 tons), which were manufactured by an Iranian piping company (Ahvaz Pipe Mills) were employed in the construction of this bridge. It took about six months to complete the bridge.

To make the Beesat Bridge, these pipes were placed into the river and oriented along the direction of the water flow. This allowed the water to

FIGURE 1.1. The Beesat Bridge structure on the southern section of the River Arvand (built in 1986, Iran).

pass by unhindered while creating the bridge structure. Pipes were stacked, starting at the bottom, until sufficient height was reached above the water level. The rows of pipes were then linked by means of earing hooks and then welded. After placing smaller diameter pipes between the large ones to make a flatter upper surface, asphalt was laid on top to cover the crevices, creating a 12-m wide drivable road surface [11,12,13].

1.2   Oil and Gas Industries

Pipes are used as means of transporting oil and natural gas between the processing and distribution centers. Examples include the pipelines crossing Africa, Asia, Europe, North America (e.g., Canada, Mexico, Puerto Rico, and the United States), South America, and Oceania. To deliver propane gas to customers in large volumes, it is more efficient for it to be converted to fluid using very high pressures.

Natural gas is transported by pipelines after it is extracted from wells: (a) gas at low pressure is transferred by pipelines with small diameters from the wells to the manufacturing facilities, where they are processed into other products; (b) gas at high pressure is transported from the manufacturing facilities to the interstate, intrastate, and international destinations (the high pressure is maintained by the pumping stations through which the gas passes); and (c) gas delivered to the main processing or distribution facilities is carried by small-diameter pipelines. The main difference between the pipes and tubing is their sizing. Pipelines are also used to transport irrigation and portable water, waste (e.g., sewer waste), slurries (e.g., coal), and chemicals (e.g., ammonia).

Natural gas transmission pipelines require high pressure for transmission. The pressure is maintained by the compressor stations located (about every 65 km to 160 km) along the way. These compressors are very powerful, outputting about 36,000 hp, a rate comparable to a large jet engine. Natural gas moves inside the pipeline at about 40 km/h (11.1 m/s). Pipelines have diameters that vary from 0.5 to 48 inches. The larger ones transfer the fluids from the processing center to the major distribution stations, while the smaller ones connect the distribution and processing centers. Transmission pipelines are usually made of steel, coated with corrosion-protection materials (e.g., coal tar enamel or light blue fusion bond epoxy) [14,15,16]. Natural gas, which comprises gases such as butane, propane, and ethane, is discretized at the processing facility, with the excess contents and contaminants such as hydrogen sulfide removed.

Usually, ethanethiol is added to the natural gas to make it smell like rotten eggs, in case it leaks, since natural gas is odorless. Storage is usually done inside waterproof underground storage facilities.

Underground pipelines are normally placed about 1.8 m (6 ft) deep below the surface. Interestingly, gas pipelines are intentionally not laid out in an exactly straight fashion; instead, gentle S-curves are added. The reason is to avoid pipe damage due to thermal expansion. While the seasonal temperature variation below the ground surface declines with depth, there is still significant variation at the typical pipe-laying depth. For example, soil temperature observations were made for oil pipeline projects in 2004–2005 in the Mackenzie River Valley in the Fort Simpson area of the Northwest Territories. These measurements recorded seasonal variations from a minimum of 2.0 °C to a maximum of 6.3 °C. While a temperature change of 4.3 °C does not appear to be large, when the thermal expansion is calculated for the tens of kilometers of the pipeline, the effect becomes significant. Using this temperature difference with the steel thermal expansion coefficient of 11.7×10^{-6} m/mK for the 10-km pipeline results in a length change of 0.5 m. If the pipes were laid in a straight line, this expansion would cause significant sideways movement, likely leading to pipe damage [17,18,19,20,21].

1.3   Organic Systems

There is a vital piping system that all humans make use of and without which they cannot survive—the human circulatory system. The human body incorporates perhaps the most complex flow system of them all, operating reliably for decades in a nearly unfailing fashion. Hemodynamics, the dynamics of the blood flow within the veins and arteries, is responsible for this operation, ensuring the transportation of the nutrients and hormones, gases such as oxygen and carbon dioxide, as well as metabolic wastes. Of course, heat transfer plays a critical role here as well. The blood flow regulates the body temperature, directing heat to the parts of the body where it is needed most, which may sometimes leave the fingers freezing as the body decides that maintaining the core temperature is more critical to the person's survival. Blood is a non-Newtonian fluid, meaning that its viscosity can change depending on the environmental conditions. The vessels are also flexible to accommodate flow variation and facilitate fluid movement. This makes the flow model in the arteries and veins both interesting and challenging.

Another example is the umbilical cord. This cord is a tube that connects mother to her baby. It has three blood vessels: (a) one vein, carrying the food and oxygen from the placenta to the baby, and (b) two arteries, carrying the waste from the baby back to the placenta. These blood vessels are cushioned and protracted by Wharton's jelly. If the cord is too long, too short, does not connect well to the placenta, or gets knotted or squeezed, these conditions cause problems. The urethra is another example, and it connects the urinary bladder to the uterus meatus. Its function is to let urine discharge from the body. Its structure is fibrous and muscular, with its length varying between 4–20 cm. The ureter tube that is about 20–30-cm long and about 3.5 cm in diameter, is made of small muscles connecting the kidney to the urinary bladder [22,23,24].

1.4   Pipe Materials

Pipes that are used to convey the fluids are made of materials such as wood, fiberglass, glass, plastics, metal (e.g., steel, copper, and aluminum), and occasionally concrete. Surface roughness is one of the factors that affects the flow regime inside the pipes. Depending on the fluid types, there are different challenges faced when designing the pipelines. For example, some of these materials (e.g., ammonia) are highly toxic. Therefore, not only the piping routes need to meet the right-of-way constraints, but also the pipe's physical, thermal, and mechanical characteristics need to comply with regulations. In addition, all the fittings such as valves, intersections, and seaming materials—with which the pipes are joined together—should be carefully selected according to the performance requirements. In Canada, interprovincial pipelines are under the supervision of the National Energy Board; the equivalent United States agency is the Federal Energy Regulatory Commission (FERC).

One may not think of wood as a suitably durable pipe material, but wooden pipelines exhibit characteristics such as a resistance to corrosion, electrolysis, and decay (rot). They are also easy to transport, especially in hard-to-reach areas, such as mountainous regions, making them a relatively easy-maintenance option for piping systems. The thick walls of wooden pipes provide good insulation for the transported substance, greatly diminishing the possibility of pipes freezing. Wood does not expand or contract easily with temperature changes, and that minimizes the need for the installation of expansion joints. Wooden pipes are made with staves and hoops, like barrels. It is believed that the redwoods found in the western United States can resist acids, insects, fungus, and weathering. In the sixteenth and seventeenth-century London, the pipes were tapered at the end and sealed by means of hot animal fat. It is reported that about 100,000 ft of the wooden pipes were installed during World War II in army camps and airfields [25,26].

Pipes running on the exterior of the structures may be exposed to the harsh environmental conditions due to the extreme temperatures, wind, and the sun's radiation, such as those found in arid climates. In these applications, the choice of the material is as vital as the design's geometry. In some aerospace applications, aluminum sheets are used for the heat pipe envelope [27]. They are used to maintain space nuclear systems within the recommended temperature range of 130–280 °C. Although aluminum is easily machinable, manufacturing the interior longitudinal grooves to increase the convective surface areas of the heat pipe envelope does not produce a strong structure for the given weight requirements. Therefore, titanium, which has a high strength-to-weight ratio, has been suggested as an aluminum substitute. These characteristics, in addition to its anticorrosive properties, make titanium a desirable material in aerospace applications. The main challenge in using this material is its machinability.

There are also other materials used for constructing pipes such as steel alloys, Inconel, and chrome-moly. Copper pipelines were used extensively through the twentieth century in residential plumbing and are still found in many older homes; however, due to copper's higher material and installation costs, it has been generally replaced by plastics such as PEX (cross-linked polyethylene). In addition to a higher installation cost, copper, being an excellent heat conductor, also can waste a notable fraction of heat when used to deliver domestic hot water, especially if a hot-water recirculation system is being used. Such energy waste may be reduced by adding insulation around the pipes; for example, in the form of closed-cell polyethylene foam semi-slit tube sleeves.

1.5   Thermal Management and Pipes

In industrial installations, it is often necessary to monitor the operating conditions of the pipelines. Instruments, such as temperature and pressure gauges, may be employed for this purpose. They can communicate by wire or wirelessly, using satellites or cellular networks, with central controllers using Supervisory Control and Data Acquisition (SCADA) systems. This information can be processed; for example, to detect leaks. Comparing the flow rate data between two different locations along the pipe can provide this information by calculating the difference between the two values.

Temperature extremes and mechanical loads must be carefully considered in pipe design to avoid failure due to accumulation of residual thermal stresses, fatigue due to thermal cycling, or exceeding the material strength. The type of load varies depending on the environment in which the pipes are operated. Conditions that may need to be addressed in pipe design are installations in earthquake-prone regions, high winds, vibrations, and fluid hammer due to the bends in the pipes. Sharp corners can cause high-stress regions within the pipe, and so pipe bend radii must be appropriately sized. Cryogenic pipes, which transport extremely cold fluids, must be carefully designed to avoid the steel structures becoming brittle when exposed to such low temperatures.

A heated or cooled fluid moving through a pipe is an important means of transporting heat to or from the system of interest; such an arrangement is used in various thermal management systems (e.g., heat exchangers). Internal or external fins are often connected to pipes to increase the heat transfer rate. An example of an effective thermal management system that can operate without using any powered fluid pumping mechanism is a heat pipe. Its reliability and effectiveness led to its use in aerospace cooling applications. A heat pipe has an array of narrow channels within it that perform a wick function. The vaporized liquid molecules travel via these channels upstream to the cool end (condenser) where they are drawn in by the capillary forces, lose the absorbed excess heat, and form a liquid, which then flows back to the warm end (evaporator) of the pipe to repeat the cooling cycle.

Teleheating, also known as *district heating*, is a method of heat distribution by means of hot water or steam. Although the pipes are insulated, the heat wastage is significant. Such piping systems are typically laid underground; stations along the pipeline routes may be added that can store heat and release it when the demand is high. This generated heat is then transferred to the users' central heating system by means of heat exchangers, isolating the heating fluid in the local system.

For some private homes and industrial spaces, heating can be done using a non-isothermal heated water flow inside the network of pipes built into the floor. These systems deliver heat by warming up the large convective surface areas of the floor slab, delivering heat by the radiation and free convection modes. Humans are quite sensitive to the radiant heat—it adds to a sense of comfort, just like standing in front of a lit fireplace does. Poorly insulated cold walls will make one feel chilled.

An HVAC duct is another example where the fluid (air in this case) flows through a channel that can be straight, bent, or split into many smaller branches. If the HVAC system is used for heating or cooling, not just ventilation, in addition to the flow rate, one also needs to be concerned that the air at appropriate temperature reaches the diffusers where it enters the intended service spaces. It must provide comfort to people or meet cooling or heating requirements of the equipment. Thus, heat transfer modeling is an important element of the HVAC system design.

Another use of thermal management is use of the *@Source-Energy* pipe system, which is essentially the same as a concrete pipe; however, it also extracts energy in the form of heat from the waste in the pipes and adjacent ground, it integrates a geothermal pipe in its concrete. In these pipes, in addition to the exterior concrete pipe, a high-density polyethylene (HDPE) conduit (the same size as that of a gas line) is also wound inside the pipe core along with the reinforcements, which is filled with a heat transfer fluid (i.e., 30% ethanol-water mixture). The exchanged energy is controlled using a heat pump [28].

1.6   Heat Transfer

The word *heat* has Germanic origins, being equivalent to contemporary Dutch *hitte* or German *heize*. In French, it is *chaleur*, which comes from the Latin *calor*, from which *calorie*, the term for energy or heat unit, was derived [29]. We can speak of the heat of the sun and the heat of fire, the heat is conducted, convected, or radiated; it may be generated, stored, or released. There is the heat of combustion or the latent heat of fusion. It can be the quantity of energy required for a certain process to occur, or it is released as a result of a process. Concepts of *heat transfer* and *heat flow*, as parts of the engineering syllabus, are taught at many educational establishments and are shared among several engineering fields.

As students acquire knowledge about the heat transfer, they may be surprised that things that appear common sense knowledge today have taken many centuries of human thought to discover. Among the ancient scientists who have contributed to these discoveries is Avicenna (Ibn Sīnā)—an Iranian (Persian) polymath in the eleventh century. Among his well-known works is *The Book of Healing (Kitāb al-Shifā')*, which is a comprehensive philosophical and scientific encyclopedia, divided into four parts of logic, natural sciences, mathematics (a quadrivium of arithmetic, geometry, astronomy, and music), and metaphysics. In his book of *Heaven and Earth*, he states that heat is generated from the motion in external things. This is to say that the thermal energy has a dynamic nature, similar to what is known as the *Dynamical Theory of Heat*. His theory on heat was reported in 1253 in the Latin text entitled *Speculum Tripartitum*. For example, when water is heated, the heat present in the water is generated with aversion on behalf of the water and then energy is *received* by the matter. Avicenna was also reportedly the first who employed an air thermometer to measure air temperature in his scientific experiments [30].

Avicenna's famous classic authoritative reference work, the *Canon of Medicine (al-Qānūn fī al-Ṭibb)* completed in 1025, consists of five books, which have been used at many western medical schools (e.g., Montpellier, France until the seventeenth century) and is still used in the East. In Book 1—essays on the basic medical and physiological principles, anatomy, regimen, and general therapeutic procedures—he states that the body parts have their own temperaments, degree of heat, and moisture. He further provides two methodically ordered lists, identifying: (a) ten body members versus their *degree of heat*, starting from the breath and ending to the skin and (b) thirteen body members from the coldest to the hottest, starting from the serious humor and ending with the skin. He also introduces the temperature equilibrium concept in the body parts, suggesting that its deficiency causes ailments in the body.

In *Book 2*—where he lists medical substances, arranged alphabetically, following an essay on their general properties—he states that drugs must not be exposed to extreme heat or cold or stored near other substances. He uses the term *innate heat* as the attrition of the blood parts, that occurs due to its circularly motion in the arteries. His theories on heat are like the thermodynamics laws, relating the temperature of multiple bodies in *equilibrium*; the *conservation* of energy for believing that the general energy is to be constant and is transformable from one form to another; and that heat transfer has a direction, flowing from a matter with the higher level to that of the lower one. Avicenna eventually founded the entropy concept, where a ground state (no energy) is defined at extremely low temperatures. Note that the *Dynamical Theory of Heat*, which was the foundation of the thermodynamics as a branch of physics, only originated in the nineteenth century.

In the West, until the invention of the thermoscope by Galileo Galilei—an Italian astronomer, physicist, and engineer in the seventeenth century—there was no numerical measure of the degree of heat. As far as heat itself, until the discoveries of the nineteenth century, scientists thought that it was a physical substance. Therefore, they associated it with the characteristics of a weightless liquid, also known as *caloric fluid* or *frigorific particles* (particles of cold). This term *(frigorific)* is attributed to Robert Boyle, a seventeenth-century British natural philosopher, who hypothesized that particles of cold are transferred between objects [31].

Today, we know that heat is the measure of the kinetic energy stored in the random motion of atomic particles in matter, and temperature describes the intensity of this motion. Historically, extensive work has been conducted on the equivalency of heat, energy, and work. Heating processes, such as quicker conduction heat transfer by aluminum compared to ceramic or plastic or convection heat transfer when boiling water, are known facts. Even young children know about temperature. They may be running a temperature when they are sick, where a thermometer is employed to measure their body temperature. This knowledge provides the ability to describe physical systems mathematically to model the reality. This modeling is either done analytically or numerically. The analytical models include mathematical relations that express the physical relations between independent variables identifying the system behavior. Numerical models are the same as the analytical models in terms of the system's behavioral representations, except that they define numerical algorithms that are applied to the analytical models. For example, they are discretized for a domain filled with elements.

Thermodynamics, fluid mechanics, and heat transfer are known collectively as thermo-fluid sciences. They find applications in just about all of nature's phenomena as well as in most of humanity's technological fields. In nature, the sun's heating of the earth's surface, atmospheric phenomena, and the movement of oceanic currents are all connected to these sciences. Human technological endeavors, such as the internal combustion engine of a car, the turbofan engine of an airliner, the floating of a ship on an ocean, the buoyancy forces that keep the airplane in the air, the frying of your morning eggs, and the heating of your house, are all connected to thermo-fluid sciences.

End Notes

[8]   K. Peszynski,, D. Perczynski, Lidia Piwecka, "Mathematical Model of a Fountain with a Water Picture in the Shape of an Hourglass," *EPJ Web of Conferences*, 213, 02064, EFM 2018, 2019; https://doi.org/10.1051/epjconf/201921302064

[9]   https://www.reviewjournal.com/local/the-strip/heres-a-behind-the-scenes-look-at-how-the-bellagio-fountains-are-maintained/

[10] https://youtu.be/kvOJ37pPKaM

[11] https://en.wikipedia.*org/wiki/First_Battle_of_al-Faw#Iranian_attack*

[12] https://www.aparat.com/v/c1JM4/کلیپ_پل_بعثت_ResaneSabz.ir
      I am negotiating with the green flag of content; seeking the untold stories of the Beesat Bridge; having become acquainted with the memories of devoted souls; I am the interpreter of the Dawn 8 passage; the land of pain, is the meeting place of the divine people; it is the turning point of all the epics. When is it possible for the traveller to return and find the lost ones of Fave? I am mesmerised by the sunset at the shores of the River Arvand and the sparrows of the South Reeds.

[13] https://www.isna.ir/news/91122114203/

[14] https://www.aboutpipelines.com/en/pipeline-101/whats-in-the-pipelines/natural-gas-lines/

[15] https://www.shell.com/about-us/major-projects/perdido.html

[16] https://en.wikipedia.org/wiki/List_of_natural_gas_pipelines

[17] https://chemicalengineeringresearchandr.weebly.com/pipeline-transport.html

[18] http://naturalgas.org/naturalgas/transport/

[19] https://www.engineeringtoolbox.com/pipes-temperature-expansion-]coefficients-d_48.html

[20] http://pstrust.org/wp-content/uploads/2015/09/2015-PST-Briefing-Paper-02-NatGasBasics.pdf

[21] https://www.builditsolar.com/Projects/Cooling/EarthTemperatures.htm

[22] https://www.britannica.com/science/umbilical-cord

[23] https://www.britannica.com/science/urethra

[24] https://www.britannica.com/science/human-renal-system/The-ureters

[25] https://www.britannica.com/technology/coal-slurry

[26] https://www150.statcan.gc.ca/t1/tbl1/en/tv.action?pid=2510005601

[27] K.N. Shukla, "Heat Pipe for Aerospace Applications—An Overview," *Journal of Electronics Cooling and Thermal Control*, 5, pp. 1–14, 2015.

[28] LAFARGE, Northern Alberta Region Drainage Systems, Concrete Pipe, 2018, Precast Stormwater & Wastewater Solution, shttps://www.lafarge.ca/sites/canada/files/atoms/files/2018_edmonton_pipe_catalogue.pdf

[29] https://en.oxforddictionaries.com/definition/heat

[30] Robert Briffault, *The Making of Humanity*, pp. 380, 2015.

[31] https://www.britannica.com/biography/Robert-Boyle

HEAT TRANSFER MODELING

Heat transfer modeling is founded on the principles of *thermody-namics*. This science focuses on the motion of particles making up the matter stimulated by heat, resulting in changes of the internal energy and heat generation (HG). In thermodynamics, the four laws are as follows:

(1) Zeroth law—two objects, each in equilibrium with a third object, are in equilibrium with one another.

(2) First law—the internal energy of objects remains constant.

(3) Second law—the entropy of the universe increases over time, meaning the changes are positive for any given system.

(4) Third law—the entropy of an object approaches zero when its temperature approaches absolute zero.

The term *object* used in this context is interchangeable with *system*, since both define a cluster of molecules with the equation of states ruling over them [32].

Interacting forces within fluids (e.g., gases, liquids, and plasmas) are governed by fluid mechanics. This is also known as the *third law of Newtonian mechanics*, sometimes called the *action-reaction law*—for every action, there is a reaction, equal in magnitude and in the opposite direction. *Fluid mechanics* is further categorized into fluid *statics* and *dynamics*, based on the temporal status of the fluid molecules. The former relates to the *stationary* systems, also known as the *steady-state* systems, and the latter to the *transient* systems. The variation of the atmospheric pressure with altitude is an example where statics applies. *Hydrostatics* is the reason for the self-leveling of tea within a cup, making liquid's surface flat. Shock absorbers, such as those used in some aircraft landing gear suspension systems, are an example of a fluid dynamics application. Depending on the type of fluids (i.e., liquids or gases), more specialized disciplines (e.g., *hydrodynamics* and *aerodynamics*) have evolved. Aerodynamics investigates the flow patterns and forces over any object moving through air.

2.1   Basic Concepts

Heat transfer is a thermo-fluid science that focuses on the transportation of heat in a continuous medium. The continuum consists of molecules and is identified by boundaries. Depending on the spacing between the particles, these entities agglomerate and form matter. The ratio of their mean free path to the characteristic length—also known as the *Knudsen number* (*Kn*)— decides if the matter is a fluid or solid. A larger space allows molecules to freely move in their environment without interacting significantly with their neighbors. If the *Kn* is larger than 1, free molecular flow is observed, where molecules can freely move to occupy the space available to them within their container (as in gases). Molecules with a smaller *Kn* (about 1) have their movement constrained to greater extent (as in fluids). Therefore, fluids (i.e., gases and liquids) are identified by the spacing between their molecules, adopting the shape of their container. In solids, the *Kn* is considerably smaller than 1 and the molecules are tightly packed.

Energy (as heat) can be transferred by mechanical interaction, requiring a medium (gas, liquid, or solid). It can also be transferred between the matter's elemental particles or by the transmission and absorption of electromagnetic waves, requiring no medium. Heating a skillet on an induction oven raises the skillet temperature due to the electromagnetic waves agitating the skillet's iron atoms; the next step is conducting this heat through the skillet particles. Holding your hand close to the skillet and feeling the heat is heating by electromagnetic interaction (*radiation* heat transfer) while touching the skillet directly is heating by mechanical interaction (*conduction* heat transfer). Taking the skillet off the heat and leaving it exposed to surrounding air makes it cool down by *convection* heat transfer.

Depending on the boundary definition, well-defined or not, a control volume or system is defined. The former can be either an *open* or *closed* system. In a closed system, energy in the form of heat crosses the boundaries. In an open system, both energy and mass can do the border crossing. This energy is either in the form of heat or work. Phase change for a specific matter is an example of a system, the boundaries of the matter change from the more defined shape (e.g., liquid or solid) to a less specific one (e.g., gas). Steam flow in a turbine is an example of a control volume, where mass with a given energy level, which is a function of its pressure and temperature, enters the turbine and leaves it at a higher level, generating work.

When a matter is exposed to heat, the matter may undergo phase changes; its state changing from solid to liquid (melting), liquid to gas (vaporization), liquid to solid (freezing), gas to liquid (condensing), solid to gas (sublimating), and gas to solid (depositing). During the phase change process, temperature remains constant; therefore, the process is *isothermal*. Temperature variation is directly associated with the average kinetic velocity of the matter's molecules while heat determines the flow of the created *spatial* or *temporal* energy. Heat transfer problems can be categorized in the following processes: (a) *isothermal*—temperature remains constant; (b) *isobaric*—pressure remains constant; (c) *isovolumetric* (or isochoric)—volume remains constant; (d) *adiabatic*—no energy is transferred; (e) *isentropic*—entropy remains constant; and (f) *isenthalpic*—enthalpy remains constant.

All thermal sciences are governed by natural physics and employ: (a) analytical, inseparable relations—derived from mathematical relations, (b) empirical relations—obtained from experimental observations, and (c) hybrid relations—predicted from correlating the two-said physical modeling approaches. These studies focus on calculating spatial fluid properties such as temperature, pressure, density, and velocity within a time domain and therefore defining temporal properties. When processing thermal science data, either interpreting or presenting them, it is possible to take advantage of certain special cases. If the system properties do not vary over time, a *steady-state* condition is achieved; otherwise, the system is *transient*.

When post-processing the results of any system study, if multiple variables are plotted against each other, by keeping one property constant, the *iso-property* (e.g., *isochoric*) contour lines or surfaces can be produced. *Isotherms* (lines of equal temperature) and *isobars* (lines of equal pressure) plotted on the weather charts are among more typical cases. Examples of less known plots are *isogeotherms* (lines of constant mean annual temperature) and *isodrosotherms* (lines of constant dew point) plotted on the weather charts.

2.2 Thermal Analysis of Systems (Components and Subcomponents)

Thermo-Fluid analysis describes behavior of systems involving heat and flow, with system properties that may depend on temperature. It may be also referred to as *heat transfer* or *flow modeling*. Thermal analysis describes the thermal response of a system as a function of *predictors* (inputs). These predictors may be material characteristics—such as thermophysical and optical properties—or process parameters—such as laser scanning speed or operating temperature.

2.2.1 Thermal Properties of Materials

The three thermophysical properties of *density* (ρ), *thermal conductivity* (k), and *specific heat capacity* (c) form the foundation of all heat transfer problems. A fourth, dependent property, *thermal diffusivity* (α), may also be used to characterize the material. Thermal diffusivity is the ratio of thermal conductivity to the product of density and specific heat capacity. Any type of mass, momentum, or energy conservation employs one or more of the first three properties.

Material properties are vital ingredients of any modeling and, depending on the modes of heat transfer or analysis type, some are more dominant than others. These properties may vary in space (*spatial*), time (*temporal*), or under environmental conditions (*environmental*). Nonconstant properties introduce nonlinearities and non-homogeneities to the physics that make the problem more challenging to tackle. An example is modeling material exposure to intense heat sources such as fire in which heat-material interaction over time is needed. For such models, thermal capacity and thermal conductivity are required. Density is another needed property that is available in most cases for being the basic material specification. Inaccuracies in material properties can sometimes lead to significant variations in thermal analysis results.

Thermal properties of materials usually involve a combination of energy (J), temperature (K), mass (kg), length (m), or time (s). Adding (or taking away) energy from a material increases (or decreases) the degree of *agitation* in the form of *translational*, *rotational*, and *vibratory* motion of the material's elementary particles; the level of this *agitation* is expressed by the material's temperature.

Specific heat capacity is the amount of energy needed to increase the temperature of material by one degree centigrade (J/kgK). The larger it is, the more capacity the matter has to hold its thermal energy; for example, for water it is 4,200 J/kgK, while for cast iron it is 460 J/kgK.

Thermal conductivity describes how much energy (J) can travel per unit time (s) per unit length (m) for a temperature gradient of one degree (°C). The larger it is, the more conductive the material is; for example, for alumina (a ceramic material) it is 27 W/mK, while for copper it is 401 W/mK, which is almost fifteen times greater than that of the ceramic material. *Thermal diffusivity* represents the combination of these thermophysical properties and is defined as the ratio of thermal conductivity to the product of density and heat capacity ($\alpha = k/\rho c$). This property describes the temporal variation of temperature within matter, the changes that occur with respect to time, when constituting the energy conservation equations.

If a material is a mixture of two or more distinct elements or compounds, it can be classified as *homogeneous, inhomogeneous,* or *heterogeneous*. In a homogeneous material, the components are mixed at such a fine level that any small macroscopic sample of the material has the same proportion of the constituents as any other. In an inhomogeneous material, taking similar samples results in sample-to-sample variation of the constituent proportions. In other words, homogeneous materials are consistent in composition and character (e.g., some metals), while inhomogeneous materials are inconsistent in composition or character due to the substantial material variations (e.g., rice pudding). Heterogeneous materials are inconsistent in composition or character for similar materials (e.g., chocolate chip cookies). If the mixture's thermal properties are not available, one way to approximate them is to use the rule of mixtures. It provides an estimate of the *equivalent* property, which is the sum of the products of the individual property value and its corresponding mass fraction.

Spatial properties can change within a geometry or specific domains within a geometry. This introduces thermal non-homogeneities in properties. The change can be: (a) spatial (e.g., varying thermal conductivity as a function of the location or direction), (b) thermal (e.g., change of specific heat capacity with temperature), or (c) temporal (e.g., changes in metabolism in a living organism).

Properties that are expressed per unit length (e.g., thermal conductivity) or length squared (e.g., elastic modulus—Pa = N/m^2) can also vary with the material direction. If such variation exists, the material is *anisotropic*. Furthermore, anisotropic materials may be transversely *isotropic* or *orthotropic*. The former has invariant properties within a plane but different properties in the direction orthogonal to this plane. Think of a thin membrane—properties within its plane are the same in all directions, but they are different in the transverse direction. Orthotropic materials have properties which differ along three orthogonal directions. For example, a sheet of rolled steel has different properties in the direction of rolling, perpendicular to the rolling direction, and transverse to the sheet plane.

When selecting materials for engineering applications, careful consideration must be given to their thermal and mechanical properties. Aluminum, for example, has a good strength-to-weight ratio (33.3 kN m/kg) and, due to the oxidized layer that is formed on its surface when exposed to air, it is almost corrosion-free. However, it does not perform that well in the high-temperature applications due to its relatively low melting point (660.2 °C) when compared to other metals. Aluminum has a high thermal conductivity (238.5 W/mK), which makes it suitable for heat transfer applications, such as heat sinks or sources. Also, it has large linear thermal expansion coefficient. Aluminum, however, has a small strength-to-weight ratio.

Titanium can be used in very high-temperature applications due to its high melting point (1,650–1,670 °C); it is harder and stronger than aluminum, but more costly. Its strength-to-weight ratio (48.9 kN m/kg) is higher than that of aluminum. This can be a benefit in structural application, but it also makes the material difficult to machine or process. Another benefit is that it has lower thermal expansion coefficient (8.9 μ/mK) compared to aluminum's (23.6 μ/m K). However, due to its low thermal conductivity (17 W/mK), titanium is a poor candidate for heat transfer applications. These examples show that there are many things to consider when selecting appropriate materials in engineering applications (Table 2.1).

TABLE 2.1. Comparison of thermomechanical properties of aluminum and titanium.

Material	Melting Point (°C)	Density (kg/m^3)	Tensile Strength (MPA)	Tensile Strength-to-Density (m^2/s^2)	Linear Thermal Expansion (μm/mK)	Thermal Conductivity (W/mK)
Aluminum	660.2	2,700	90	33	23.6	238.5
Titanium	1,650-1,670	4,500	220	49	8.9	17

2.2.2 Static versus Dynamic

There are two types of thermo-fluid analysis: (a) transient and (b) stationary. The former case involves temporal system characteristics (properties change with time), while the latter involves no temporal dependence. Occasionally, the system's instantaneous behavior depends on the past performance; such a system is referred to as a *dynamic system*. However, that is not necessarily the case for all transient systems. In other words, a system can be time-dependent static but not necessarily dynamic. A static system is *memoryless* while a dynamic system has an "elephant" memory. In a memoryless system, the output is the function of the input. For example, $f(t) = te^{at}$ is a static system, which requires a certain time variable (t) to identify the value of the function $f(t)$ at that instance (t). No previous data is required to define this function; this is a memoryless system $(f(5)|_{a=0.5} = 60.91)$, even though the result depends on time (transient). However, $f'(t) = te^{at} + f'(t-a)$ is a system that has a memory, since it requires a certain time variable to not only identify the value of the function $f'(t)$ at that particular instance (t), but also the value of the function $f'(t)$ at a s prior to that instance— $f'(5)|_{a=0.5} = 103.61$. Functions $f(t)$ and $f'(t)$ are presented in Figure 2.1.

As the next step, curves were fitted to the same functions $(f(t)$ and $f'(t))$, with the results presented in Figure 2.2. The two fitted functions $(y(x)$ and $y'(x))$ are third-degree polynomials and do not identify if the functions are static or dynamic. Therefore, functions alone do not determine the dynamic nature of a system; other characteristics are needed to determine that.

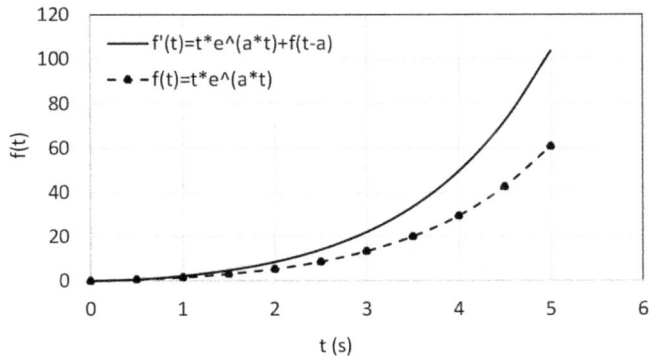

FIGURE 2.1. Function $f(t)$ versus the t representing static (dotted line) and dynamic (solid line) systems (a = 0.5).

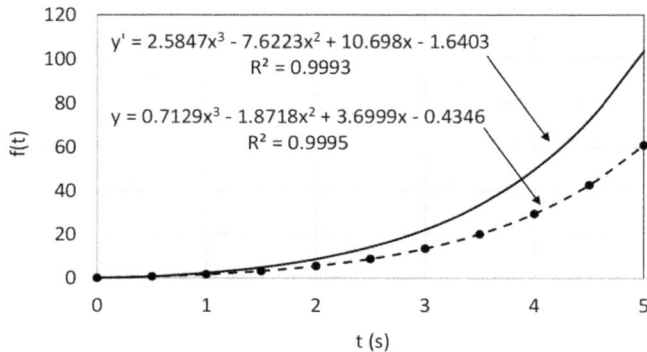

FIGURE 2.2. Fitted function $f(t)$ versus the t representing static and dynamic systems ($a = 0.5$).

Let us consider adjusting the temperature of a room using a thermostat. Inputting the desired temperature to the thermostat control panel is inherently not time-dependent; you may set whatever temperature level you wish at any time. However, reaching the desired room temperature after adjusting the thermostat is a transient process in which temperature increases (or decreases) with time. Given the real-time ambient conditions, the input temperature is translated to an electrical signal, which is then communicated to the boiler to generate more warm air, or the air conditioner to generate more cool air. The heated flow is then transferred through the in-floor pipes, radiators, or ducts by means of conduction, radiation, or convection heat transfer modes to the thermally controlled zones. The temperature of the environment changes on a continuous basis until it reaches the set value.

A thermometer is an intervening device, ensuring the desired temperature is achieved by measuring the temperature at each heating or cooling step. Eventually, the thermal transient process becomes steady at the set value. At this point, the behavior of the system does not change with time anymore. In most cases, it is assumed that the system has completed several temporal processes before reaching the steady-state condition; the process events depend upon the previous events and therefore heating process is a dynamic system while inputting temperature is a static process.

For a transient analysis, where the temporal variation is desired, the time predictor is considered either as an additional coordinate to the three spatial ones or as a separate variable where it influences the thermophysical properties—expressed in terms of the temporal variation of the boundary conditions. An example is the definition of the volumetric heat generation term for the case of a laser contour welding process, where the profile of the heat source changes along the x-, y-, and z-coordinates and varies with time (since the beam is scanning the part). The heat source can be applied cyclically—turning on and off—to study the effect of the heating and cooling (after the heat source is no longer active). The time-varying heat source, either in the form of heat generation inside the geometry or boundary conditions applied to the internal or external borders, follows similar rules [33].

2.2.3   Energy Balance

In thermo-fluid systems, the conservation of energy principle states that the energy should be conserved in all subcomponents. Their total energy should be zero, demonstrating that the balance of energy has been reached. Conservation of energy requires that the total energy inputted into and generated within the system is the same as the total outputted from and stored by the system. Figure 2.3 shows schematically the general form of the energy balance for a continuum, Equation (1).

$$\dot{E}_{in} + \dot{E}_{generated} = \dot{E}_{out} + \dot{E}_{storage} \qquad (1)$$

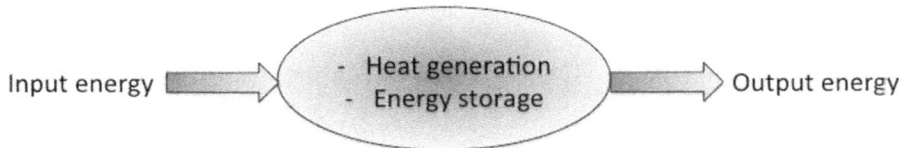

FIGURE 2.3. Energy balance diagram for a continuum (e.g., a parcel of air).

The conservation of energy requirement means that the energy balance is to be complied with for any small and identifiable portion (element) of the material that satisfies the continuity of mass, energy, and momentum. Energy can enter and leave the continuum; however, the boundaries remain constant. The continuum is identified by its size, mass, and thermophysical properties. Thermophysical properties may be *temporal* (transient—change with time), *spatial* (non-homogeneous—change with direction and location within the geometry), *temperature-dependent*, or *constant*.

Energy is defined in different forms inside this environment. It is either in the form of heat entering or leaving the continuum by conduction, convection, or radiation modes of heat transfer, changes of internal energy (or energy storage), or heat generation ($\dot{E}_{generated}$) inside the continuum. The heat leaving the continuum by conduction is the same as the heat entering the continuum by conduction plus the spatial variations over the length of travel, and it is time-independent (steady-state), Equation (2).

$$\vec{q}_{out}(x+dx, y+dy, z+dz) = \vec{q}_{in}(x,y,z) + \nabla(\vec{q}_{in}(x,y,z)) \tag{2}$$

The internal energy is time-dependent (transient); it represents the variation of rate of energy storage ($\dot{E}_{storage}$ in W) expressed in the following form that includes specific heat capacity of the continuum (c_p in J/kgK), mass (m in kg), volume (dV in m³), temperature (T in K), and time (t in s), Equation (3).

$$\dot{E}_{storage}(x,y,z) = mc_p \frac{dT(x,y,z)}{dt} \tag{3}$$

The heat generation can be time- and location-dependent (transient and temporal, respectively). It represents the heat generated inside the heat-conducting medium. It can be expressed in units of energy rate per unit volume—Equation (4)—where $\dot{E}_{generated}$ is heat generation (W) and $\dot{q}$ is rate of energy generated per unit volume (W/m³). Internal energy ($\dot{E}_{internal}$) is the energy storage and energy generation terms, Equation (5). Substituting Equations (2) to (5) into Equation (1) results in Equation (6).

$$\dot{E}_{generated}(x,y,z) = \dot{q}dV = \dot{q}dxdydz \tag{4}$$

$$\dot{E}_{internal} = -mc_p \frac{dT(x,y,z)}{dt} + \dot{q}dV \tag{5}$$

$$\dot{q}dV + \nabla(\vec{q}_{in}(x,y,z)) = mc_p \frac{dT(x,y,z)}{dt} \tag{6}$$

The energy balance for the case studies presented in this work is set for a *system* where mass does not enter or leave the system boundary—only energy in the form of heat and work does. There are some cases in which mass crosses the boundary or there is no net mass transfer (i.e., the inlet mass is the same as that of the outlet); therefore, there is no mass transfer—this is a *control volume* (versus the system) problem. The boundary of a system may expand or contract. For a control volume, however, both energy and mass may enter and leave the boundaries, which do not expand or contract. Your body as a source of the sensible heat you experience transported by sweat, or rush of blood and tears, is a control volume, with the possibility for organic fluids entering and leaving the body parts. Each body part has a set boundary that essentially does not change, though it may expand or contract. Examples of the processes including mass transport are evaporation, precipitation, and distillation.

Geometry in which heat transfer takes place can be defined by the *Cartesian* coordinate system in one (1D), two (2D), or three (3D) dimensions. For some 3D shapes, the *cylindrical* or *spherical* coordinate system can facilitate the modeling task. There are cases where the model can be simplified by reducing the number of dimensions. One case is where the length of the plane transverse to heat transfer direction is large compared to the other dimensions, including the dimension along which heat is transferred. In this case, the heat transfer along the transverse direction can be ignored. This is where a 3D model can be simplified to a 2D model. Another case is an axisymmetric model, meaning that the model (i.e., properties, boundary and initial conditions, and state variables) are the same about a symmetry axis. In this case, a 3D model can be simplified to a 2D axisymmetric model. A pipe exposed to a uniform transient heat flux on its exterior surface is an example of such model.

2.3   Modes of Heat Transfer

Dependent variables are the driving forces for defined physics. When modeling heat transfer in *solids* and *fluids*, temperature is the dependent variable. This is analogous to pressure being the dependent variable when modeling fluid flow. Heat is transferred from the point with the higher temperature to that with the lower one. Fluid moves from the point with higher pressure to that with the lower one. Heat or fluid movement continues until all points reach an equilibrium state, meaning that their temperature or pressure equalizes.

The dependent variable is to be measurable so that the derivative may be calculated. For heat flow to be determined, temperature is the state variable. The variation of temperature throughout matter—either in the form of solid or fluid—is either time-dependent (temporal) or space-dependent (spatial). The gradient of temperature (i.e., spatial variation) results in heat conduction—from a region with a higher temperature to that with a lower temperature. The rate at which this equalization takes place is proportional to the thermal diffusivity and spatial derivative of temperature. As you may recall, thermal diffusivity is the ratio of heat conductivity to the product of density and specific heat capacity. This property is the characteristic of the material, affecting temperature change over time (i e., the transient temperature).

The mechanism of heat transfer depends on the medium in which the heat is being transferred. Heat transfer is achieved primarily by the mechanisms of conduction and radiation. For conduction to happen, either in its pure or subsidiary forms (such as convection), molecules need to be present. While in the radiation form, electromagnetic waves are the energy-transmitting agents, and no intervening molecules are needed. This is how the sun's radiant energy reaches the earth's atmosphere and passes through the atmospheric layers to be absorbed by the planet's surface.

For solids, in which molecules are near each other, the conduction mode of heat transfer is dominant. For molecules flowing in the form of a fluid (i.e., liquid or gas), heat transfer takes place by means of advection, which is the combination of the convection and conduction due to the fluid flow and the solid surface they may come in contact with. When a gas comes in contact with a solid, a hybrid heat transfer mechanism takes place—a combination of the conduction and convection heat transfer modes, both in the solid and liquid as well as their interface. Additionally, the momentum of the fluid bulk transfers some of the energy in the form of heat.

Flow profile and velocity affect the heat transfer mechanism. This is particularly the case for conjugate heat transfer models, where the combination of fluid flow and solid heat transfer are included in a hybrid thermal heat transfer model. When a continuum *nonslip* flow passes over a wall (i.e., any solid boundary), the magnitude of the flow velocity adjacent to the wall will be zero, while for a *slip* flow this value is not zero, since the fluid can slide relative to the wall. The parabolic velocity profile associated with the continuum flow passing over the wall changes to a linear profile in a *free molecular* flow, meaning that the flow velocity, starting at a nonzero magnitude at the wall, changes (increases) linearly with increasing distance from the wall.

2.3.1 Conduction Heat Transfer

For solids, where molecules are in close contact with each other, conduction is the main mode of heat transfer. For molecules within fluids (i.e., liquid or gas), heat transfer takes place by means of advection, which is a combination of convection and conduction. Since heat is transferred though the internal energy of randomly colliding molecules, this mode of heat transfer is available for all three phases of material (i.e., solid, liquid, and gas). Obviously, the more the molecules interact, the more efficient heat transfer mechanism becomes.

Consider, as an example, the case of in-floor heating, where electric heating wires are embedded into the floor in a raster pattern. When walking barefoot on the heated floor, its heat is transferred to your feet, assuming that the floor's temperature is higher than that of your body. Another mechanism of heat transfer is the convection by spontaneous change of density for the air at the proximity of the horizontal warm surface. With no in-floor heating, your bare feet will normally feel the cold of the floor surface. This is because the heat from your sole is transferred to the floor. High thermal conductivity of the 0.05-m-thick ceramic floor tile (1.84 W/mK) compared to your skin (0.37 W/mk) are the responsible factors; however, thermal diffusivity of the ceramic is larger than that of the skin (Silicon carbide thermal diffusivity is about 1.1×10^{-6} m^2/s compared to that of the human skin that is about 9.8×10^{-8} m^2/s) [34, 35, 36, 37, 38]. This means that the heat diffusion for the ceramic floor tile is about 11 times that of the human skin, causing the cool sensation—Table 2.2.

TABLE 2.2. Comparison of thermophysical properties of some materials.

Material	Thermal Conductivity (W/mK)	Heat Capacity (J/kgK)	Density (kg/m^3)	Thermal Diffusivity (m^2/s)	Thickness (m)	Thermal Resistivity (m^{2}K/W)
Ceramic Floor Tile	1.84	840	2,000	1.10E-6	0.05	0.03
Human Skin	0.37	3,391	1,109	9.84E-8	0.002	0.005

For a derivative (e.g., heat) of a property (e.g., temperature) to be transported, a state variable must be defined that is responsible for the transportation of the derivative; it can be either measured or calculated. For the heat flow (response) to be determined, temperature (predictor) as the state variable is to be employed. Variation of temperature throughout the matter, either in the form of the solid or fluid, is either time- or space-dependent. Recall that gradient of temperature causes heat transfer to occur from the region with higher temperature to that with the lower one. Material thermal characteristics, such as thermal conductivity and heat capacity, as well as physical properties, such as density, affect this energy transfer, over the defined domain (space) and time, introducing temporal-spatial thermal characteristics.

In addition to the analytical relations and numerical approaches to solve conduction heat transfer modes, there are also diagrams suitable for specific scenarios that correlate thermal variables. An example is a Heisler-Gröber chart, consisting of three sets of charts presenting the temperature distributions inside a sphere with known radius, a semi-infinite slab with known thickness, and a cylinder with a known radius. There are limitations to be considered when using these diagrams. These limitations are due to the assumptions used to create the diagrams: (a) the initial temperature should be constant; (b) the environmental conditions (namely temperature) should remain unchanged; (c) the convection heat transfer coefficient should not vary as a function of temperature; and (d) the rate of heat generation inside the part is zero.

To obtain these diagrams, the exact solutions from the Fourier transformation (infinite slab and sphere) and the Bessel functions (infinite cylinder) are simplified, including only the first terms. There are sets of diagrams identified for each geometry that show the following: (a) the dimensionless temperature at the center of the geometry assuming an imposed temperature on the boundary of the surface as a function of the *Fourier number* $(Fo = \alpha t/r_0^2)$; (b) dimensionless temperature distribution inside the geometry as a function of the inverse *Biot number* $(Bi = hr_0/k)$, given the dimensionless radius or the thickness of the geometry; and (c) dimensionless thermal energy (heat), which is a function of the product of the *Biot number* squared by the *Fourier number* (Bi^2Fo), assuming that constant temperature boundary conditions were applied [39].

As mentioned earlier, temperature difference is the driving force for the movement of heat energy. This heat flow rate (Q in $W = J/s$) depends on the heat conduction coefficient (k in W/mK), which is the proportionality factor, area of the body normal to the heat flow (A in m²), and temperature change (dT in K) with respect to the distance (dx in m). This is described by the Fourier equation: $Q = -\nabla(kAT)$.

Fourier law is applicable to the heat conduction mode and it states that energy transfer is proportional to the gradient of the temperature with respect to the direction (coordinate system) along which it flows $\left(\dfrac{dT}{dx}, \dfrac{dT}{dy}, \dfrac{dT}{dy}\right)$ as well as the areas of the body perpendicular to the directions of heat transfer (A_x, A_y, A_z), and thermal conductivity (k_x, k_y, k_z) along the heat transfer direction. Thermal conductivity depends on the material and demonstrates how fast the molecules get agitated, showing signs of increased activity as the temperature increases. Metals in general have higher thermal conductivities compared to nonmetals. The closer the molecules are to one another, the easier it is for them to transfer their motion, which corresponds to the thermal energy.

The heat transfer flow follows the temperature gradient; $\vec{i}, \vec{j},$ and $\vec{k}$ represent unit vectors along the (x, y, z) in the Cartesian coordinate system.

Temperature gradient is a vector, having magnitude and direction. For three-dimensional space, each of the vector's three components is obtained by calculating the derivatives of temperature with respect to the dimension component (coordinate). Thus, the vector $\left(\dfrac{dT}{dx}, \dfrac{dT}{dy}, \dfrac{dT}{dy}\right)$ is obtained in the Cartesian coordinate system. These derivatives take on different forms for the cylindrical and spherical coordinate systems, which will be discussed later in this section.

The area that is normal to the heat transfer gradient vector becomes important when calculating the total power passing through a plane. For example, determining the heat transfer along the x-coordinate involves the y-z plane: the y-z plane is perpendicular to the heat transfer direction and is therefore the surface to which the heat flux is applied. Equation (7) presents the general form of changes of energy due to the conduction heat transfer in the Cartesian coordinate system in a steady-state case. Note that heat generation and change of internal energies are not included in Equation (7) and that $A_x = dydz$, $A_y = dxdz$, and $A_z = dxdy$.

$$Q_{\text{cond}(x,y,z)} = -\nabla(AkT(x,y,z))$$

$$= -A_x\left(k_x\frac{dT}{dx}\right)\vec{i} - A_y\left(k_y\frac{dT}{dy}\right)\vec{j} - A_z\left(k_z\frac{dT}{dz}\right)\vec{k} \qquad (7)$$

Similar to electric current (I) that flows inside a resistor from the areas of higher voltage (V) to those of lower ones inside a circuit ($V = RI$)—with the voltage being the driving force—heat flows from the regions of higher temperature (T_s) to those of the lower ones ($T_{s'}$), with temperature difference acting as the driving force. Therefore, it is possible to simulate the heat flow as electric current and assign the denominator term presented in equation (8) the *thermal resistance*.

$$Q_{\text{cond}} = \frac{(T_s - T_{s'})}{\dfrac{L}{kA}} \qquad (8)$$

Conductive thermal resistance is then presented by Equation (9), where L is the length of the thermal layer (m), k is homogeneous thermal conductivity (W/mK), and A is area normal to heat transfer direction (m^2). The dimension of conductive thermal resistance (R) is K/W.

$$R_{\text{th cond-cartesian}} = \frac{L}{kA} \qquad (9)$$

Depending on the geometry to be modeled, the Cartesian, cylindrical, or spherical coordinate system may be employed. For instance, a cylindrical shape, such as a pipe, is better represented by the cylindrical coordinate system. A ball is more accurately modeled using the spherical coordinate system. This assists with capturing the geometry irregularities such as bends, curves, and corners.

Changes of energy due to the conduction heat transfer in the cylindrical coordinate system are presented by Equation (10). Heat generation and the change of the internal energies are not incorporated; $\vec{i}, \vec{j}$, and $\vec{k}$ represent unit vectors along (r, θ, z); $x = r\cos\theta$, $y = r\sin\theta$, $z = z$, $r = \sqrt{x^2 + y^2}$, and $\tan\theta = y/x$.

$$Q_{\text{cond}(r,\theta,z)} = -\nabla\left(AkT(r,\theta,z)\right)$$
$$= -A_r\left(k_r\frac{dT}{dr}\right)\vec{i} - A_\theta\left(k_\theta\frac{dT}{rd\theta}\right)\vec{j} - A_z\left(k_z\frac{dT}{dz}\right)\vec{k} \tag{10}$$

Changes of energy due to the conduction heat transfer in the spherical coordinate system are presented by Equation (11). Heat generation and the change of the internal energies are not incorporated; $\vec{i}, \vec{j}$, and $\vec{k}$ represent unit vectors along (r, θ, φ); $x = r\sin\varphi\cos\theta$, $y = r\sin\varphi\sin\theta$, $z = r\cos\varphi$, and $\tan\theta = y/x$, $r = \sqrt{x^2 + y^2 + z^2}$, and $\varphi = \arccos(z/r)$.

$$Q_{\text{cond}(r,\theta,\varphi)} = -\nabla\left(AkT(r,\theta,\varphi)\right) \tag{11}$$
$$= -A_r\left(k_r\frac{dT}{dr}\right)\vec{i} - A_\theta\left(k_\theta\frac{dT}{rd\theta}\right)\vec{j} - A_\varphi\left(k_\varphi\frac{dT}{r\sin\theta d\varphi}\right)\vec{k}$$

The thermal resistance analogy is also applicable to the cylindrical and spherical coordinate systems; therefore, Equations (10) and (11) are transformed into Equations (12) and (13), respectively, where r_1 and r_2 are the internal and external radii of the geometries, and L is cylindrical length.

$$Q_{\text{cond}} = \frac{(T_s - T_{s'})}{\ln\left(\frac{r_2}{r_1}\right)\Big/2\pi kL} \tag{12}$$

$$Q_{\text{cond}} = \frac{(T_s - T_{s'})}{(r_2 - r_1)/4\pi kr_1 r_2} \tag{13}$$

Conductive thermal resistances are then presented by Equations (14) and (15), where k is the homogeneous thermal conductivity, and A is area normal to the heat transfer direction.

$$R_{\text{th cond-cylindrical}} = \frac{\ln(r_2/r_1)}{2\pi kL} \tag{14}$$

$$R_{\text{th cond-spherical}} = \frac{(r_2 - r_1)}{4\pi k r_1 r_2} \qquad (15)$$

One may derive Equations (10) and (11) from their equivalents in the Cartesian coordinate system, Equation (7). To achieve this, the following transformations are applied. The equivalent of the Cartesian coordinate system (x, y, z) in the cylindrical coordinate system (r, θ, z) is $(r\cos\theta, r\sin\theta, z)$ which implies that $\tan\theta = y/x$. The equivalent of the Cartesian coordinate system (x, y, z) in the spherical coordinate system (r, θ, φ) is $(r\cos\theta\sin\varphi, r\sin\theta\sin\varphi, r\cos\varphi)$, which implies that $\tan\theta = y/x$ and $\arccos\varphi = z/r$. Furthermore, $\frac{dT}{dx, dy, dz} = \frac{dT}{dr}\frac{dr}{dx, dy, dz} + \frac{dT}{d\theta}\frac{d\theta}{dx, dy, dz} + \frac{dT}{d\varphi}\frac{d\varphi}{dx, dy, dz}$. Note that the denominator expression dx, dy, and dz determines along which dimension component (x, y, z) the gradient is applied.

2.3.2 Convection Heat Transfer

The convection heat transfer mode occurs between a solid and fluid. It is related to the temperature difference between the solid surface in contact with the fluid and the bulk temperature of the fluid surrounding it. Occasionally, the bulk temperature is assumed equal to the average temperature between the wall surface temperature and the flow temperature at a distant location, if this temperature difference is considerable. The factors that make this proportionality an equality are the area of the surface, and a proportionality coefficient called the *convection heat transfer coefficient*. This coefficient depends on the bulk flow characteristics, such as its velocity. Surface characteristics, such as roughness or surface orientation, also affect this coefficient, which is expected to be greater for a vertical surface (due to the gravity effects and more pronounced at higher temperatures) versus the horizontal surface under similar conditions.

For a motionless fluid, this coefficient may be obtained from experimental observations. For example, for a horizontal wall adjacent to an air volume, this coefficient is about 5 W/m²K, while for a vertical wall, this value is about 10 W/m²K. The larger the convection heat transfer coefficient is, the larger the heat transfer magnitude from the solid surface to the fluid environment is. The reason is that heat transfer is facilitated by the flow in the proximity of the vertical surface, where the fluid can move freely due to its buoyancy. Since the cold fluid is denser, it moves downward, while the warm fluid moves upward, creating flow in the vicinity of the wall, which promotes heat transfer and would be represented by the higher heat transfer convection coefficient value. The horizontal surface generally exhibits a lower heat transfer rate.

The mechanism of convection heat transfer that occurs only due to the natural buoyancy of the fluid is called *free convection*. In this case, when the solid surface comes in contact with the fluid, no additional mechanism exists to facilitate heat transfer. In other words, heat transfer in this case is done spontaneously and that is why it is also known as *natural convection*. An example is warming a room by hot-water radiators. The fluid takes advantage of the generated buoyancy forces due to the variation of the density due to the temperature variations inside its bulk (e.g., flow in the vicinity of a vertical radiator). In these cases, the *Grashof Number* (Gr), showing the ratio of the buoyant to viscous forces becomes important.

If there are external sources, such as a fan (e.g., air conditioning), or a fluid pump, where the fluid is moved around artificially in the desired direction, the *forced convection* mechanism is applicable. An example is heating a room by means of an electric heater with a built-in fan. Forced convection may increase the convection heat transfer coefficient by a factor of ten to about 100 W/m²K or more.

Newton's law of cooling describes the convection heat transfer; this law states that energy transfer is proportional to the temperature difference between the surface temperature (T_s), surrounding fluid temperature (T_b), area of the exposed surface (A), and proportionality constant (h_c), which is also known as the *convection heat transfer coefficient*. Newton's law of cooling can be expressed by Equation (16), where Q_{conv} is heat transfer by convection.

$$Q_{\text{conv}} = Ah_c\left(T_s - T_b\right) \tag{16}$$

The surrounding fluid temperature (T_s), which is often considered the bulk fluid temperature, can be obtained experimentally by averaging and logarithmic mean relations. Convection heat transfer coefficient (h_c) can be predicted using the *Nusselt number* $(h_c L/k)$, where k is the thermal conductivity and L is characteristic length. Therefore, the *Nusselt number* shows the ratio of the convective to conductive forces.

The heat flow may be simulated as an electric current, with temperature difference playing the driving force. The temperature difference causes heat to transfer from the regions with higher temperatures to the ones with lower temperatures. Convective heat flow can be simulated in a similar fashion to the electric current. Equation (17) is another form of Equation (16), which shows the relation between the heat flow and temperature difference as a function of the convective surface areas and the convection heat transfer coefficient. By assigning the denominator term presented in Equation (17) as the thermal resistance $\left(R_{\text{th conv}} = \dfrac{1}{Ah_c} \right)$, Equation (17) is obtained.

$$Q_{\text{conv}} = \frac{(T_s - T_b)}{\dfrac{1}{Ah_c}} \tag{17}$$

For example, hot-water baseboard heaters rely, to a large extent, on the convective heat transfer mode to effectively deliver heat to the room. These heater types are also known as *radiators*. The term was introduced in a patent filed in 1834 by Mr. Olmsted, an inventor from Connecticut, who proposed adding a heat exchanger to a stove to improve room heating. Perhaps a *convector* would be a more appropriate name, since most of the heat transfer occurs by convection mode.

Today's typical hot-water baseboard heaters have copper pipes with closely spaced transversely mounted thin aluminum fins attached to them. Heat from the hot water circulating through the pipes, typically at about 70–80 °C, is conducted through the pipe and the fins where it warms the surrounding air. The warmer, less dense air rises due to the buoyancy forces, setting up convective air circulation. That is why blocking the air path from below or above will reduce such a heater's efficiency.

2.3.3 Radiation Heat Transfer

Thermal radiation was observed and reported throughout the history by scientists, including horticulturists. The *caldarium* is the hottest area of the greenhouse, and its existence was known in the 1700s. It was reported in a letter written in 1745 by Linnaeus to his student Samuel Nauclér, emphasizing that the temperatures inside the orangery at the University of Uppsala Botanical Garden had reached 30 °C, which was well above the desired temperatures of 20 °C in summer and 15 °C in winter. Linnaeus hypothesized that this temperature increase was due to the thermal radiation received by the windows angled so that they were exposed to the Sun's maximum radiation.

Radiation is the mode of heat transfer that does not require a physical medium for heat to propagate. In this mode, the energy is transferred by electromagnetic waves radiated by one body and absorbed by another. All objects at a temperature higher than absolute zero emit thermal radiation. One example of this phenomenon is the radiation emitted from the sun and received by the earth. This radiation is emitted in a broad range of wavelengths but, because some of that is absorbed by the atmosphere, only part of that broad spectrum reaches the earth's surface and the intensity is reduced (to about 1,000 W/m² on average), making the sun's radiation tolerable for the earth's residents. With radiation heating, non-colliding photons transfer the electromagnetic radiation versus the colliding molecules in the conduction heat transfer mode.

Being an electromagnetic wave, solar energy travels through space at the speed of light (3×10^8 m/s). To understand the concept of electromagnetic energy waves, imagine throwing a stone into a still lake—the wave ripples, radiating in all directions along the water surface from the point where the stone hits. If you imagine being stationary over any point by which the wave passes, you can measure how many waves pass that point per second—this gives you the frequency. If you freeze the motion for an instant and measure the distance between the wave crests, you will get the wavelength. Measuring the speed at which the wave crests move gives the propagation speed, which is equal to the product of the frequency and wavelength.

Emissivity is the percent of the incoming radiative energy that leaves a surface. A highly polished surface, such as a mirror, is highly reflective, not absorbing any of the incoming energy. A perfect mirror is fully reflective; it has zero absorptivity and emissivity. For example, to improve the efficiency of in-floor radiant heat systems, it is recommended to have unpolished floor surfaces due to their higher emissivity and lower reflectivity [40].

Depending on the size of a surface and how it is situated with respect to other surfaces, its radiant energy is distributed to the external entities (surfaces). An object that emits whatever energy it receives is known as a *black body* and has an emissivity of one. For this body, the emission and absorption of light are equivalent through Kirchhoff's law, which describes how the radiative energy is emitted as a function of the wavelength.

A black body radiates energy in all directions in equal fashion, so the radiation intensity is both independent of the direction (*diffuse*) and wavelength (*gray*). When modeling the radiation mode of heat transfer, one can think in terms of the surfaces and the media. Radiative energy can be emitted by a surface or medium. It can also be absorbed by these. In a model, any component can be designated as opaque (and thus not able to transmit radiation).

Surfaces can absorb or emit. The absorption is a function of the wavelength of the radiation and the incident angle. Emission can be *diffuse* (multidirectional) or *specular* (when the incoming radiation is reflected without scattering). The medium between the surfaces can be: (a) completely transmitting to radiation (like air or a vacuum), (b) partially absorbing and retransmitting to radiation, (c) absorbing and scattering to radiation, and (d) opaque. To model the radiation, one needs to calculate the radiative energy reaching the surface as well as leaving the surface. The simplest case is that of a surface facing the ambient (surroundings). If the ambient is cooler than the surface, the surface will lose heat, and vice versa.

Things get more complicated when there are surfaces that can *see* one another. Consider, for example, a hollow brick-shaped block. Figure 2.4 illustrates such a block by a 2D rectangle. A block will have six surfaces. Each surface can either face the interior or exterior surfaces. Four surfaces, marked by letters from "*a*" to "*d*," are identified in the figure. An external point heat source (like the Sun) is also shown.

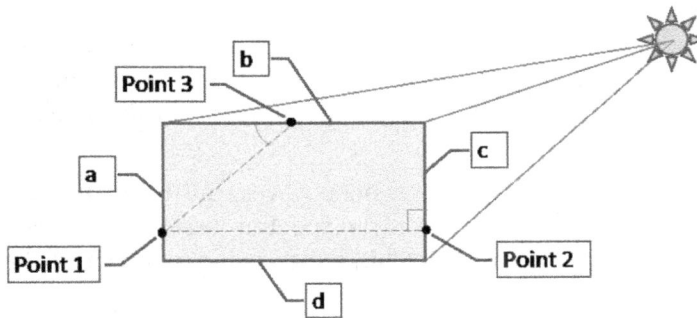

FIGURE 2.4. An illustration showing the concept of the view factor in radiation heat transfer.

Some surfaces will be visible to this radiation source ("*b*" and "*c*" exterior), and some will not (both sides of "*a*" and "*d*," the interior of "*b*" and "*c*"). Also, each point on a surface can see some surfaces but not others. The exterior "*a*," "*b*," "*c*," and "*d*" surfaces in this example cannot see each other. They are on a convex surface where this is always the case, like the exterior of a sphere. The interiors of these four surfaces can all see each other, which is the case for concave surfaces (like the interior of an ellipse). For more complex shapes and a greater number of objects, there will also be shadowing to account for.

If a surface is visible from any point on another surface (such as *Point 1* on the interior of "*a*" in the figure), the radiation it receives from the surfaces that it *sees* will also depend on the angle between the line from this point to the point on that surface. This is accounted for in a *view factor* calculation. The view factor is the percent of the energy radiated (sent out), which is received by the other object. Thus, *Point 1* sees *less* of the infinitesimal surface patch at *Point 3* than at *Point 2*, since the incident angle is smaller for *Point 3*. It is for the same reason that there are seasons on the earth: the tilted earth axis means each hemisphere will see larger or smaller incidence angles between the earth's surface and the sun during the year.

For the molecules with large mean free paths, like what is seen in space as the sun's electromagnetic radiation reaches the earth's atmosphere, passes through its layers, and is absorbed at its surface, radiation is the main mode of heat transfer. In this method, the electromagnetic waves do not require a medium to be transferred; all they need is matter with a temperature greater than the absolute zero. The charged particles of this matter (i.e., protons and electrons), moving in random motions (both speed and direction), interact with each other, generate electric and magnetic fields due to the charge acceleration and dipole oscillation, respectively, which are coupled with one other, creating electromagnetic fields, and photons are emitted as the result. This generated electromagnetic energy is released to the photon's environment and does not require a medium to propagate; therefore, it is the dominant heat transfer mechanism in the vacuum.

The unobstructed electromagnetic waves can travel far and when they do, they may be absorbed given the spectral-directional characteristics of the obstructing medium. If this hindrance is independent of the radiation wavelength, the medium is *gray* and if it is independent of the radiation direction, it is a *diffusive* medium. To calculate how much heat is lost or gained by the surface, one needs to integrate over all the other visible surfaces. This means the larger the surface area of a receiving body is, the higher percentage it receives of the total energy sent from the emitting body.

Reflectivity, *absorptivity*, and *transmissivity* of matter, also known as *optical properties*, are among the spectral-directional characteristics, defining the percent of the electromagnetic energy reflected, absorbed, or transmitted through the medium. The summation of these optical properties is one hundred percent of the total energy that reaches the object, Equation (18). If these properties are integrated over the wavelength or direction, the spectral-directional property changes to *directional* (superscript, e.g., 'α') or *spectral* (subscript λ, e.g., α_λ) properties, associated with the diffuse or gray bodies, respectively. Note that a diffuse body emits the energy isotropically, independent of the direction, while a black body emits the energy isotropically for all wavelengths.

$$\alpha'_\lambda + \rho'_\lambda + \tau'_\lambda = 1 \qquad (18)$$

Emissivity is the energy that is emitted from the surface; based on Kirchhoff's reciprocity law, it is the same as the absorptivity of the matter ($\alpha_\lambda' = \varepsilon_\lambda'$). This law states that if object 1 sees object 2, then object 2 sees object 1 (Figure 2.5). The portion of the energy that is emitted by object 1 and received by object 2 is called the *view factor* (also known as the *shape factor* or *configuration factor*, Equation (19). Note that θ_1 and θ_2 are the angles that the surface normal unit vectors (n_1, n_2) make with the line (S) connecting the two surfaces (dA_1, dA_2)—Figure 2.5.

Given that the total energy emitted from object 1 is received by the surrounding matter, which is visible to object 1 (e.g., objects 2 and 3), the total ratio of these energies is one—Equation (20). If object 1 is not able to see itself, its view factor is zero—Equation (21), where i is the surface identifier.

$$F_{1 \to 2} = \frac{1}{A_1} \int\limits_{A_1} \int\limits_{A_2} \frac{\cos\theta_1 \cos\theta_2}{\pi S^2} dA_1 dA_2 \tag{19}$$

$$\sum_{j=1}^{n} F_{s_i \to s_j} = 1 \tag{20}$$

$\therefore$ e.g., $\qquad\qquad F_{1 \to 1} + F_{1 \to 2} + F_{1 \to 3} = 1$

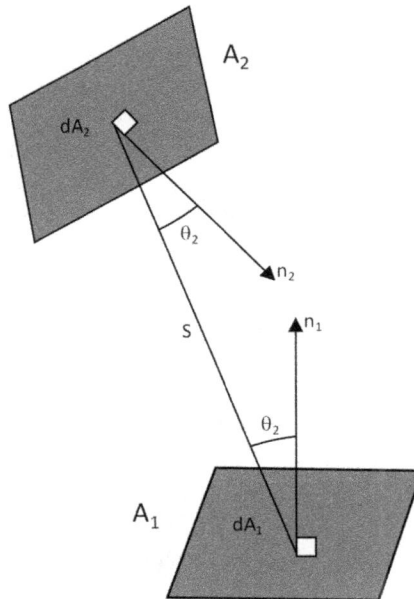

FIGURE 2.5. View factor between the two surfaces seeing one another.

$$\sum_{i=1}^{n} F_{S_i \to S_i} = 0 \tag{21}$$

∴ e.g.,

$$F_{1 \to 1} = 0$$

The thermal radiation reciprocity concept again shows that the ratio of the portions of the energy emitted from object 1 to 2, and vice versa, are related to their surface areas—Equations (22) and (19).

$$A_1 F_{1 \to 2} = A_2 F_{2 \to 1} \tag{22}$$

Highly reflective surfaces, such as mirrors, have a very low emissivity or absorptivity; they are not capable of either absorbing or emitting the radiative energy. Matter can only emit what it can absorb and that is why these values cannot exceed one hundred percent.

Human skin is an almost perfect emitter. Assuming a total body surface area of 2 m^2, and a temperature of about 37 °C, it can be estimated to emit radiated energy at a rate of about 212 W. Additional heat loss is due to convection to the environment, and can be estimated as 116 W. This is obtained by assuming the natural heat transfer convection coefficient of 3.4 W/m^2K, with the environment at 20 °C. This adds up to a total heat loss from human skin of about 327 W. This number will be affected by the clothing characteristics (e.g., surface, color, and material). If the body continues transferring energy to its environment, its temperature will keep decreasing exponentially until it equalizes with that of its environment. Therefore, the human sensible and latent heats (*metabolic heat*) are responsible for compensating for the heat loss. Average metabolic energy (*E*) depends on factors such as gender, age (*Age*), weight (*W*), level of activity, time spent in order to complete the activity (*t*), and heart rate (*HR*), which can vary from 120 W (28.66 cal/s, e.g., working at the computer) to about 430 W (102.7 cal/s, e.g., athletic exercises) [41,42,43,44,45,46,47].

Daily basal metabolic rates (*BMR* in J) for females (BMR_{Female}) and males (BMR_{Male}) based on their age (*Age* in years), height (*H* in cm), and weight (*W* in kg) based on the Harris–Benedict BMR formula are given by Equations (23) and (24) [17]. These relations are basal metabolic rates at rest. The metabolic rate, which is the amount of energy required for the body to function and perform activities such as sitting, breathing, and sleeping, decreases with age, while it increases with weight (Figure 2.6). Female and male metabolic energy (*BMR*) burns based on their level of activity and time spent on the activity are given by Equations (25) and (26).

The level of activity is accounted for in these equations via the heart rate (*HR* in beats per minute). Increasing the heart rate raises the energy use per unit time, with greater increase for males. In these relations, *t* is the time in minutes and *E* is the energy burned in Joules. To obtain the results in calories, the energy values are divided by 4.18. The energy burn rate increases with increasing body weight for males and decreases for females (Figure 2.7) [49].

$$BMR_{Female} = -4.676\,Age + 9.563\,W + 1.850\,H + 655 \tag{23}$$

$$BMR_{Male} = -6.755\,Age + 13.75\,W + 5.003\,H + 66.5 \tag{24}$$

$$E_{Female} = (0.074\,Age - 0.1263\,W + 0.4472\,HR - 20.4022)t \tag{25}$$

$$E_{Male} = (0.2017\,Age + 0.1988\,W + 0.6309\,HR - 55.0969)t \tag{26}$$

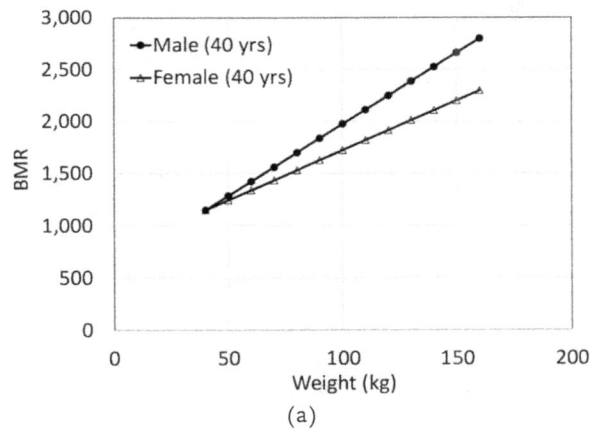

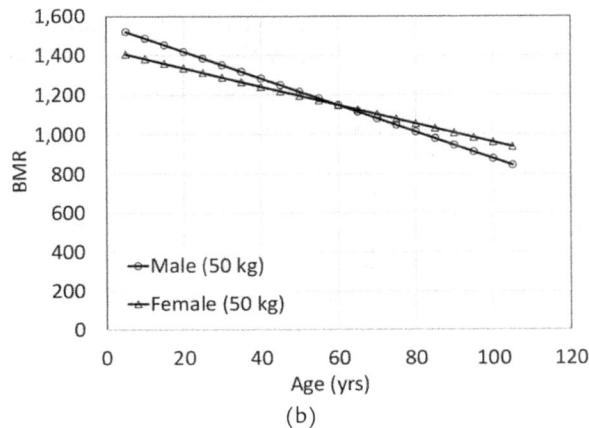

FIGURE 2.6. BMR for a human adult (*H* = 160 cm) as a function of: (a) Weight, (b) Age.

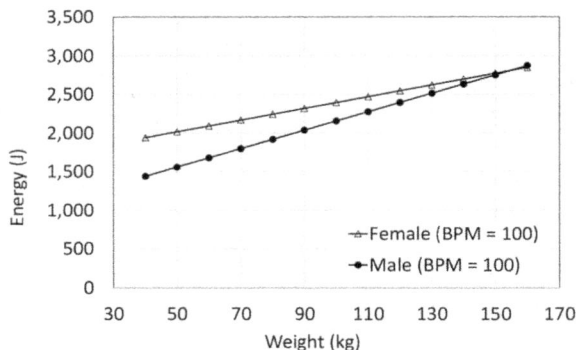

(a)

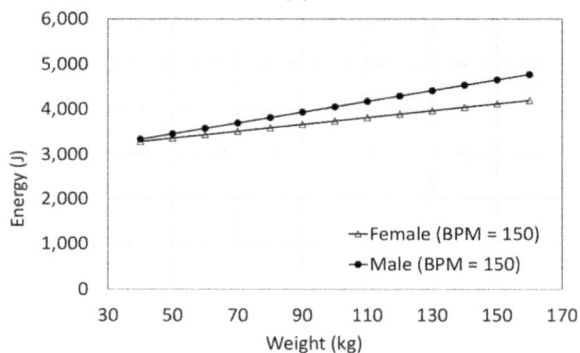

(b)

FIGURE 2.7. Hourly energy burned for a human adult as a function of weight
(*Age* = 40 years, *H* = 160 cm): (a) 100 BPM, (b) 150 BPM.

The refractive index is a material optical property that is defined as the ratio of the speed of light in vacuum to that in the material. In general, the denser the environment is, the higher this index is. This index determines the degree to which the light rays bend (i.e., refract) as they pass from one medium to another. Snell's law uses the indices of refraction for the two media to describe the relationship between the angles of incidence and refraction. The refractive index also depends on the light wavelength and changes by a few percent over the visible spectrum, becoming smaller with increasing wavelength. This dependence leads to display of a multicolored light spectrum when a white light passes through a transparent prism and the same mechanism also creates rainbows when sunlight refracts through water droplets.

The whole spectrum of electromagnetic radiation ranges from the shortest wavelengths of ionizing radiation, and continues to visible, microwaves, and radio waves. The near ultraviolet and near, medium, and far infrared waves form visible light. Gamma rays, hard and soft X-rays, and extreme ultraviolet waves form the ionizing radiation with energy levels as high as one million electro-volts compared to the waves in the visible range (one electro-volt). Longer wave radiation has much lower energy levels—as low as 10^{-15} electro-volts.

Due to its dual wave-particle nature, electromagnetic radiation can be characterized both in terms of wavelength/frequency and in terms of the photon energy. Frequency (f in 1/s or Hz) is inversely related to wavelength by $f = c/\lambda$, where c is the speed of light (2.99,792,458 m/s), and l is wavelength (m). Photon energy is proportional to the frequency and is expressed by Planck's law ($f = E/h$), where E is the photon energy in J, and h is Planck's constant (6.626,068,960 × 10^{-34} Js = 4.135,667,330×10^{-15} eVs).

Since an object's optical properties may vary as a function of the radiation wavelength, one cannot determine if the material is absorptive or emitting simply by its visual appearance. Detailed data on the materials' spectral-directional properties are required. For example, although a white surface has a low emissivity and absorptivity in the visible range, the opposite is true in the infrared range. A black surface is highly absorptive and emitting in both ranges. Glass windows pass the light in the visible and near infrared ranges but do not transmit well in the mid and far-infrared ranges. Therefore, we are able to see through the glass by letting in the visible light; however, glass does not let the light emitted from objects at room temperature escape, maintaining the temperature inside the greenhouse and gradually increasing it (the *greenhouse effect*).

The effect of thermal radiation is not limited to increasing matter's temperature; it can also apply a very small force to an object and therefore create momentum that may change trajectory of a spacecraft. This effect caused a problem for the Pioneer 10 and 11 spacecraft. They were launched in 1972 and 1973, respectively, to investigate Jupiter and Saturn's solar wind, passing through the asteroid belt. The two spacecraft were among the first five human made objects to reach the escape velocity (the velocity required for an object to escape the gravity of the Earth). For both spacecraft, the asymmetric thermal radiation due to the exposure of one side to the sun generated minute forces on the surfaces exposed and momentum as a result, affecting the spacecraft's trajectory. It is believed that if the distance between the satellite and the planet (in case of Pioneer 10) was increased to three times radius of the planet, this drift would not have occurred. This deviation from the trajectory due to the thermal radiation is now also known as the *Pioneer anomaly*.

In addition to the color and surface roughness, the material properties such as degree of crystallinity and the molecular bonding method also affect the way the materials interact with light. An example is adding pigmentations, fillers (glass fibers), and other additives (e.g., carbon black) to thermoplastic materials that are to be welded using laser transmission welding (LTW) process. In this process, a laser light passes through the first part to be joined and is absorbed by the second part, thus generating heat at the interface that melts the polymer and forms a joint. Thermoplastics have very low absorption in the near-infrared part of the spectrum used by the typical joining lasers. Therefore, for joining to occur, the natural polymer needs to be modified using an absorbing additive, such as the most used carbon black.

If the matter is perfectly absorbing at all wavelengths and directions of light incidence, it is called a *black body* $(\alpha'_\lambda = \varepsilon'_\lambda = 1)$. For matter to act as a black body, it must be at thermal equilibrium, meaning that there is no variation in temperature through the matter and therefore no heat transfer or thermal energy flow exists. This condition follows the zeroth law of thermodynamics, meaning that the temperature within the matter does not change spatially or temporally. It is to be noted that in a system that is thermodynamically in equilibrium, mass transfer is also negligible in addition to the energy transfer in the form of heat and work. However, there are states of equilibrium in the matter, where permeable or non-permeable portions of it undergo equilibrium processes and as a result, the system's total entropy increases. This is explained by the second law of thermodynamics and emphasizes the irreversibility of the system.

The spectral-thermal radiance of the body, $B(\lambda, T)$, is the total energy that leaves the surface of the body in the form of radiation per unit frequency, angle, and area. It follows Planck's relation, Equation (27), where T is the absolute temperature. For a body at any particular temperature above absolute zero, the spectral radiance has a distribution curve that is similar to a bell (Gaussian) curve, though it is not symmetrical. The peak value of this curve decreases as the absolute temperature decreases and the peak position of this curve shifts toward the longer wavelength or lower frequency.

$$B(\lambda,T) = \frac{2hc^2}{\lambda^5} \frac{1}{e^{\frac{hc}{\lambda K_B T}} - 1} \tag{27}$$

Planck's law describes the spectral density of the blackbody radiation as a function of temperature. The assumption is that the total radiation emitted from a black body equals the one that it receives, and it does not vary with the direction of the beam. To find the wavelength associated with the maximum temperature reached, the derivative of this relation with respect to wavelength is taken. This results in a relation that shows that temperature (T) and maximum wavelength (λ_{max}) are inversely related; this relation is also known as *Wien's Displacement Law*, Equation (28), where b is a proportionality constant, also known as *Wien's displacement constant* (2.897,772,917 × 10⁻³ mK). Since the frequency is the inverse of the wavelength, the peak frequency is the inverse of the peak wavelength ($f = c/\lambda$) and therefore, Equation (28) may be rewritten in the form of Equation (29) and so the absolute temperature of a radiating body is linearly dependent on the frequency at which it emits the thermal radiation. In other words, from Wien's displacement law, using the wavelength at the peak, the temperature can be inferred. Note that for relatively low temperatures, the radiation is emitted at long (infrared) wavelengths and is therefore not visible to the human eye.

$$\lambda_{max} = b/T \qquad (28)$$

$$f_{max} = (c/b)T \qquad (29)$$

A plot of the temperature versus the distance from the Sun is presented in Figure 2.8 for different planets. For example, Venus has a mean surface temperature of about 464 °C, while Mercury's mean surface temperature is about 167 °C. For comparison, the mean surface temperature of the Earth

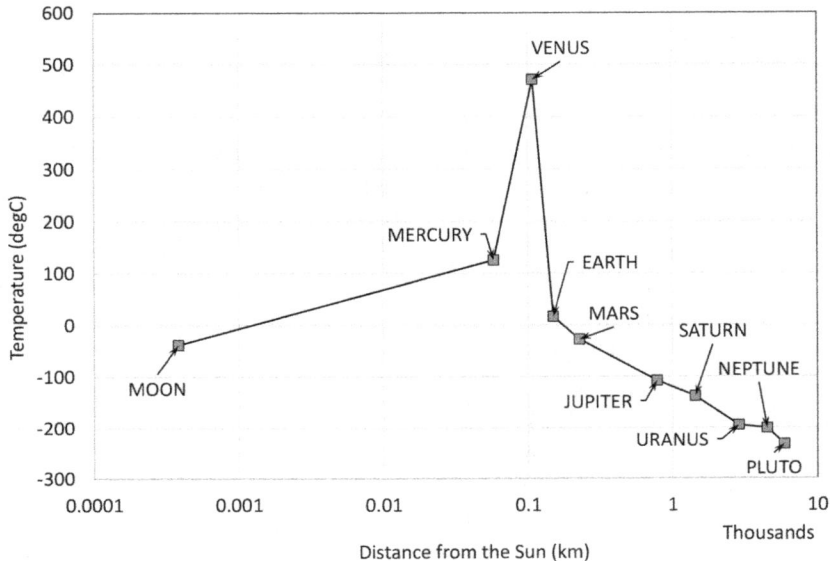

FIGURE 2.8. Mean surface temperature versus the distance from the Sun.

is about 15 °C. One interesting observation is that while Mercury is closer to the Sun than Venus, its mean surface temperature is much lower. This is due to the presence of a thick atmosphere on Venus; Mercury has almost no atmosphere. It also has the greatest daily variation of temperature in the solar system: between −180 °C at night and 430 °C during the day [50,51].

A cavity can behave as a black body (Figure 2.9). A pinhole cavity functions as a light trap; as the light passes through its opening, it hits the opposite surface, and then it continues bouncing back within this cavity until its energy is fully absorbed. The walls of the cavity are assumed to be opaque to the incoming radiation beam, meaning that it will not allow any light to escape.

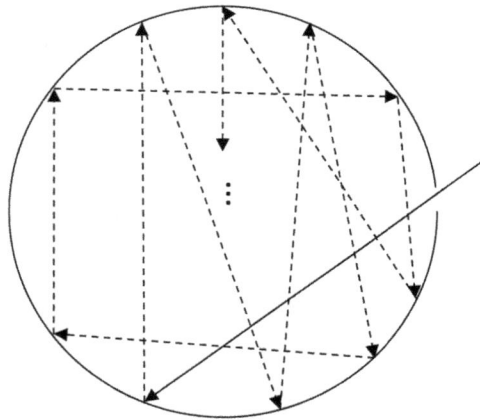

FIGURE 2.9. Spectral radiance inside a cavity.

Stephan-Boltzmann's law for thermal radiation is expressed by Equation (30), which shows that the energy transfer is proportional to the difference between the surface temperature of the emitting object (T_s in K) raised to the fourth power and that of the environment or receiving body ($T_{s'}$ in K) also raised to the fourth power. The radiation is also proportional to the area of the emitting body (A in m²), emissivity (ε, dimensionless) of the emitting body, and Stephan-Boltzmann's constant ($\sigma = 5.670,374,419 \times 10^{-8}$ W/m²K⁴). The emissivity property is a proportionality constant that describes how good a body is at emitting thermal radiation as determined by its optical and surface properties, and can vary from 0 to 1. Equation (30) presents the radiative heat (Q_{rad} in W) that flows from the emitting body (s) to the receiving body (s'). Note that this energy also can be expressed in terms of heat flux (q in W/m²). The view factor ($F_{s \to s'}$) is to be taken into consideration if the radiation from the emitting body partially reaches the receiving body. Note that the reciprocity relation that was presented earlier is applicable in this scenario. $F_{s \to s'}$ is the percent of the energy that leaves the surface of the emitting body (s) and reaches the surface of the receiving object (s'). The total spectrum (all wavelengths and directions) of emitted energy, integrated over the entire spectrum, expressed by Equation (30) represents the black body radiation, where $\alpha = \epsilon = 1$.

$$Q_{\text{rad}\,s \to s'} = F_{s \to s'} \sigma \epsilon A \left(T_s^4 - T_{s'}^4 \right) \tag{30}$$

Note that it is also possible to discretize the difference of the temperatures of the emitting and receiving bodies raised to the fourth power to the binomial form of difference of the temperatures squared and continue this process until the temperature difference expression is obtained as a single term versus the rest of the parameters—Equation (31). The multiplier of the temperature difference in this scenario is the equivalent of the convection heat transfer coefficient and is known as the *radiative convection heat transfer coefficient* (h_r)—Equation (32). Equation (31) then can be simplified to (33).

$$Q_{\text{rad}\,s \to s'} = F_{s \to s'} \sigma \epsilon A \left(T_s^2 + T_{s'}^2 \right) \left(T_s + T_{s'} \right) \left(T_s - T_{s'} \right) \tag{31}$$

$$h_r = F_{s \to s'} \sigma \epsilon A \left(T_s^2 + T_{s'}^2 \right) \left(T_s + T_{s'} \right) \tag{32}$$

$$Q_{\text{rad}\,s \to s'} = h_r (T_s - T_{s'}) \tag{33}$$

The heat flow may be simulated as an electric current, with thermal resistance defined by Equation (34) and difference of the fourth powers of temperature playing the driving force—Equation (35).

$$R_{\text{th rad}} = \frac{1 - \epsilon_1}{\sigma A_1 \epsilon_1} + \frac{1}{\sigma F_{s \to s'} \epsilon A} + \frac{1 - \epsilon_2}{\sigma A_2 \epsilon_2} \tag{34}$$

$$Q_{\text{rad}\,s \to s'} = \frac{\left(T_s^4 - T_{s'}^4 \right)}{\dfrac{1 - \epsilon_1}{\sigma A_1 \epsilon_1} + \dfrac{1}{\sigma F_{s \to s'} \epsilon A} + \dfrac{1 - \epsilon_2}{\sigma A_2 \epsilon_2}} \tag{35}$$

2.3.4 Thermal Management

Electro-mechanical and biological systems almost always generate heat during their operation. The heat may be generated through mechanical means such as friction between the subcomponents, electrical currents, electromagnetic fields, biological functions such as sensible heat, and localized heat sources. In most cases, the heat generated needs to be dissipated as effectively and efficiently as possible. The discipline concerned with dissipating heat from working systems is called *thermal management*. Thermal management may be accomplished using any of the following methods: (a) adding extended surfaces, also known as *fins*; (b) introducing cooling channels; (c) implementing additional mechanical systems such as fans and heat pipes; (d) interfacing the parts so that the contact areas are increased for efficient heat transfer by using thermal patty, oil, or thermal tape at the adjacent surfaces; (e) changing the object's geometry, such as its thickness; and (f) varying material properties.

Other thermal management approaches take advantage of the combination of the above methods. These may include, for example, using a cold plate at the interface of the heated objects, cooling flow, electrostatic fluid acceleration that creates flow without use of moving parts, and synthetic jet air cooling that involves ejection and suction of the flow across an opening resulting in zero flow mass balance. Some highly advanced techniques involve using phase change materials, capable of storing and releasing large amounts of heat when phase change occurs, and synthetic diamond cooling sinks for their high thermal and low electrical conductivities in applications such as high-power laser diodes, transistors, and semi-conductor technologies, where the use of thermally conductive materials such as copper can result in substantial variations in the electrical or magnetic fields and as the result reduction of system efficiency (e.g., linear induction motors—LIM).

Convection cooling by means of an oil pump in an aircraft engine is an example of thermal management. The oil dissipates the generated heat away from the heated parts such as cylinder head, in addition to acting as cleaner, lubricant, and sealant. In this application, there are two factors to consider carefully to achieve efficient heat transfer. One is the design of the system (e.g., pump location inside the cooling circuit) and the type of the cooling agent (i.e., oil). Oil has higher vaporization temperature with respect to water (above 100 °C) at atmospheric pressure and therefore is a better heat sink in absorbing the heat and as a result a more efficient coolant. This means that oil can be used for thermal management applications where the maximum temperature exceeds that of the boiling water.

Thermal management techniques are employed to ensure the ice does not form or is separated from the aircraft wing's leading edge and therefore does not let it progress by agglomerating the ice crystals to the wing mid-sections. One of these methods is to incorporate resistive heating elements into the leading edge of the horizontal stabilizers or other control surfaces.

To ensure the temperature is maintained at the desired level, a temperature sensor that measures the temperature along with an overheat sensor, which gets activated for temperatures about 154 °C (310 °F), are designed into the temperature control system [52]. This method is also known as *resistive de-icing* method. Another thermal management method to prevent ice from forming is taking a small fraction of the hot air generated within the compressor of turbofan or turboprop aircraft engines (known as the *bleed air*) and directing it toward the control surfaces.

Fire management is achieved by variety of methods that focus on different corners of the combustion triangle by: (a) removing the fuel source; (b) suppressing the flame so that the airflow is eliminated; (c) cooling the fire so that the combination of the fuel-oxygen cannot reach the *flash point* required for its perpetuation; or (d) adding a *fire retardant* to the mixture so that the chemical reaction is delayed. Use of materials such as intumescent paint or tape in hard-to-reach areas or where space is limited is among the fire management methods that delay spreading the fire to the adjoining areas.

Another method of thermal management is in radiative applications. This is mainly achieved by selecting materials with suitable emissivities. The higher the emissivity is, the more broadband energy is absorbed by the object. There are cases in which the emissivity of the surface is large while the absorptivity is small. An example is white paint, with large emissivity of about 0.93 and a low absorptivity of about 0.16. Therefore, the roofs of some houses in warm-arid regions are painted white—this provides effective thermal management. In the same way, the interior of a white car should be cooler than a black one if both are left parked outdoors on a sunny summer day.

2.4 Governing Equations

Experimental correlations have been the basis for many thermo-fluid formulae. In this approach, tests are carried out to investigate the influence of change of a single parameter or number of them on a control volume or system. The parameters can either be thermophysical properties of the materials such as heat capacity and thermal conductivity or temperature-induced ones such as stress, creep and oxidation life, magnetic fields, and phase change. In a complex system such as a heat exchanger, water temperature, pressure, and velocity are the determining factors for heat transfer mechanisms, and its efficiency as well as flow regimes.

Theoretical relations derived from experiments show an approximate relationship between two or more parameters; for example, they identify that these parameters are directly or inversely related. The correlating factors can be (a) material-dependent, such as the conductivity in Fourier's law, or (b) process-dependent, such as the convection heat transfer coefficient in Newton's law of cooling. The correlated value obtained from the former case defines a thermophysical property of the material while the experimental setup or processes influence that of the latter.

In some cases, the correlation value is a constant parameter, which may be of general significance in physics. An example of such a parameter is Stefan-Boltzmann constant that relates radiated electromagnetic energy to the object's temperature in Stefan-Boltzmann's law of thermal radiation. There are cases in which no exact mathematical relations can be achieved by fitting an experimental relationship into a theory; this is the definition for an empirical relationship. Examples include the release of magnetic energy during a solar flare, heat transfer in external flows, and shear stress in non-Newtonian fluids. In these cases, different equations may be applicable to different conditions. For example, the *Reynolds number* is employed as a criterium when setting up laminar and turbulent flow models, the latter flow type being capable of addressing flow disruptions and eddies.

Figure 2.10 shows schematically the general form of energy conservation diagram for the Cartesian coordinate system. The figure shows an infinitesimal cube with dimensions dx, dy, and dz. Heat flux q is shown entering or leaving the cube's faces. For example, along the x-coordinate, the spatial variation of heat flux is represented by the gradient dq/dx. It includes all modes of heat transfer. The radiation and convection terms shown are applied in the form of boundary conditions. As in other disciplines, such as the balance of forces in solid mechanics, the energy conservation law can be expressed separately along each x- y- and z-coordinates—Equation (36). A matrix can then be created that is a linear combination of the energy conservation in three dimensions in addition to the time component. T, q, k, dx, dy, and dz, and t are the temperature, (K), heat flux (W/m²), thermal conductivity (W/mK), infinitesimal distances along the x-, y-, and z-coordinates, and time (s), respectively. The heat flux defined by Equation (36) is proportional to the temperature gradient (dT/dx in K/m), where the conductivity (k in W/mK) is the proportionality constant. Equation (5), presented in section 2.2.3, shows that the rate of energy storage is a function of the variation of the internal energy ($\dot{E}_{internal}$ in W) over time and energy generated inside the material due to any heat source or sink ($\dot{E}_{generated}$ in W)—m is mass (kg), and c_p is specific heat capacity (J/kgK).

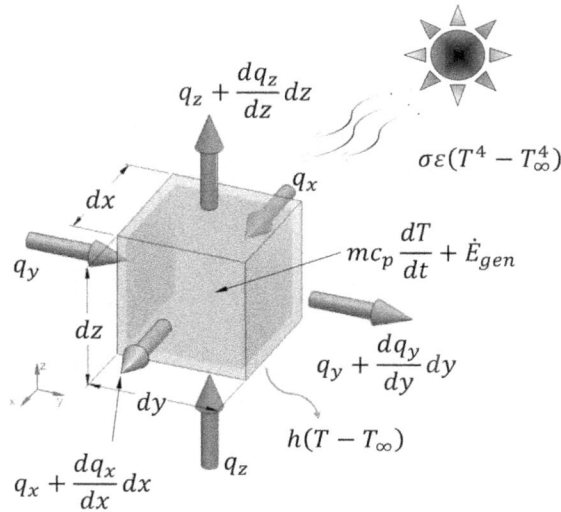

FIGURE 2.10. General form of energy conservation diagram in the Cartesian coordinate system.

The outgoing energy is the incoming energy plus the variations of the energy along the length, where the energy is transported, expressed in the form of the derivative of the energy in the direction of the energy transportation. This energy balance applies along each of the three coordinates (x, y, and z). This spatial variation is shown by the wave equation applicable to the conduction heat transfer mode, Equation (36). Note the terms dx, dy, and dz are the block dimensions along the three said coordinates.

$$q(x+dx,y+dy,z+dz) = q(x,y,z) + \frac{dq(x,y,z)}{(dx,dy,dz)}(dx,dy,dz) \tag{36}$$

For a one-dimensional coordinate system, assuming the thermal variations occur along the x-coordinate, the wave equation, representing the heat flux rate can be simplified to Equation (37). Similarly, along the y-and z-coordinate systems, the heat flux rate can be expressed in terms of Equations (38) and (39), respectively. Equations (40), (41), and (42) are the equivalents of the said equations at the coordinate (dx,dy,dz) from the original coordinate $(0,0,0)$.

$$q_y = -k_x\left(\frac{dT}{dx}\right) \tag{37}$$

$$q_y = -k_y\left(\frac{dT}{dy}\right) \tag{38}$$

$$q_z = -k_z \left(\frac{dT}{dz} \right) \tag{39}$$

$$q(x+dx) = q(x) + \frac{dq(x)}{dx} dx = -k_x \left(\frac{dT}{dx} \right) + \frac{d}{dx} \left(-k_x \frac{dT}{dx} \right) dx \tag{40}$$

$$q(y+dy) = q(y) + \frac{dq(y)}{dy} dy = -k_y \left(\frac{dT}{dy} \right) + \frac{d}{dy} \left(-k_y \frac{dT}{dy} \right) dy \tag{41}$$

$$q(z+dz) = q(z) + \frac{dq(z)}{dz} dz = -k_z \left(\frac{dT}{dz} \right) + \frac{d}{dz} \left(-k_z \frac{dT}{dz} \right) dz \tag{42}$$

Substituting aforementioned relations into the energy balance (Figure 2.10) in Equations (6) and (7) results in Equation (43), which after simplification results in the heat diffusion equation presented by Equation (44), demonstrating that spatial and temporal temperature profiles are related to the change of internal energy and heat generation within the material in the Cartesian coordinate system. $\dot{q}$ is the volumetric heat generation (W/m^3). Note that (A_x, A_y, A_z) are areas perpendicular to the heat flow direction along the x-, y-, and z-coordinates $(dydz, dxdz, dxdy)$. dV is the volume of the block whose mass is $m = \rho dV = \rho dxdydz$.

$$A_x \left(\frac{dq_x}{dx} \right) dx + A_y \left(\frac{dq_y}{dy} \right) dy + A_z \left(\frac{dq_z}{dz} \right) dz + \dot{q}dV = m c_p \frac{dT(x,y,z)}{dt} \tag{43}$$

$$\frac{d}{dx} \left(k_x \frac{dT}{dx} \right) + \frac{d}{dy} \left(k_y \frac{dT}{dy} \right) + \frac{d}{dz} \left(k_z \frac{dT}{dz} \right) + \dot{q} = \rho c_p \frac{dT}{dt} \tag{44}$$

To balance the energy in the cylindrical coordinate system (Figure 2.11), Equations (6) and (10) are combined, resulting in the heat diffusion—Equation (45). Similarly, for the spherical coordinate system (Figure 2.12), after balancing the energy and combining Equations (6) and (11), the heat diffusion equation is obtained—Equation (46).

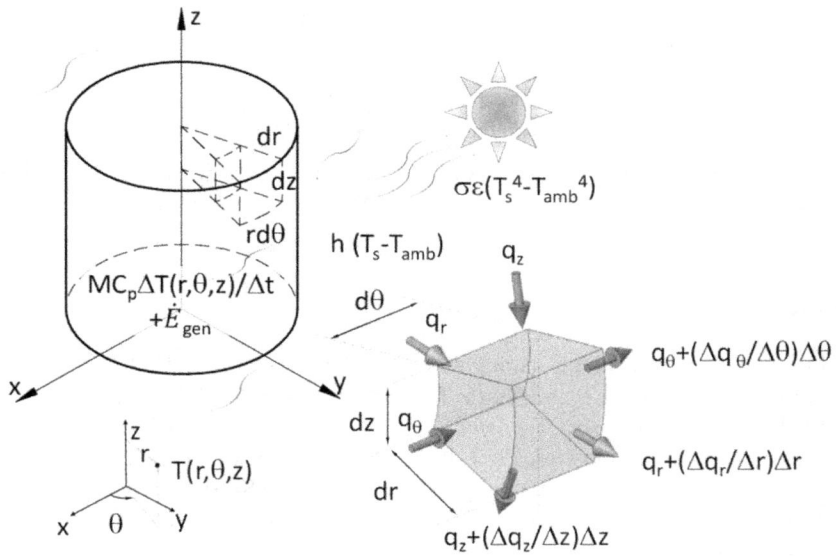

FIGURE 2.11. General form of energy conservation diagram in the cylindrical coordinate system

$$\frac{1}{r}\frac{\partial}{\partial r}\left(k_r r\frac{\partial T}{\partial r}\right) + \frac{1}{r^2}\frac{\partial}{\partial \theta}\left(k_\theta\frac{\partial T}{\partial \theta}\right) + \frac{\partial}{\partial z}\left(k_z\frac{\partial T}{\partial z}\right) + \dot{q}_{\text{gen}} = \rho C_p\frac{dT}{dt} \qquad (45)$$

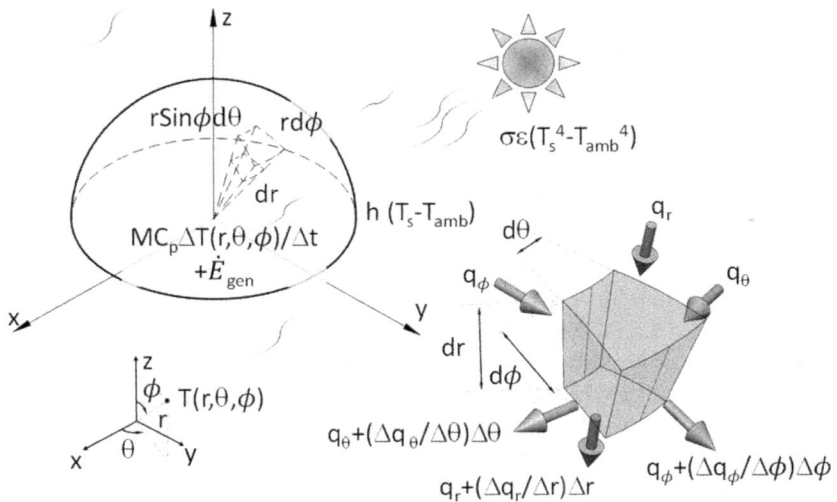

FIGURE 2.12. General form of energy conservation diagram in the spherical coordinate system

$$\frac{1}{r^2}\frac{\partial}{\partial r}\left(k_r r^2 \frac{\partial T}{\partial r}\right) + \frac{1}{r^2 \sin^2 \varphi}\frac{\partial}{\partial \theta}\left(k_\theta \frac{\partial T}{\partial \theta}\right)$$

$$+ \frac{1}{r^2 \sin^2 \varphi}\frac{\partial}{\partial \varphi}\left(k_\varphi \sin\varphi \frac{\partial T}{\partial \varphi}\right) + \dot{q}_{\text{gen}} = \rho\, C_p\, \frac{dT}{dt} \tag{46}$$

The upcoming chapters present both analytical, numerical, and finite element analysis approaches to solve the thermal partial differential equations (heat and wave problems) presented above.

End Notes

[32] https://web.stanford.edu/~cantwell/AA283_Course_Material/AA283_Course_Notes/AA283_Aircraft_and_Rocket_Propulsion_Ch_09_BJ_Cantwell.pdf

[33] Layla S. Mayboudi, *Heat Transfer and Thermal Modelling of Laser Transmission Welding of Thermoplastics*, PhD dissertation, Queen's University, 2009.

[34] https://www.ijser.org/researchpaper/THE-EFFECTS-OF-FLOORING-MATERIAL-ON-THERMAL-COMFORT-IN-A-COMPARATIVE-MANNER-Ceramic-tile-and-wood-flooring.pdf

[35] Myron L. Cohen, "Measurement of the Thermal Properties of Human Skin, A Review," *The Journal of Investigative Dermatology*, 69(3), pp. 333-338, 1977, https://www.jidonline.org/article/S0022-202X(15)45119-X/pdf

[36] https://itis.swiss/virtual-population/tissue-properties/database/thermal-conductivity/

[37] https://itis.swiss/virtual-population/tissue-properties/database/heat-capacity/

[38] https://itis.swiss/virtual-population/tissue-properties/database/density/

[39] https://en.wikipedia.org/wiki/Heisler_chart

[40] REF https://en.wikipedia.org/wiki/Underfloor_heating

[41] https://www.verywellfit.com/walking-calories-burned-by-miles-38871544

[42] http://fitnowtraining.com/2012/01/formula-for-calories-burned/

[43] https://en.wikipedia.org/wiki/Talk%3AHeart_rate

[44] http://taggedwiki.zubiaga.org/new_content/5044ca257bc71037f69dd996514f9b35

[45] https://www3.nd.edu/~nsl/Lectures/mphysics/Medical%20Physics/Part%20I.%20Physics%20of%20the%20Body/Chapter%202.%20Energy%20Household%20of%20the%20Body/2.3%20Heat%20losses%20of%20the%20body/Heat%20losses%20of%20the%20body.pdf

[46] http://www.calories-calculator.net/Calculator_Formulars.html

[47] https://www.engineeringtoolbox.com/metabolic-heat-persons-d_706.html

[48] J. Arthur Harris and Francis G. Benedict, "A Biometric Study of Human Basal Metabolism," *Proc Natl Acad Sci U S A*, Dec; 4(12), pp. 370–373, 1918.

[49] http://www.shapesense.com/fitness-exercise/calculators/heart-rate-based-calorie-burn-calculator.shtml

[50] https://solarsystem.nasa.gov/resources/681/solar-system-temperatures/

[51] https://coolcosmos.ipac.caltech.edu/ask/168-What-is-the-temperature-on-the-Moon-

[52] http://navyflightmanuals.tpub.com/P-861/Wing-Leading-Edge-Anti-Ice-System-78.htm

CHAPTER 3

FINITE ELEMENT ANALYSIS

The Finite Element Method (FEM) is a numerical technique in which the geometry is divided into a finite number of small pieces called *elements*. One advantage of defining such elements is that it enables the division of regions into smaller regions that more accurately represent the associated physics. Element size and shape may vary by region, depending on the physics they represent. Each element can have its own distinct properties. Elements are in contact with the adjacent elements.

Solving the FEM problems consists of solving m conservation equations (m is the number of nodes) when there is only one field variable. For each node, an equation is written for each field variable (e.g., temperature in heat transfer models), as a function of the data of the surrounding nodes, to find the value of the variable at the given node. The field can be defined in 1D, 2D, or 3D spaces. For example, if there are eight nodes with a single field variable (e.g., x displacement), eight equations are required; if there are eight nodes with two field variables (e.g., x and y displacements), sixteen equations are required (Figure 3.1).

Each node requires its own boundary and initial conditions. From algebra, you may recall that if you attempt to solve an equation with two independent variables, to obtain a unique solution, you need to solve it in combination with a second linearly independent equation that includes at least one of these two independent variables. Expanding the equation from 2 to m state variables requires m linearly independent equations. The same concept applies to solving the FEM equations.

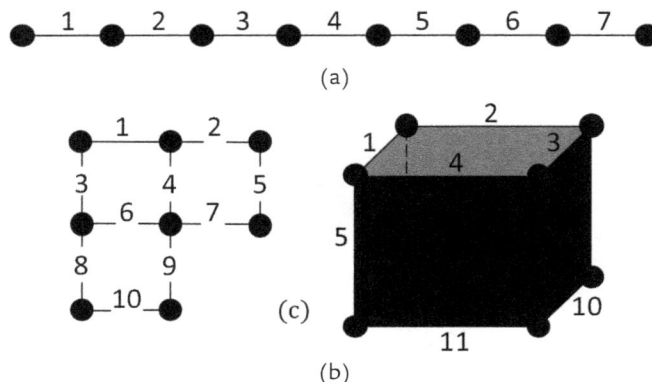

FIGURE 3.1. Element and nodes: (a) 1D, (b) 2D, and (c) 3D.

When analyzing thermo-fluid numerical models, either using the Finite Difference Method (FDM) or FEM, the conservation of energy principle must be applied to all elements or nodes. For nodes, the total energy of zero confirms that the balance of the energy at each node has been met, meaning that the total nodal incoming energy equals the total outgoing energy. Since an element occupies a line, area, or volume, as determined by its spatial dimension, the balance of energy should still be satisfied; however, in this case, the total elemental incoming energy should be equal to the total outgoing energy.

3.1   Geometry

The number of dimensions to be used in setting up the physics geometry depends on the model shape, boundary conditions, and computational resources (e.g., time and machine). The dimensions can start at zero for the simplest cases and progress to one (1D), two (2D), and three dimensions (3D) as complexity increases. The zero-dimension approach, also known as the *lumped capacity technique*, assumes that the temperature is spatially uniform throughout the model. In a 1D numerical analysis, one coordinate is required to identify the position of a point and heat is transferred in only one direction (e.g., the *x*-coordinate), meaning that heat transfer along the remaining coordinates, which form a plane, is ignored or heat is integrated over the remaining plane. One advantage of 1D numerical analyses is that they allow comparison with the simplified analytical solutions, thus enabling validation of the numerical analysis. In a 2D numerical analysis, two coordinates are needed to identify the position of a point and heat is transferred in two directions (e.g., the *x*- and *y*-coordinates). In other words, heat transfer transverse to the active *Work Plane* is ignored or the heat is integrated over the third dimension of the geometry. In a 3D numerical analysis, the most comprehensive approach, three coordinates are needed to represent the position of a point within the geometry (x-, y-, and z-coordinates) and heat is transferred in all three directions.

In cases where the geometry, material properties, and boundary conditions have axial symmetry, one can reduce the model by one dimension. Thus, for example, a cylinder has axial symmetry, and so this 3D shape can be represented by a 2D axisymmetric model without any loss of fidelity. A 2D shape, like a flat ring, can be replaced by an equivalent 1D axisymmetric model.

Symmetry about a plane can be also used to reduce the model size. For a geometrical shape, such *reflectional* symmetry can exist in 3D space about one, two, or three planes. Again, if the boundary conditions are also symmetrical, the model can be reduced to one-half, one-quarter, or one-eighth of the original size, respectively. A similar concept applies to 2D space, where reflectional symmetry can exist about one or two lines.

Another type of symmetry that can be taken advantage of is *rotational* symmetry. Here, the model can be represented by rotating a particular shape m time about an axis, giving an *m-fold* symmetry. Thus, a three-petal shamrock flower can be considered to have a threefold symmetry, while a four-leaf clover has a fourfold symmetry. Such models can then be reduced by modeling only the repeating element.

Some shapes will have multiple symmetries. You can decide which one will be most advantageous to use. For example, a hexagonal nut (ignoring threads) has reflectional symmetry about the three principal planes in addition to a sixfold rotational symmetry (Figure 3.2). Here you can reduce

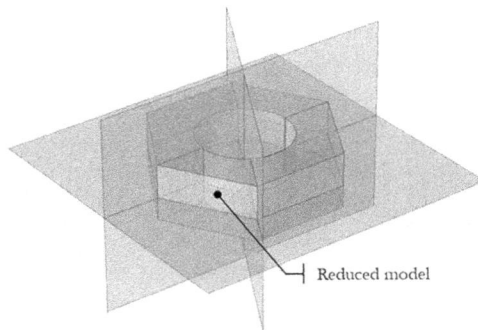

Reduced model

FIGURE 3.2. Hexagonal nut shape with symmetry planes.

the model to one-twelfth the size by utilizing the sixfold symmetry together with the reflectional symmetry about the horizontal plane, as shown in the figure. The extra up-front time spent to identify these geometry characteristics is effort well spent, since it forms the foundation of all subsequent steps, saving time and computational resources.

3.2 Material Properties

Material properties are important for the development of good quality models. One should try to obtain the most accurate material properties possible to assure accurate model predictions. However, obtaining accurate property values is sometimes challenging. Thus, an analyst should be aware of which properties have greater impact on the solution. The relative importance of different material properties may be determined by the thermo-fluid regime, mode of heat transfer, or analysis type. One can use sensitivity analysis methods to determine the effect of uncertainty in any property on the desired model output.

Material properties may vary in space (spatial), time (temporal), or environmental conditions (environmental). Nonconstant properties may introduce nonlinearities and non-homogeneities to the physics, making the problem more challenging. To describe temperature-dependent material properties, an FEA tool may employ analytical or piecewise functions. A property definition table contains a list of combinations of known temperature-property value pairs. For temperature values between those listed, interpolation functions are used; this can be a linear or a higher order function. If the temperatures in the solution exceed the limits of the range of temperatures for which the property values are given, one can choose to either extrapolate linearly or to keep the values constant, equal to the value of the nearest extreme point.

Let us review next the material property settings which may be required as inputs for a physics model set up in a typical FEM software tool. Usually, such tools have a built-in library of materials, which may be expandable with optional add-ons. Thus, if the material you need for your model is available within one of these sources, simply selecting it defines common inputs such as density, specific heat capacity, and thermal conductivity. If needed, any predefined properties may be changed, missing properties can be added, or a completely new material may be defined from scratch. For example, thermal conductivity may be defined as an isotropic, diagonal, symmetric, and anisotropic property.

When setting up a radiation problem, wavelength-dependent surface properties can be selected, which are either constant, depending on the solar and ambient conditions, or have multiple spectral bands and hence are wavelength-dependent. In most cases for transparent media, the refractive index needs to be defined. The *refractive index* of a medium is the ratio of the speed of light in a vacuum to that of the medium and is therefore always more than one. For water, this value is 1.33, meaning that light travels 33 percent faster in a vacuum than in water. For air, the refractive index is close to one. A transparent medium needs to be defined for a domain enclosed by diffuse surfaces that face each other.

The *surface-to-surface* radiation method is used to model cases where heat transfer by conduction, convection, and radiation are present in combination with radiation from internal or external surfaces. To model this phenomenon, one needs to define several settings. First, the method is selected as direct area integration, hemicube, or ray shooting. In the direct area integration method, the radiation between surfaces is calculated directly, not considering the obstructing (shadowing) surfaces, eliminating the surfaces that do not face each other. In the hemicube method, shadowing effects are included. The ray shooting method calculates the view factors given the wavelength and direction. To complete these settings, the radiation integration order, radiation resolution, tolerance, and maximum number of adaptations are set. Solution techniques include setting up the surface radiosity that can be linear, quadratic, cubic, quartic, or quantic. Surface *radiosity* or *radiant intensity* is the amount of radiation flux emitted from the surface as a function of the radiation wavelength.

3.3 Analysis Types

Any set of solution settings for a model may be referred to as a *study*. Analysis type selection specifies whether the study will be time-independent (i.e., *stationary*) or time-dependent (i.e., *transient*). A stationary study does not mean that the actual modeled physical system never changes over time, but that the analyst is interested in finding out what happens after the system has reached a steady-state condition. This is the state of the system at some theoretically infinite time. In a time-dependent study, the analyst is interested in the state of a system as time passes. If the study is run over a sufficiently long period of time, a steady-state condition may be reached, as well. For example, temperature may not change any further for a given fixed rate of heat input in a thermal problem. A steady-state condition may be reached only if the model boundary conditions are constant. Thus, if the model is exposed to a heat input that increases linearly over time, a steady-state temperature distribution cannot be reached.

Selecting the analysis type may also depend on the objective of the analysis. If an analyst is interested in studying the thermal response of a train underframe to fire to make sure it complies with the fire test code for rail transportation vehicles (ASTM E2061 or NFPA 130), she should study the time response for the first 15 min of the exposure time by performing a transient analysis to obtain the temperature history over that time.

For a heat exchanger, the analyst is interested in evaluating the spatial thermal performance, which can be done by plotting the temperature profile along a specific path (e.g., the liquid cooling channel), after the heat exchanger has been operating for some time and temperatures have stabilized. Therefore, a steady-state analysis is appropriate in this case.

3.4   Boundary and Initial Conditions

Just as material properties are important to accurately represent the modeled system, the boundary and initial conditions are important to correctly describe the conditions to which the modeled system is exposed. For heat transfer problems, setting the initial conditions means defining the temperature from which the solution starts (e.g., the room temperature of 20 °C can be a default starting point). Boundary conditions may be defined as insulated (a default condition that is automatically applied), temperature, heat flux, convective, or radiative. These boundary conditions are defined for the nodes (1D models); edges or points (2D models); and domains, surfaces, edges, or points (3D models).

3.5   Mesh Size and Time Step

FEM involves dividing the geometry into small elements and solving the energy and mass governing equations for each element and for the number of time steps or iterations required to reach the specified analysis time (for transient problems) or steady-state (for stationary problems). The number of iterations required for a solution to converge depends on the initial conditions to start the solution, and it may increase or decrease depending on the residuals. Residuals are the estimates of the difference between the calculated and desired values. The temporal and spatial steps can be controlled when setting up the analysis. Spatial step is related to the mesh size, which may vary within the geometry. The temporal (time) step is varied by the solver as the solution progresses.

The choice of the element size for meshing in FEM is similar to the choice for image resolution. If the image pixels are large relative to the detail in the picture that the analyst would like to see, they are not going to get a clear image of these details. Thus, a smaller pixel size is needed. However, if the analyst just wants to get an overall impression of an image, the analyst may increase the pixel size, reducing the total number of pixels (or elements in FEM). When meshing, unlike in images, you can vary your pixel (element) size throughout the model. For example, intense heating processes, such as laser welding, require fine detail resolution around the exposed regions, where temperature is changing rapidly in space and time, which can be achieved by local reduction of the element size and time step.

Assume that one decides on a mesh size. The next step is to make sure the element size produces converging results that are reasonable. One way to achieve this is to change the element size from larger to smaller values and review the variation of the numerical results (i.e., sensitivity analysis). When this variation is reduced below some appropriate lower limit, no further reduction in element size is required.

When a meshed model is solved, there are two types of errors: (a) round-off and (b) truncation. The former occurs when one decides to round the number to the closest value, using only the desired number of decimals. The latter case is when one decides to keep only a specific number of decimals. A simple example is to represent 14.557123 as 14.56, 14.55, or 14.557. The first two examples show the same number when it is either rounded off or truncated with two figures after the decimal; the third example could be either rounded off or truncated to the same number when three figures after the decimal are employed. There is a balance between the two errors, especially where they are accumulated due to the increased number of numerical equations, which is the case if the number of elements is increased. They usually show an opposite trend—decreasing versus the increasing for the roundoff and truncation errors. Time step and mesh sensitivity analyses provide good compromises. Due to the accumulation of the computational errors with the decreasing element size, after converging to the most accurate solution, the solution may begin diverging (i.e., deviating from the exact solution).

3.6   Solution Control and Convergence

Conservation laws should be satisfied when solving equations for heat transfer of any type. Dependent variables (e.g., temperature) are calculated using independent variables (e.g., thermal conductivity) as well as initial values. Note that the *independent* variables are the inputs to the models while *dependent* variables are the results given the independent variables. The equations are solved, and the residuals are obtained. The *residuals* are the actual sum difference from the zero-sum case. For example, for the energy conservation law to be valid, the total energy entering an element should equal the total energy leaving an element, including the energy storage and energy generated within the element. The vector summation of all the terms should be zero (error) and therefore any nonzero value is the residual error.

Zero residuals are not normally possible, and so a small nonzero tolerance value needs to be used so that the program uses that as the acceptable criteria and stops further iterations. For instance, if a user sets a 10^{-5} tolerance value for a solid heat transfer analysis problem, most probably they will be happy with the results if the solution is reached within reasonable time. However, if the user were to employ the same tolerance for a flow problem, there is a good chance that the analysis may require an excessive number of iterations, leading to very long solution time (convergence) or in some cases to not converging at all.

Figure 3.3 is an example of a convergence plot for a single-parameter time-dependent analysis. It shows the reciprocal of step size versus the time for a transient analysis using a logarithmic vertical scale. Thus, larger time

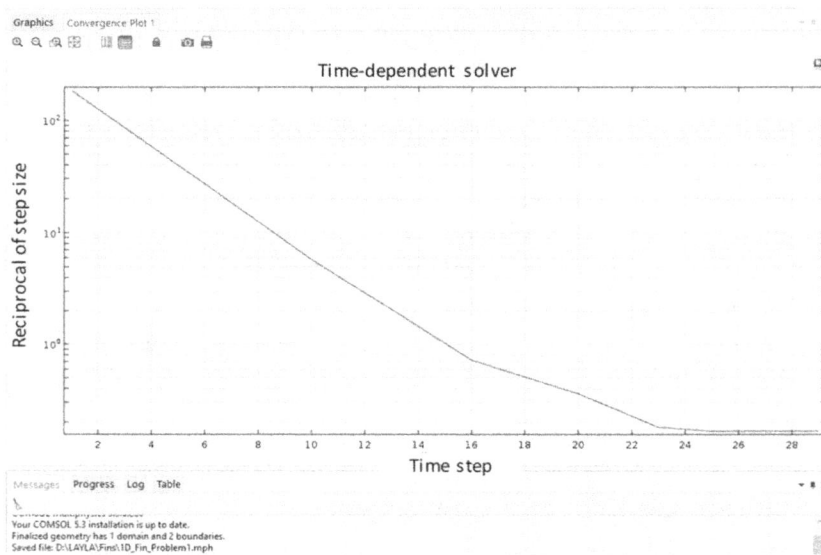

FIGURE 3.3. Example of a convergence plot for a 3D analysis for a heat transfer model.

steps are made as the solution progresses. At each solution step, the solver estimates the next time step required to obtain accurate solution. Although Figure 3.3 is generated in COMSOL Multiphysics, similar convergence plot data can be also obtained from Partial Differential Equation (PDE) solvers such as those implemented in the MATLAB environment.

CHAPTER 4

AN INTRODUCTION TO *MATLAB*

MathWorks offers two product families that can be employed for mathematical modeling: MATLAB and Simulink®. With Simulink, one can model a system consisting of multiple sub-systems and investigate the effect of the individual sub-systems on the overall performance. A system's behavior, such as the thermal response to the individual components after varying key variables, can be investigated in these models. To interact with the models, flow diagrams are created that are visual representations of the modeled system. Like other modeling tools, this approach leads to creation of smart prototypes, resulting in cost savings during the design process as well as during the rest of the product's lifecycle.

MATLAB can be employed to investigate a system or its sub-systems in detail. This is accomplished by introducing mathematical models, developing algorithms, providing numerical solutions to the models, analyzing the data using visual tools, and generating outputs such as diagrams and tables. The last step of output generation can be either carried out within MATLAB or by exporting data to a third-party tool such as tecplot or Microsoft Excel.

The MATLAB software package comprises the core application and a set of the optional toolboxes dedicated to a variety of specialized applications. The toolboxes most relevant to thermal modeling can be found within the *Math and Optimization* product family. Within this family, there are six toolboxes available: *Curve Fitting, Optimization, Global Optimization, Symbolic Math, Mapping,* and *Partial Differential Equation Toolboxes* [1,2].

The rest of this chapter will introduce the basic MATLAB environment; this should be of value to those who want to learn how to use the MATLAB tool in general. The chapter closes with several code examples that highlight best programming practices within the MATLAB environment. Techniques for using MATLAB specifically for thermal modeling will be introduced in Chapter 5.

4.1 Desktop

The MATLAB application desktop consists of several panels. By default, the *Current Folder* and *Details* panels are on the left, the *Editor* and the *Command Window* panels are in the center, and the *Workspace* is on the right side. A command toolstrip with several tabs is found on top. There are also two toolbars: *Quick Access* and *Current Folder*. Arrangement of panels can be customized and saved using the *Layout* tool found in the *Environment* group of the HOME toolstrip tab. The two toolbars and many other features can be customized via the *Preferences* found in the same group.

Looking at the MATLAB desktop by starting on the left side, the *Current Folder* panel is found that shows the list of files in the folder indicated by and selected via the *Current Folder* toolbar located just below the toolstrip (Figure 4.1). The *Details* panel below the *Current Folder* panel shows the relevant information for the selected file—variables for the *.mat* file and functions for the *.m* file.

The MATLAB *Command Window* (lower center) is where the user enters command lines, sees the text output of the running script, and responds to the text prompts. It has a powerful *Command History* feature that all the MATLAB users should learn to utilize for improved efficiency. It is accessed by pressing the *Up* arrow on the keyboard and that can be then navigated with *Up/Down* arrows. For example, a typical workflow may be to go back to one of the previously executed commands, bring it back to the cursor (>>), revise, and execute it again, without having to retype everything. The *Command Window* can be cleared of the commands printed on the screen by a *clc* command, returning the cursor to the top line within the window, but without clearing the command history.

Any commands executed via the command line can also be entered into the MATLAB script (*.m*) file. One can also select (with a shift key) any number of the commands from history and save them to a script file. The script files may be called later as the input to the MATLAB program, edited or run, generating data and diagrams. The *.m script files are normally edited using the built-in editor that opens the *Editor* panel automatically when a script file is opened (located in the top-middle in Figure 4.1). The script files are stored as plain ASCII text, and thus can be edited with any text editor (such as the Windows Notepad).

Finally, on the right, the *Workspace* panel shows contents of the current *Workspace*. The variables that have been imported into or created within MATLAB are stored in the *Workspace* memory. To view or edit these variables, one may either employ the *Workspace* panel or the *Command Window* (Figure 4.1). All the *Workspace* variables can be saved into a compressed *.m at file using the *save* command, and then restored by using the *load* command. The *Workspace* data files can also be loaded as input to the *.m files.

The *Workspace* variables can be deleted in bulk by the *clear* command, with command options allowing for selective deletion as well. The *clear*

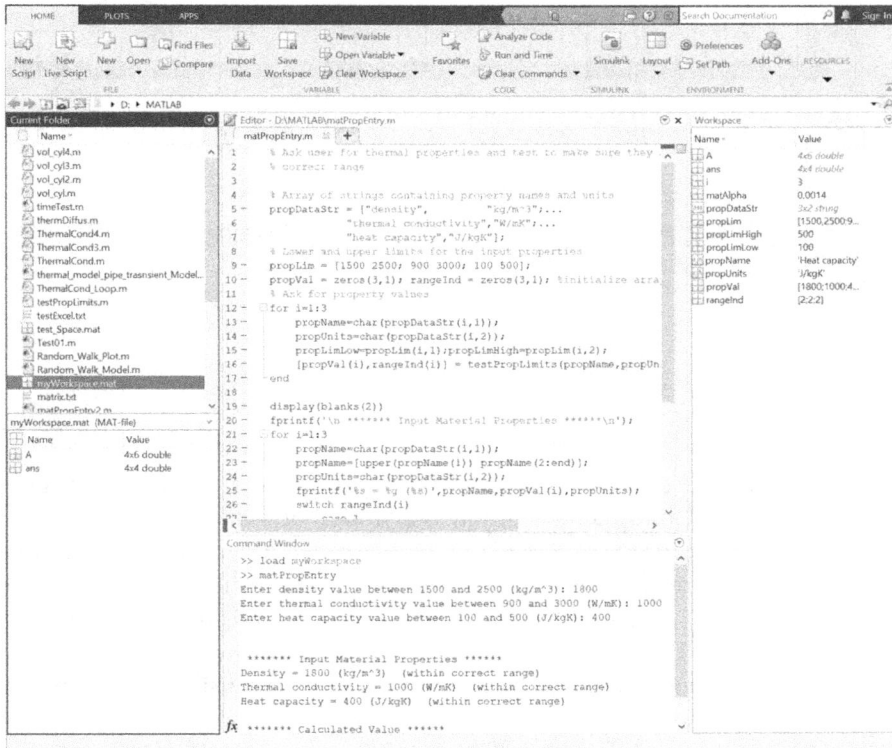

FIGURE 4.1. The MATLAB *HOME* toolstrip with the *Workspace* and *Command Window* panels.

variable_name command deletes the *variable_name* from the *Workspace*. The wildcard (*) character can be used to clear all variables containing the specified sequence of characters. For example, *clear abc** will delete from the *Workspace* all the variables with names starting with *abc*.

Different toolstrips can be selected by choosing among several tabs. For example, under the *HOME* tab, available tools are *New Script*, *Find Files*, *Set Path*, and *Help* (Figure 4.2). Other tab menus are (a) *PLOTS*, where plots can be generated with selected styles; (b) *APPS*, where the MATLAB-compatible applications can be imported and used; (c) *EDITOR*, where *.m* file-related commands can be carried out; (d) *PUBLISH*, where work can be published and formatted in a custom style; and (e) *VIEW*, where the number of the *Editor* window panels and their method of display are selected. If the selected components are not available (e.g., variables as input to create plots) the command tools on the related menu are grayed out (i.e., *PLOTS* in Figure 4.3 is grayed out because no compatible variable has been selected in the *Workspace* panel).

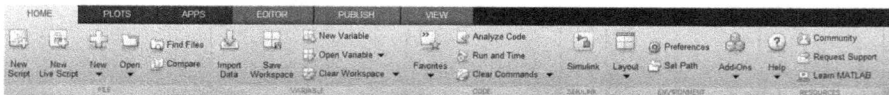

FIGURE 4.2. The MATLAB *HOME* toolstrip.

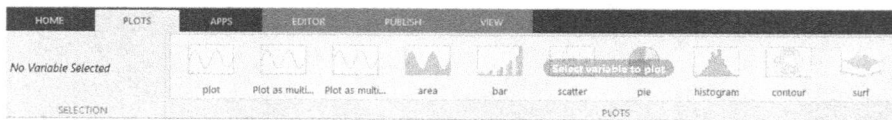

FIGURE 4.3. Grayed out *PLOTS* toolstrip.

4.2   Variables

In MATLAB, variables are not declared at the start of the program; they are created automatically upon assignment of a value. If this variable is assigned a set of numbers (a vector or a matrix) of a particular size, then an array of the corresponding dimensions is created in the program memory. Expanding this array by adding more elements requires redefining its size internally. Thus, it may be more computationally efficient to create a variable with an array of zeroes of appropriate size using a built-in *zeros* function discussed in Section 4.3.4, for example, before entering the loop, where the array size is repeatedly expanded.

Variable data types include numbers, characters, and strings, and logical and structural arrays. There are also several built-in constants (e.g., pi for $\pi = 3.1415$). To display in the *Command Window*, a simple list of variables available in the *Workspace*, the *who* command is used. To obtain more detailed information, such as name, size, bytes, and class, the *whos* command is used (Figure 4.4).

```
>> who

Your variables are:

T              k              qy             test2          thermalModelS
ans            msh            qz             test2test2     thermalresults
cp             q              rho            test3
i              qx             test           test_2

>> whos
  Name                Size              Bytes  Class

  T                   9411x1            75288  double
  ans                 1x1                   8  double
  cp                  1x1                   8  double
  i                   1x1                   8  double
  k                   1x1                   8  double
  msh                 1x1                   8  pde.FEMesh
  q                   1x1                   8  double
  qx                  9411x1            75288  double
  qy                  9411x1            75288  double
  qz                  9411x1            75288  double
  rho                 1x1                   8  double
  test                1x1                   8  double
  test2               1x1                   8  double
  test2test2          1x1                   8  double
  test3               1x1                   8  double
  test_2              1x1                   8  double
  thermalModelS       1x1                   8  pde.ThermalModel
  thermalresults      1x1              341709  pde.SteadyStateThermalResults
```

FIGURE 4.4. The MATLAB *who* and *whos* commands.

4.2.1 Numeric Variables

By default, all numeric variables in MATLAB are stored as double-precision (8 byte/64 bit) floating-point values and are identified by the data type (class) of *double*. To convert a number to a single-precision (4 byte/32 bit) value, the *single* function is used (Figure 4.5). Also, one can convert the floating-point values into integer variables, signed and unsigned, of length from 1 to 8 bytes. If the data can be represented by integers, large volumes of data can be handled at faster speeds by storing them and operating on them as integers. Operations combining integers and double variables return integers. Conversion is done by a group of functions such as; for example, the *int16* command is used to convert to a signed 16-bit integer or *unit 16* to convert to an unsigned 16-bit integer.

Numerical values, built-in numbers, or mathematical constants such as π can be displayed in the *Command Window* with more or fewer digits by entering, respectively, the commands *format long* (15 decimal places) and *format short* (4 decimal places)—Figure 4.5.

```
>> b=single(5)

b =

    single

      5

>> c=double(5)

c =

      5

>> whos
  Name      Size            Bytes  Class     Attributes

  b         1x1                 4  single
  c         1x1                 8  double
```

FIGURE 4.5. *Single* versus the *double* precision.

4.2.2 Character Vectors and Strings

Text can be stored in MATLAB either as a character vector or a string. A character vector is created by enclosing text in single quotes, such as 'Heat Transfer'. Prior to MATLAB R2016b, this was the only way that sequences of text characters were stored. After that release, a new variable type of string was introduced to facilitate handling of longer text segments. Strings are created by enclosing text in double quotes, such as 'Heat Transfer'. One can think of a character vector as a sequence of character codes stored in a linear array. For a string, a single text segment enclosed in double quotes is treated as an element of an array; this array can then contain any number sof string elements.

Character vectors can be concatenated using square brackets; this method does not work for strings and a + sign must be used instead. In the example (Figure 4.6), note that the *myCharVector* size is 1 × 13 whereas the *myString* size is 1 × 1. To create a character vector that includes a numerical value, a *mum2str* function needs to be used (Figure 4.6).

```
>> myCharVector=['Heat',' ','Transfer']      >> size(myCharVector)

myCharVector =                                ans =

    'Heat Transfer'                               1      13

>> myString = "Heat" + " " + "Transfer"      >> size(myString)

myString =                                    ans =

    "Heat Transfer"                               1      1

            >> matDensity=500;
            >> myCharVector=['Density = ',num2str(matDensity)]

        myCharVector =

            'Density = 500'
```

FIGURE 4.6. *Character* vectors versus the *strings*.

In MATLAB, characters are stored using Unicode format with UTF-16 encoding that can represent over 1 million distinct codes. The first 128 symbols of this code use the same encoding as the ASCII character set, where each character is stored as an unsigned 7-bit integer. Character vectors are then just arrays of these integers that are identified internally as character sets and thus displayed as such; they can be converted to their corresponding ASCII code values. Thus, *double('test')* will return a numeric array of [116 101 115 116]. One can also use *uint32* to convert the characters to unsigned 32-bit integers instead of double-precision numbers. If a numerical operation is carried out on the character vector elements (such as addition or subtraction), the array is automatically converted to numerical values. Since the encoding is in alphabetical order, one can manipulate characters by addition or subtraction if so desired. For example, upper-case characters are encoded by integers that are 32 bits smaller than the lower-case ones. Thus, if one assigns *a = 'test'*, the function *char(a − 32)* will return *"TEST."* Here the *char* function converts numbers to the corresponding characters.

There are numerous other character vector and string manipulating functions in MATLAB. For example, the *blanks(j)* function is used to create a string with j blank spaces. The *deblank(string_name)* function is used to remove the trailing blank spaces of strings. The *strtrim(string_name)* function is used to remove both leading and trailing white space. The *upper(string_name)* and *lower(string_name)* functions are used to convert characters, respectively, to uppercase or lowercase (Figure 4.7).

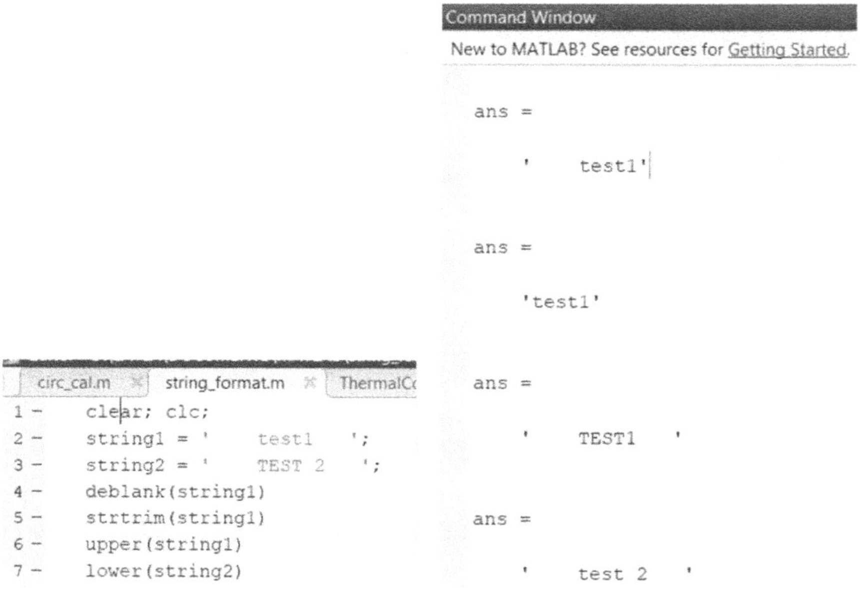

FIGURE 4.7. The *deblank, strtrim, upper,* and *lower* text manipulation: (a) Commands, (b) Outputs.

4.2.3   Logical Variables

A variable of logical data type is created because of evaluation of a logical expression involving relational operators (e.g., <,>, = =), a logical test function (e.g., *isnumeric*), or type conversion from a numeric variable using *logical* function. For example, logical test functions *isnumeric, isfloat* and *isinteger* are useful for identifying the numeric value type and return a logical *true* or *false*. In the example below, 10 random integers from 1 to 100 are generated and placed in the array *myNumbers* (Figure 4.8). Each element of this array is then evaluated with a logical expression to test if it is greater than 50 and a logical array *over* 50 is created as an output of this evaluation. This logical array can then be used as an index to extract from *myNumbers* only those values that are greater than 50. Use of matrix indexing is further discussed in Section 4.4.1.

```
>> myNumbers=randi(100,1,10)

myNumbers =

      46      9     23     92     16     83     54    100      8     45

>> over50=myNumbers>50

over50 =

  1×10 logical array

   0   0   0   1   0   1   1   1   0   0

>> myNumbers(over50)

ans =

      92     83     54    100
```

FIGURE 4.8. Use of a *logical* array as an index of a numeric array.

4.2.4 Variable Names

When working with variables in MATLAB, the following five points should be remembered:

(1) Variable names must start with a letter, not a digit (e.g., 2*test* is not a valid variable name; however, *test* 2 is correct).

(2) No spaces can exist between the variable characters (e.g., *test* 2 is an incorrect variable name).

(3) Variable names are case sensitive (e.g., *test* 2 is different from *Test* 2).

(4) Function names should not be used when assigning variable names (e.g., *pde*, which is a partial differential equation demo function, should not be used as a variable name); the *which* command can be used to test if a name is associated with any function.

(5) If a mathematical operation result is not assigned to a variable name by an equal sign (=), the operation result is stored in the built-in variable *ans*, which is the short form for *answer*; the current value of the *ans* variable can be used in the subsequent expressions by entering *ans* at the command line.

4.3 Creating Matrices

Matrices are fundamental to MATLAB. The name of the software itself is derived from *MATrix LABoratory*. It is important to be able to create and manipulate matrices and vectors. The latter are essentially a special case of a matrix, with only one dimension greater than one. Vectors can be combined to form matrices.

4.3.1 Manual Matrix Creation

To create a row vector, a sequence of numbers is separated by spaces and enclosed within square brackets; to create a column vector, the row vector may be transposed by using a single quotation mark appended to the right of the closing bracket or placed after the vector variable name (Figure 4.9). A second quotation mark appended performs another transpose, thus returning the vector to its original orientation. Another way to create a column vector is by adding a semicolon (;) after each number to create a new row. When defining matrices in general, rows are separated by a semicolon (;). Within each row, elements are separated by a blank space or a comma (,).

```
>> b=[1 2 3 4]                    >> b''

b =                               ans =

     1     2     3     4              1     2     3     4

>> b'                             >> a = [1; 2; 3; 4]

ans =                             a =

     1                                1
     2                                2
     3                                3
     4                                4
```

FIGURE 4.9. Defining row and column vectors and using the transpose operator.

4.3.2 Generation of Vectors with Equally-Spaced Values

In many applications, a vector comprising a sequence of equally spaced numbers is needed. There are several ways to obtain such a vector:

(1) The most common approach is to specify the start and end points as well as the fixed increment with the use of the colon (:) operator. For example, if vector A is defined as $A = 1:2:10$, the resultant vector is

$A = [1, 3, 5, 7, 9]$. The increment between the start (1) and end (10) points is 2. This increment can be negative as well. For example, if $B = 10{:}-2{:}1$, the resultant vector is $B = [10, 8, 6, 4, 2]$.

(2) Another method is by using the *linspace* function. Its advantage relative to the colon (:) operator is that one directly specifies the number of values to be generated for the array. With the colon operator, the numbers are generated by incrementing until the end value is exceeded. For example, to generate 5 equally spaced numbers between 5 and 90, one can execute $C = $ linpace$(5, 90, 5)$. It results in $C = [5, 26.25, 47.5, 68.75, 90]$.

(3) A related function called *logspace* distributes logarithmically the specified number of values between the two end-points $10{\wedge}a$ and $10{\wedge}b$, where a and b are the exponents given as input; for example, for $D = $ logpace$(0, 2, 5)$, the resultant vector ranges from $1E0$ and $1E2$: $D = [1, 3.162, 10, 31.62, 100]$.

(4) Matrices and vectors can be concatenated. For example, the above C and D vectors can be concatenated and create vector $E = [C, D] = [5, 26.25, 47.5, 68.75, 90, 1, 3.162, 10, 31.62, 100]$.

4.3.3 Random Number Matrices

In some applications, sets of random numbers, either real or integer, need to be generated. Random real numbers in the interval $[0, 1]$ are generated by the *rand* function (e.g., $rand = 0.1216$, $rand*100 = 82.5853$). Random integer numbers are generated in the interval starting at 1 and ending at the value given in the *randi* function argument (e.g., $randi(5) = 3$, $randi(10) = 8$). One can also specify lower and upper bounds for the random integers as in, for example, $randi([50, 100]) = 70$. Matrices of random numbers can be generated as well (Figure 4.10). $randi(10,5)$ generates a 5×5 matrix with random integers from 1 to 10; $randi([10\,50], 5)$ generates a 5×5 matrix with random integers ranging from 10 to 50.

```
>> randi(10,5)                    >> randi([10 50],5)

ans =                             ans =

    10    1    1    7    6            19   16   40   40   33
    10    2    8    6    7            17   38   28   19   34
     9    5    3    5    6            14   32   11   26   47
     6    8   10   10    8            40   31   35   41   35
     1    4    1    4   10            28   46   30   30   40
```

FIGURE 4.10. Random integer generating function.

To obtain a non-square matrix made of random double-precision values, the number of rows (i) and columns (j) are specified: $rand(i, j) = A_{i \times j}$. Example below shows generation of 3×4 matrix of integers between 5 and 20 (left) and a 2×3 matrix of double-precision values between 0 and 1 (right) (Figure 4.11).

```
>> randi([5,20],3,4)              >> rand(2,3)

ans =                             ans =

       7     5     8    18            0.0642    0.7095    0.9327
      10    20    10     8            0.9599    0.2798    0.5807
      17    19     5    14
```

FIGURE 4.11. Random integer and real variables.

To create normally distributed values, the *randn* command may be used (e.g., *randn* = 0.2884). The distribution of these random numbers should tend to the average value of zero and standard deviation of one. About 68.7% of the generated values are expected to be within one standard deviation ($-\sigma < x < \sigma$).

4.3.4 Special Matrices

There are several special matrices that can be generated, which are often needed in matrix operations. These include matrices of *zeros* or *ones*, identity, and diagonal matrices (Figure 4.12). These matrices can be made by specifying the number of rows (i) and columns (j), as shown in Equation (47).

$$A_{i \times j} = zeros(i, j), A_{i \times j} = ones(i, j) \text{ and } A_{i \times i} = eye(i) \tag{47}$$

```
>> zeros(2,3)          >> ones(3,4)              >> eye(3)

ans =                  ans =                     ans =

      0    0    0          1    1    1    1          1    0    0
      0    0    0          1    1    1    1          0    1    0
                           1    1    1    1          0    0    1
```

FIGURE 4.12. The *zeroes*, *ones*, and identity matrices.

A diagonal matrix (i.e., one with non-zero elements along the diagonal only) can be generated by specifying a vector as input to the *diag* function. If a matrix is specified as input to the same function, a vector is produced equal to the specified diagonal. Thus, if *A* is a matrix, the *diag(A, X)* function outputs a column vector obtained from the elements of the *X*-th diagonal of *A*, and *diag*(A) = *diag*(A, 0) is the main diagonal.

For example, Figure 4.13 shows first a 4 × 6 matrix *A* created from random integers varying from -20 to 20. Its four-element main diagonal vector is extracted with the *diag*(A) function. Using this extracted vector as input to the same *diag* function produces a 4 × 4 diagonal matrix with the specified vector values along its diagonal and zeroes elsewhere.

```
>> A=randi(40,4,6)-20

A =

     9      7     19     11     16    -14
    11    -13     -6     -9     19     -9
    -8    -15      4      1      2     14
     8      0    -11      8    -14     -9

>> diag(A)'

ans =

     9    -13      4      8

>> diag(diag(A))

ans =

     9      0      0      0
     0    -13      0      0
     0      0      4      0
     0      0      0      8
```

FIGURE 4.13. Extracting the diagonal values from a matrix and creating a diagonal matrix.

4.4 Operating on Matrices

4.4.1 Matrix Indexing

In MATLAB, it is very useful to be able to access any element of a matrix, either to obtain its value, to test its value, or to assign a value to it. An element's location within a matrix is known as its index, and thus this process is referred to as *indexing*. The most used technique is indexing by element positions; the second way is by using a single index; and the third approach is indexing with logical values.

Each element of a two-dimensional matrix is identified by its row and column numbers, as shown in Equation (48). The first index (i) identifies the row and second index (j) the column. When *indexing by position*, one can then use these indices to reference any matrix element.

$$A(i, j) = a_{ij} \qquad (48)$$

One can reference individual elements, ranges of rows and columns—$A(3{:}4, 1{:}2)$, selected rows/columns by listing them in square brackets, or entire rows/columns by using a colon (:) (Figure 4.14). Using a keyword *end* one can reference the last row/column as shown in the last example in Figure 4.14 that selects the last two elements of the third row.

```
>> A

A =

      9      7     19     11     16    -14
     11    -13     -6     -9     19     -9
     -8    -15      4      1      2     14
      8      0    -11      8    -14     -9

>> A(3,5)

ans =

      2
```

```
>> A(3:4,1:2)

ans =

     -8    -15
      8      0

>> A([1 3],:)

ans =

      9      7     19     11     16    -14
     -8    -15      4      1      2     14

>> A(3,end-1:end)

ans =

      2     14
```

FIGURE 4.14. Matrix indexing by position.

If only *a single index* is used in a two-dimensional array, it references the array values as if they were all listed in a one-dimensional vector going down each column, from left to right. For example, $A(7)$ in matrix A shown in Figure 4.14 is equal to -15. In the same matrix, $A(3{:}5)$ evaluates to $[-8\ 8\ 7]$.

Indexing with logical values allows one to select elements based on logical tests. Applying a logical test to a matrix results in a logical array of the same dimensions with 1/0 values indicating, where the test evaluation was *true* or *false*. This array can then be used to reference the elements, where the test evaluated to *true*. For example, to find all the elements in A that are greater than 12 or smaller than -12, the logical expression

$(A < -12 \mid A > 12)$ is evaluated, a matrix *indA* of logical values is obtained and is then used as index into A (Figure 4.15, left). One can also assign value to the index-selected elements as shown in Figure 4.15 (right), where value of 12 is assigned to all elements that were greater than 12. Finally, if one needs a list of index locations for all elements, where the condition evaluates to *true*, the *find* function can be employed. For example, find $(A < -12)$ returns [6 7 20 21]. These are single indices into A, with element at index 7, for example, equal -15.

```
>> indA=(A<-12 | A>12)

indA =

  4×6 logical array

    0    0    1    0    1    1
    0    1    0    0    1    0
    0    1    0    0    0    1
    0    0    0    0    1    0

>> A(indA)'

ans =

   -13   -15    19    16    19   -14   -14    14
```

```
>> A2=A;A2(A>12)=12

A2 =

     9     7    12    11    12   -14
    11   -13    -6    -9    12    -9
    -8   -15     4     1     2    12
     8     0   -11     8   -14    -9
```

FIGURE 4.15. Matrix indexing with logical values.

4.4.2 Arithmetic Operators

MATLAB arithmetic operators include the standard ones such as $+$, $-$, $*$, $/$, $\wedge$ for addition, subtraction, multiplication, division, and exponentiation. When these are employed between scalars, regular mathematical rules are followed that the reader will be familiar with. With MATLAB, however, one can also operate either between combinations of scalars and matrices or between matrices only. Regarding the former case, to facilitate dealing with matrices, MATLAB in some cases carries out operations that would not be allowed if strict mathematical rules were applied. For example, it would not be correct to write a mathematical expression, where scalar a is added to two-dimensional matrix B since they do not have matching dimensions. However, MATLAB assumes that you mean to add a to every element of B and thus will compute the expression $a + b$ without an error message. Any valid mathematical expression will, of course, also work in MATLAB. $a*B$, $B*a$, B/a all produce the expected results. a/B gives an error message that matrix dimensions must agree. With a square matrix D, one can write $a*D^{-1}$, which multiplies scalar a and a matrix inverse of D (Figure 4.16).

```
>> B=A(1:3,1:2)        >> a=2

B =                    a =              >> B*a
                                                            >> B*C
      9     7                 2        ans =
     11   -13                                               ans =
     -8   -15          >> a+B                  18    14
                                               22   -26              36    277   -189
>> C=A(1:2,4:6)        ans =                  -16   -30             238    -71    -37
                                                        >> D=A(1:2,5:6)   47   -413    247
C =                           11     9   >> B/a                      D =
                              13   -11                                   >> B'*C'
     11    16   -14           -6   -13   ans =                   16   -14
     -9    19    -9                                              19    -9   ans =
                      >> a*B              4.5000    3.5000
>> B'+C                                   5.5000   -6.5000    >> a*D^-1      387    200
                      ans =             -4.0000   -7.5000                    79   -175
ans =                                                         ans =
                              18    14   >> a/B                             >> B'*C
     20    27   -22           22   -26   Error using  /             -0.1475    0.2295   Error using  *
     -2     6   -24          -16   -30   Matrix dimensions must agree.  -0.3115    0.2623
```

FIGURE 4.16. Use of arithmetic operators for scalars and matrices.

When carrying out operations between matrices, two different operation types exist. First are the regular arithmetic operations, such as those mentioned above. In MATLAB, they are called *matrix operations*. They are carried out following the linear algebra rules regarding the matching dimensions. To add or subtract matrices, they must have the same number of rows and columns, as shown in Equation (49).

$$C_{i \times j} = A_{i \times j} + B_{i \times j}$$

$$\therefore \qquad C(i,j) = A(i,j) + B(i,j) \qquad (49)$$

$$\therefore \qquad c_{ij} = a_{ij} + b_{ij}$$

To multiply two matrices (Figure 4.16, right), the number of columns in the first matrix should be the same as the number of rows in the second matrix, as given in Equation (50).

$$C_{i \times k} = A_{i \times j} \times B_{j \times k}$$

$$\therefore \qquad C(i,k) = A(i,j) \times B(j,k)$$

$$\therefore \qquad c_{ik} = \sum_{j=1}^{n} a_{ij} \cdot b_{jk} \qquad (50)$$

The second operation type is special to MATLAB. These operations are done with the element-by-element operators and are called *array operations*. They are coded to allow fast computation and thus speed up (by 5 to 10 times) and simplify certain operations that would otherwise require execution of computationally expensive loops. Element-by-element operators are indicated by adding a period before a regular arithmetic operator. As addition/subtraction are by definition element-by-element, the addition of a period in front of them does not make sense and is not permitted. Period can be placed in front of other operators to produce their element-by-element equivalents: (.*, ./, .^). These must be applied between matrices of the same dimensions or between matrix and a scalar (Figure 4.17).

```
>> B'.*C

ans =

    99    176    112
   -63   -247    135

>> B'./C                              >> C.^2

ans =                                 ans =

    0.8182    0.6875    0.5714          121    256    196
   -0.7778   -0.6842    1.6667           81    361     81
```

FIGURE 4.17. Use of element-by-element operators for scalars and matrices.

4.4.3 Relational Operators

The relational operators are used to identify if the two expressions are equal (= =), not equal (~=), or to compare their values (<, <=, >, >=). Logical operators are used in logical expressions and include *or* (|), *and* (&), and *not* (~). The functional forms of these can be used instead (e.g., *or*, *and*, or *not* functions such as *or*(A,B)). There are also *short-circuiting* versions of *and* (&&) and *or* (||). In the former case, if the first operand evaluates to *false*, the *false* result is returned without evaluation of the second operand; in the latter case, if the first operand is *true*, a *true* result is returned without evaluation of the second operand. This should speed up the code execution for very large data sets. Several examples of relational operator use are given in Figure 4.18.

```
                              >> B'<C

                              ans =
                                                    >> indC=(C>10)&(C<15)
>> B'                           2×3 logical array
                                                    indC =
ans =                           1    1    0
                                0    1    1           2×3 logical array
    9    11    -8
    7   -13   -15             >> abs(C)>15            1    0    0
                                                      0    0    0
>> C                          ans =
                                                    >> C(indC)
C =                             2×3 logical array
                                                    ans =
   11    16   -14               0    1    0
   -9    19    -9               0    1    0           11
```

FIGURE 4.18. Use of relational operators for scalars and matrices.

4.4.4 Matrix Reshaping and Rearrangement

Matrices are collections of entities (such as numbers) organized into arrays of one or two or more dimensions. MATLAB provides several tools for rearranging the entities within the arrays and for changing the number and length of the dimensions. An example of a commonly used rearrangement is the transpose (') operator introduced in earlier sections. Another example is the use of the *reshape* function to change how a set of numbers is organized. In the example given here, the objective is to create a 3 × 4 matrix, containing integers from 1 to 12 and that are to be incremented row-wise, starting with 1 in the (1, 1) element. This is done in a single line using the *reshape* function (Figure 4.19). First, a row vector is created containing 12 integers from 1 to 12; it is then reshaped into a 4 × 3 matrix, where the numbers are incremented column-wise; the matrix is then transposed to create the desired 3 × 4 matrix, where the numbers are incremented row-wise.

```
>> BB=reshape(1:12,[4 3])'

BB =

     1     2     3     4
     5     6     7     8
     9    10    11    12
```

FIGURE 4.19. Use of the *reshape* function.

One can rearrange array elements with the *sort* function. In its simplest form, the function will sort a one-dimensional numerical vector *a* in ascending order with *sort*(*a*); for a 2D numerical array, *sort*(*A*) will sort each column of *A* in the ascending order. Entering *sort*(*A*, 2) will sort rows in the ascending order; entering *sort*(*A*, 'descend') will sort columns in the descending order. From the release R2017a, string arrays can be sorted, as well. In the example in Figure 4.20, the first column of matrix *A* is sorted in the ascending order and then matrix *C* is created with rows rearranged to follow the same sorted order. Array indices in vector *iB* obtained by the sorting action are used in *A*(*iB*,:) to create the new sorted *C* matrix.

```
>> A=randi(20,4,6)

A =

     3    11     3     1    10     3
    18     9     4    19     7    16
    13     2     5    19    19     8
     8     5     9    10     8     5

>> [B,iB]=sort(A(:,1));iB'

ans =

     1     4     3     2
```

```
>> C=A(iB,:)

C =

     3    11     3     1    10     3
     8     5     9    10     8     5
    13     2     5    19    19     8
    18     9     4    19     7    16
```

FIGURE 4.20. Use of the *sort* function.

Another rearrangement type is to reverse the element order and it is carried out using the *flip* function. It has similar input options to the *sort* function. Figure 4.21 shows how the matrix A used above is flipped by having all its columns reverse their order; in the next example, row 4 of matrix A is reversed; finally, the order of a character vector MATLAB is reversed to produce text written backwards.

```
>> flip(A)

ans =

     8     5     9    10     8     5
    13     2     5    19    19     8
    18     9     4    19     7    16
     3    11     3     1    10     3

>> flip(A(4,:))

ans =

     5     8    10     9     5     8
```

```
>> flip('MATLAB')

ans =

    'BALTAM'
```

FIGURE 4.21. Use of the *flip* function.

4.4.5   Extracting Information about Matrices

The most basic information about the matrix is its dimensions. These are obtained by using the *size* command. The number of matrix rows and columns thus obtained can be employed within the code (Figure 4.22). A related *length* function would typically be applied to a row or column vector to find the number of elements it contains. For matrices, the command returns the greater of the number of rows or columns (i.e., the longest matrix dimension).

```
>> length(C(1,:))

ans =

        3

>> size(C)          >> length(C)

ans =            ans =

    2    3            3

>> size(B)          >> length(B)

ans =            ans =

    3    2            3
```

```
C =

    11    16   -14
    -9    19    -9

>> B

B =

     9     7
    11   -13
    -8   -15
```

FIGURE 4.22. Use of the *size* and *length* commands for vectors and matrices.

The maximum value for each column of matrix A is obtained by using the $max(A)$ function. Similarly, minimum for each column is obtained by the $min(A)$ function. If the same functions are applied to a vector, its maximum or minimum is obtained. Thus, the $max(max(A))$ function shows the overall maximum value of the matrix, see Equation (51).

$$B_{1 \times 1} = max(max(A_{i \times j})) = max(max((a_{ij}))$$

and
$$B_{1 \times 1} = min(min(A_{i \times j})) = min(min(a_{ij})) \tag{51}$$

Starting from the R2018b release, one can also use the $max(A, [], \text{'}all\text{'})$ function to obtain the maximum of all values. With this function, the dimension can be specified along which the results are produced. If the maximum value for each column is needed, the $max(A, [], 1)$ function can be used; to get the maximum for each row, $max(A, [], 2)$ function can be used (Figure 4.23). Providing two matrices of equal dimensions as the input to the max or min function returns a matrix of the same size containing the larger or smaller element value in an element-by-element comparison.

Product and sum of various subsets of matrix elements can be obtained with the functions structured like the max or min. For matrix A, executing the $prod(A)$ or $sum(A)$ function returns a row vector containing, respectively, products or sums of elements along each column. Equation (52) shows the product and sum over j.

$$B_j = prod(A_{i \times j}) = \prod_{i=1}^{m}(a_{ij}) \quad \text{and} \quad B_j = sum(A_{i \times j}) = \sum_{i=1}^{m}(a_{ij}) \tag{52}$$

```
                              >> max(C,[],'all')

C =                           ans =

      11     16    -14            19
      -9     19     -9
                              >> max(C,[],1)
>> max(C)
                              ans =
ans =
                                  11     19     -9
      11     19     -9
                              >> max(C,[],2)          >> max(B',C)
>> max(max(C))
                              ans =                   ans =
ans =
                                  16                      11     16     -8
      19                          19                       7     19     -9
```

FIGURE 4.23. Use of the max function for vectors and matrices.

If products or sums along the rows are required, the $prod(A, 2)$ and $sum(A, 2)$ functions can be employed, which result in vectors with m rows. Note that, unlike for the *max* or *min* functions, the empty matrix ($[\,]$) is not required as input for these functions.

The overall product or sum, as for the *max* or *min* functions, can be obtained by applying them twice, as shown in Equation (53). Alternatively, from R2018b, one can also use expressions such as the $sum(A, \text{'all'})$ function.

$$B_{1 \times 1} = prod(prod(A_{i \times j})) = \prod_{i=1}^{m} \prod_{j=1}^{n} (a_{ij}) \quad \text{and}$$

$$B_{1 \times 1} = sum(sum(A_{i \times j})) = \sum_{i=1}^{m} \sum_{j=1}^{n} (a_{ij})$$

(53)

4.4.6 Matrix Inverse

The calculation of a matrix inverse is an important concept in linear algebra, and it is closely related to the task of solving a system of linear equations. Using MATLAB, one can directly calculate an inverse of a square matrix by either raising it to -1 power or using the *inv* function (Figure 4.24). However, knowing the mathematics behind the inverse calculation helps to understand and troubleshoot the results if issues arise. For example, commanding MATLAB to determine the inverse of matrix

$C_{4\times4}$ in Figure 4.25 results in *Inf* (infinite) matrix components. Further investigation shows that the determinant of matrix $C_{4\times4}$ is zero. You can ensure the matrix is not ill-conditioned by calculating its condition number using the *cond(C)* function. If the condition number of a matrix is significantly greater than 1, the matrix inverse will be very sensitive to very small errors in the input matrix element values; an infinite condition number corresponds to a non-invertible matrix. Another method to calculate the matrix X in $AX = B$, is to directly obtain it by dividing matrix B by matrix A $(X = B\backslash A)$. Note the use of the backslash (\) in this expression.

```
>> C=[1 0.5 2 0.25;-5 0 -1 0.25;.25 1 0.5 0;0.75 -0.25 1.5 -0.5]

C =

    1.0000    0.5000    2.0000    0.2500
   -5.0000         0   -1.0000    0.2500
    0.2500    1.0000    0.5000         0
    0.7500   -0.2500    1.5000   -0.5000

>> C^-1

ans =

   -0.0163   -0.2222   -0.0217   -0.1192
   -0.1951    0.0000    1.0732   -0.0976
    0.3984    0.1111   -0.1355    0.2547
    1.2683    0.0000   -0.9756   -1.3659
```

FIGURE 4.24. Calculating the inverse function.

```
>> C=[1 0.5 2 0.25;-0.5 0 -1 0.25;.25 1 0.5 0;0.75 -0.25 1.5 -0.5]      >> inv(C)
                                                                         Warning: Matrix is singular to working precision.
C =

    1.0000    0.5000    2.0000    0.2500                                 ans =
   -0.5000         0   -1.0000    0.2500
    0.2500    1.0000    0.5000         0                                   Inf   Inf   Inf   Inf
    0.7500   -0.2500    1.5000   -0.5000                                   Inf   Inf   Inf   Inf
                                                                          Inf   Inf   Inf   Inf
>> C^-1                                                                    Inf   Inf   Inf   Inf
Warning: Matrix is singular to working precision.
> In mpower>integerMpower (line 50)
In ^ (line 49)


ans =

   Inf   Inf   Inf   Inf
   Inf   Inf   Inf   Inf
   Inf   Inf   Inf   Inf
   Inf   Inf   Inf   Inf
```

FIGURE 4.25. Indeterminate inverse matrix.

The analytical method to obtain an inverse of a square matrix A is to divide the adjugate of A, *adj(A)* by the determinant of A. The adjugate in turn is a transpose of a cofactor matrix of A. For example, for a 2 × 2 matrix, Equation (54) is applicable.

$$A_{2\times2} = \begin{bmatrix} a & b \\ c & d \end{bmatrix}$$

$$\therefore \quad A^{-1}{}_{2\times2} = \frac{1}{det(A_{2\times2})} \begin{bmatrix} d & -b \\ -c & a \end{bmatrix} \quad (54)$$

This equation shows that if the determinant is zero, the matrix is not invertible.

Transposing a matrix is when the elements at certain rows and column (i, j) are switched with the elements at the columns and rows (j, i). See Equation (55).

$$A_{i\times j} = a(i, j)$$

$$\therefore \quad A'_{i\times j} = a'(i, j) = a(j, i) \quad (55)$$

The calculation of the inverse is one method to find a solution of a system of linear equations. Assume there is an equation $AX = B$, where $A_{i\times j}$, $B_{i\times k}$ are matrices consisting of known elements and X is the variable matrix (unknown). To find matrix $X_{j\times k}$, one method is to use the inverse, as shown in Equation (56).

$$AX = B$$

$$A^{-1}AX = A^{-1}B \quad (56)$$

$$\therefore \quad IX = A^{-1}B$$

$$\therefore \quad X = A^{-1}B$$

Note that for $AX = B$ to be valid: (a) the number of columns of matrix A should be the same as the number of rows of matrix X; (b) the number of rows of matrix A should be the same as the number of rows of matrix B; and (c) the number of columns of matrix X should be the same as the number of columns of matrix B; $A_{i\times j} X_{j\times k} = B_{i\times k}$. Furthermore, the determinant of matrix $A_{i\times j}$ should not be zero. To find the determinant of a 2 × 2 matrix, the following relation is used, Equation (57).

$$A_{2\times2} = \begin{bmatrix} a & b \\ c & d \end{bmatrix}$$

$$\therefore \quad det(A_{2\times2}) = \begin{vmatrix} a & b \\ c & d \end{vmatrix} = ad - bc \quad (57)$$

To find the determinant of a 3 × 3 matrix, Equation (58) is used.

$$A_{3 \times 3} = \begin{bmatrix} a & b & c \\ d & e & f \\ g & h & i \end{bmatrix}$$

$$\therefore \qquad det(A_{3 \times 3}) = a \begin{vmatrix} e & f \\ h & i \end{vmatrix} - b \begin{vmatrix} d & f \\ g & i \end{vmatrix} + c \begin{vmatrix} d & e \\ g & h \end{vmatrix} \tag{58}$$

$$A_{3 \times 3} = \begin{bmatrix} a & b & c \\ d & e & f \\ g & h & i \end{bmatrix} \therefore det(A_{3 \times 3}) = a(ei - fh) - b(di - fg) + c(dh - eg) \tag{59}$$

4.4.7 Systems of Linear Equations

In the MATLAB environment, it is possible to use built-in tools to solve systems of linear equations in which the number of variables is the same as the number of linearly independent equations. In general, programing languages (like C or Fortran) would need to have dedicated code written employing multiple loops (e.g., *for*, *while*) to implement elimination techniques, such as the Gauss-Seidel method, iteratively. However, in the MATLAB environment, this can be achieved by a built-in linear solver function (*linsolv*). It solves the equation $AX = B$, where X is the state variable vector and is unknown. Note that, as mentioned earlier, for the systems of linear equations to have a definite solution, the determinant for the unknown variables multiplier matrix A should not be zero—$|A| \sim\, = 0$ to satisfy Equation (56)—$X = A^{-1}B$. If the system of equations has a definite solution (Figure 4.26), the output will appear in the *Command Window*; otherwise, a warning message will be shown: *Matrix is singular to working precision* (Figure 4.27). Note that it is possible to solve a single system of equations, Equation (60) and Figure 4.26, or multiple systems of equations, Equation (61) and Figure 4.28— $A_{m \times n} X_{n \times \kappa} = B_{m \times \kappa}$.

$$A = \begin{bmatrix} 1 & 2 & 3; 2 & 3 & 4; 3 & 4 & 1 \end{bmatrix} \quad and \quad B = \begin{bmatrix} 6 & 4 & 1 \end{bmatrix}$$

$$A_{3 \times 3} X_{3 \times 1} = B_{3 \times 1}$$

$$\begin{cases} x + 2y + 3z = 6 \\ 2x + 3y + 4z = 4 \\ 3x + 4y + z = 1 \end{cases} \therefore \begin{cases} x = -9.75 \\ y = 7.50 \\ z = 0.25 \end{cases} \tag{60}$$

```
Command Window

A =

        1        2        3
        2        3        4
        3        4        1


B =

        6        4        1


ans =

    -9.7500
     7.5000
     0.2500
```

```
linear_equation.m   tictoc.m
1 -    clear all; clc
2 -    A=[1 2 3;2 3 4;3 4 1]
3 -    B=[6 4 1]
4 -    linsolve(A,B')
```

(a)   (b)

FIGURE 4.26. Solving a single system of linearly independent equations: (a) Script, (b) Solution.

```
Command Window

A =

        1        4        7
       10       14       18
       20       25       30


B =

        1        4        7

>> linsolve(A,B')
Warning: Matrix is singular to working precision.

ans =

      NaN
     -Inf
      Inf
```

FIGURE 4.27. Solving a system of linearly dependent equations.

$$AA = [1 \ 2 \ 3; 2 \ 3 \ 4; 3 \ 4 \ 1] \quad \text{and} \quad B = [6 \ 4 \ 1; 5 \ 3 \ 0; 4 \ 2 \ -1]$$

$$A_{3 \times 3} X_{3 \times 3} = B_{3 \times 3}$$

$$
\begin{cases} x_1 + 2y_1 + 3z_1 = 6 \\ 2x_1 + 3y_1 + 4z_1 = 4 \\ 3x_1 + 4y_1 + z_1 = 1 \end{cases} \quad \therefore \begin{cases} x_1 = -8 \\ y_1 = 7 \\ z_1 = 0 \end{cases}
$$

$$
\begin{cases} x_2 + 2y_2 + 3z_2 = 5 \\ 2x_2 + 3y_2 + 4z_2 = 3 \\ 3x_2 + 4y_2 + z_2 = 0 \end{cases} \quad \therefore \begin{cases} x_2 = -6 \\ y_2 = 5 \\ z_2 = 0 \end{cases} \tag{61}
$$

$$
\begin{cases} x_3 + 2y_3 + 3z_3 = 4 \\ 2x_3 + 3y_3 + 4z_3 = 2 \\ 3x_3 + 4y_3 + z_3 = -1 \end{cases} \quad \therefore \begin{cases} x_3 = -3 \\ y_3 = 2 \\ z_3 = 0 \end{cases}
$$

```
Command Window

   AA =

         1      2      3
         2      3      4
         3      4      1


   BB =

         6      4      1
         5      3      0
         4      2     -1
```

```
linear_equation.m    tictoc.m    am
1 -   clear all; clc
2 -   AA=[1 2 3;2 3 4;3 4 1]
3 -   BB=[6 4 1;5 3 0;4 2 -1]
4 -   linsolve(AA,BB)
```

```
   ans =

      -8.0000     -6.0000     -3.0000
       7.0000      5.0000      2.0000
      -0.0000     -0.0000     -0.0000
```

(a)   (b)

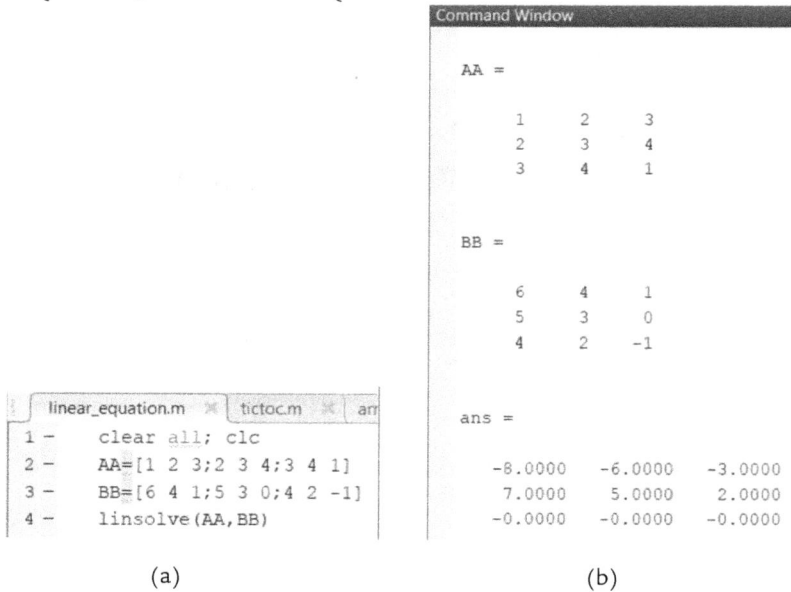

FIGURE 4.28. Solving multiple systems of linearly independent equations: (a) Script, (b) Solution.

4.5 Built-in Functions

There is a very large set of built-in functions available in the MATLAB base installation, plus many more via the add-on toolboxes. The best way to learn about them is by using the MATLAB help facility, available as a separate window via the *Quick Access* toolbar (top-right of the Desktop window) or via the *Home* toolstrip. One can also enter the *help* command followed by the function name within the *Command Window*. Many functions have multiple pages of help text. Typing the *more on* command at the command line displays this text one page at a time. Press the *enter* key to advance by one line, press space bar to advance by one page, or type *q* to exit to command line. Typing the more off command disables this feature.

Some functions can be called in two different ways: either using the command syntax or the function syntax. If no output from the function needs to be obtained, a simpler command syntax can be used. In this case, the function input argument is added after the function name and is always treated as a character vector, e.g., *load myWorkspace.mat*. With the function syntax, the function name is followed by parentheses within which arguments are listed. These can be variable names or values, e.g., *load('myWorkspace. mat')*. Thus, if you need to pass input to a function via a variable, then a function syntax needs to be used. In the example (Figure 4.29), variable *myWSFileName* is defined to be equal to character vector *'myWorkspace'*. This variable can then be used as input to the *load* command in its function form but not in its command form, as the error message shows.

```
>> myWSFileName='myWorkspace';
>> load (myWSFileName)
>> load myWSFileName
Error using load
Unable to read file 'myWSFileName'. No such file or directory.

>> load myWorkspace
>> load ('myWorkspace')
```

FIGURE 4.29. Executing built-in functions using the command or function syntax.

Here are several functions that carry out operations that will be useful in subsequent chapters:

(1) Floating-point numbers may be rounded up, down, or rounded to the nearest decimal, using the *ceil, floor*, and *round* functions. For instance, $ceil(11.4) = 12, floor(11.4) = 11, round(11.499) = 11$, and $round(11.5) = 12$. Note the difference between the first two functions and the last one.

(2) Module operation results in the remainder of a division operation; for example, $mod(50, 3) = 2$.

(3) One can convert radians to degrees and vice versa using functions in following examples: $radtodeg(3.14) = 179.9087$ and $degtorad(180) = 3.1416$.

(4) Trigonometric operations such as *sine* and *cosine* can be implemented using the functions in the form of $sin(\alpha)$ or $sin(\beta)$, where α and β are in radian and degrees, respectively. Trigonometric functions like these are used extensively when defining cyclic boundary conditions.

(5) *isa(obj, ClassName)* identifies if the object belongs to the specified class category; *ClassName* can be, for example, *double, single, logical,* and *char*. *isa(obj, ClassCategory)* identifies if the object belongs to the specified class category; *ClassCategory* can be numeric, float, or integer. The result of the *is a* function *isa* logical *true* or *false*.

(6) *isnan(a)* is another logical function that tests the input for a particular property. There are over seventy *is*∗ functions that test their input for things like whether it is an empty matrix, an integer, or a string. The *isnan(a)* function tests whether each element of the input array is *not-a-number (NaN)* and returns an array of the same size containing corresponding elements that have logical *true* values if they are, and *false* if they are not. A *NaN* means that this element is neither a real nor a complex number. For example, if one attempts to calculate a 0/0, a *NaN* results. However, attempting to calculate 1/0 results in infinity *(inf)*, and not in a *NaN* (Figure 4.30).

```
Command Window

>> A = 0./[0 1 2 3 4]

A =

    NaN     0     0     0     0

>> isnan(A)

ans =

  1×5 logical array

    1     0     0     0     0

>> B = 1./[0 1 2 3 4]

B =

        Inf    1.0000    0.5000    0.3333    0.2500

>> isnan(B)

ans =

  1×5 logical array

    0     0     0     0     0
```

FIGURE 4.30. Identifying the *NaN* variables within an array.

4.6   Scripts

Any sequence of the MATLAB commands can be saved to a script file (*.m). Scripts are just text files and so it is possible to create and edit them with any text editor outside the MATLAB environment. However, the MATLAB editor offers many additional features that aid in creation, editing, and debugging of scripts. One way to create a new script within the MATLAB environment is via *HOME > New Script* (Figure 4.31); another way is via *EDITOR > New > Script*.

Figure 4.31 shows a sample script that calculates the face perimeter, area, and volume of a cube. When a script is opened, the *EDITOR* tab in the toolstrip is activated (Figure 4.32). The script is saved under the name *Cube.m* in the current folder. The output for the script shown in Figure 4.32 is presented in Figure 4.33. Note that the script is essentially the same as the one shown in Figure 4.31; however, the semicolon (;) after the last line (the formula to calculate the *face_perimeter*) is omitted. As it is seen, the only visible output variable in the *Command Window* is the *face_perimeter*. The rest of the variables (*volume* and *face_area*) are not shown in the *Command Window*. Normally, a semicolon (;) is added to the end of each command line to suppress the output to the *Command Window* when running a script.

A script can be executed either by clicking on the *Run* command in the *EDITOR* menu or by simply typing the script's name (without the extension) on the command line. If any script or function name are entered on the command line or if a function external to the script is called from within it, MATLAB needs to know where this script or function are located. First, it looks in the current folder identified by the *Current Folder* toolbar. If the

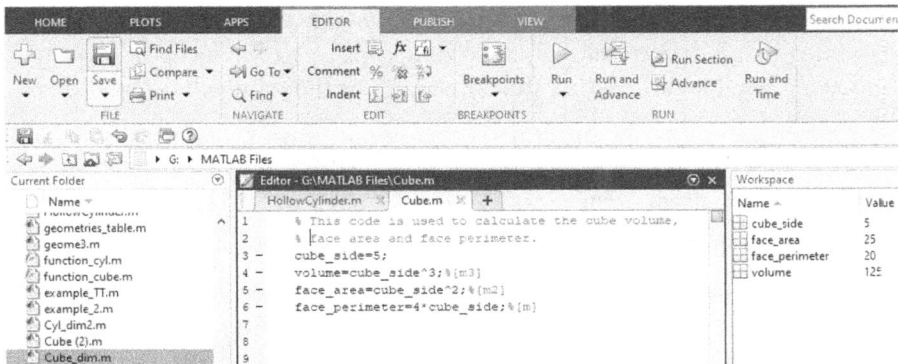

FIGURE 4.31. Creating a script file.

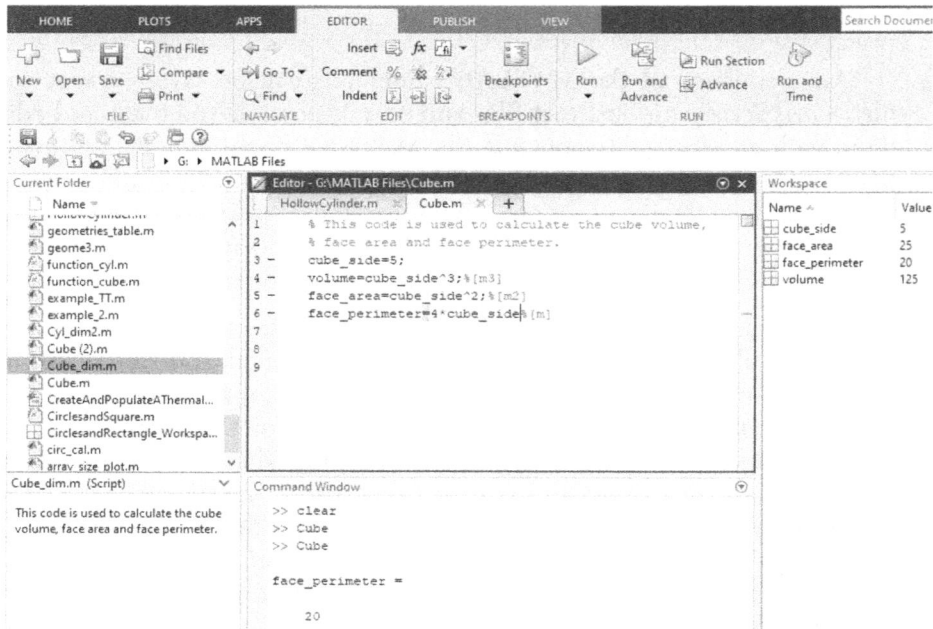

FIGURE 4.32. Saving a script file and running the script.

script/function is not found there, then the MATLAB path is searched. The path contains a sequential list of folder locations, where the program searches for any script/function name it is trying to execute. The search is carried out from the top of the list until the match is found. One can use the *Set Path* tool in the *HOME* menu tab to add new locations or change the list sequence.

When trying to execute a script or function, it is often helpful to find out where this script or function is located to make sure that the right one is being used. To obtain this information, use the *which* command followed by the script/function name that you are looking for. A folder location is returned, allowing you to confirm that the correct script/function will be executed.

Note that all script and function names are case-sensitive. For example, if one entered the script name starting with a lower-case letter, *cube*, MATLAB sends an error message:

Cannot find the exact (case – sensitive) match for 'cube'.

Then, it suggests the following: The closest match is: *Cube in D:\MATLAB\ Cube.m* and *Did you mean: >> Cube*. If the user agrees by pressing the *Enter* key, the suggested function will be run (Figure 4.33).

All variables assigned values within the executed script (whether they are displayed or not within the *Command Window*) remain in the *Workspace* (Figure 4.32 and Figure 4.33). It is possible to include comments within the script, which is a good practice that will pay off whether you are looking at this script in the future or someone else is trying to understand what you have done. Comment text can be added anywhere on the line; any text on the line after the comment (%) operator will be treated as a comment and thus not executed (Figure 4.32). In the MATLAB editor, the information after the comment operator will be highlighted with green color.

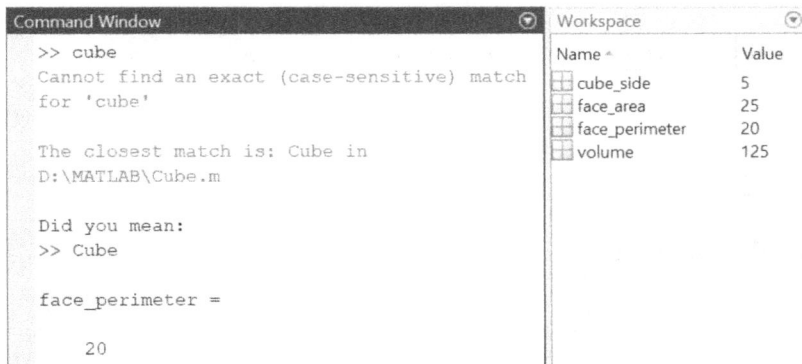

FIGURE 4.33. The *Command Window* showing the script name entry and its output.

4.7 Input-Output Techniques

In the example provided in Figure 4.32, the input variable *cube_side* is defined within the script, and so no user input is needed. Instead of pre assigning a variable value in a script, one can also ask for the user input. This is accomplished using the *input* function (Figure 4.34), which includes as its own input argument a character array that prompts the user to enter the requested numerical value or text. In the latter case, an *s* parameter is included after the prompt text. In the example code *Cube_dim* (Figure 4.34), the user is prompted with *"Enter the cube side length:"* Note that single quotes must be used, as the prompt text is a character vector. In the example shown in Figure 4.35, the user is to input the cylinder's *radius* and *height*; the *volume* and *total_area* are calculated based on the input value.

```
  Cube_dim.m  ×  +
1      % This code is used to calculate the cube volume, face area and face perimeter.
2  -   cube_side=input('Enter the cube side length:   ');
3  -   volume=cube_side^3;%[m3]
4  -   face_area=cube_side^2;%[m2]
5  -   face_perimeter=4*cube_side%[m]
6
```

Current Folder	Command Window	Workspace
Name ^	>>	Name ^ / Value
cylinder_vol.m	>> Cube_dim	cube_side 10
example_TT.m	Enter the cube side length: 10	face_area 100
matPropEntry.m		face_perimeter 40
matrix.txt	face_perimeter =	volume 1000
myWorkspace.mat		
Cube_dim.m (Script)	40	

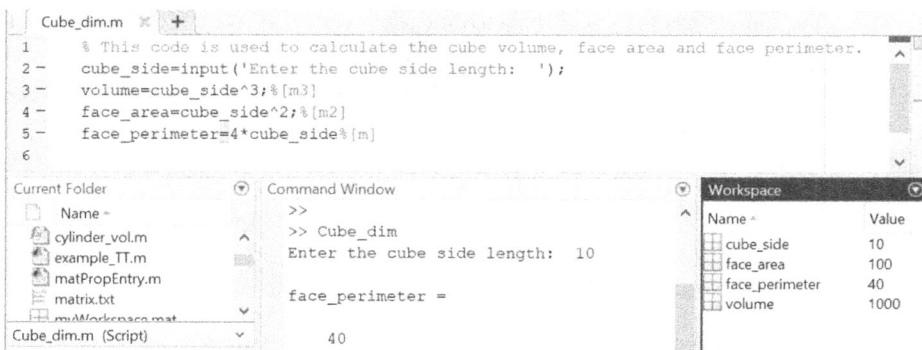

FIGURE 4.34. Asking for a single input in a cube parameter calculation script.

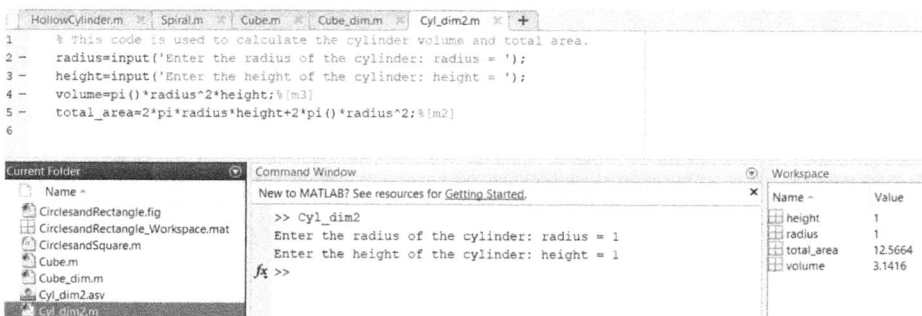

```
  HollowCylinder.m   Spiral.m   Cube.m   Cube_dim.m   Cyl_dim2.m  ×  +
1      % This code is used to calculate the cylinder volume and total area.
2  -   radius=input('Enter the radius of the cylinder: radius = ');
3  -   height=input('Enter the height of the cylinder: height = ');
4  -   volume=pi()*radius^2*height;%[m3]
5  -   total_area=2*pi*radius*height+2*pi()*radius^2;%[m2]
6
```

Current Folder	Command Window	Workspace
Name ^	New to MATLAB? See resources for Getting Started.	Name ^ / Value
CirclesandRectangle.fig	>> Cyl_dim2	height 1
CirclesandRectangle_Workspace.mat	Enter the radius of the cylinder: radius = 1	radius 1
CirclesandSquare.m	Enter the height of the cylinder: height = 1	total_area 12.5664
Cube.m	fx >>	volume 3.1416
Cube_dim.m		
Cyl_dim2.asv		
Cyl_dim2.m		

FIGURE 4.35. Asking for multiple inputs in a cylinder parameter calculation script.

The simplest way to view the value of a variable is to type its name while omitting the semicolon (;) at the end of the line. This produces an output, showing the variable name and value. For double-precision arrays or scalars, only the name and value are displayed. For other types, such as logical, for example, the array dimensions and the variable type are displayed as well. Equivalently, one can also use the *display* command with the variable or value one would like to print to the *Command Window* given as input. For a variable/matrix, it will show the same information as leaving out the semicolon (;) would. However, it may be better programing practice to explicitly state the intent of displaying the variable value as opposed to just leaving out the semicolon (;). For example, in Figure 4.36, *display*(*blanks*(2)) is used to add a blank line to the display. Another option is to use the *disp*(*variable*) function; it shows the value associated with the variable, but omits display of the variable name and any other information.

In some cases, it is needed to control more precisely the formatting of the displayed numbers and text. This can be accomplished by using the

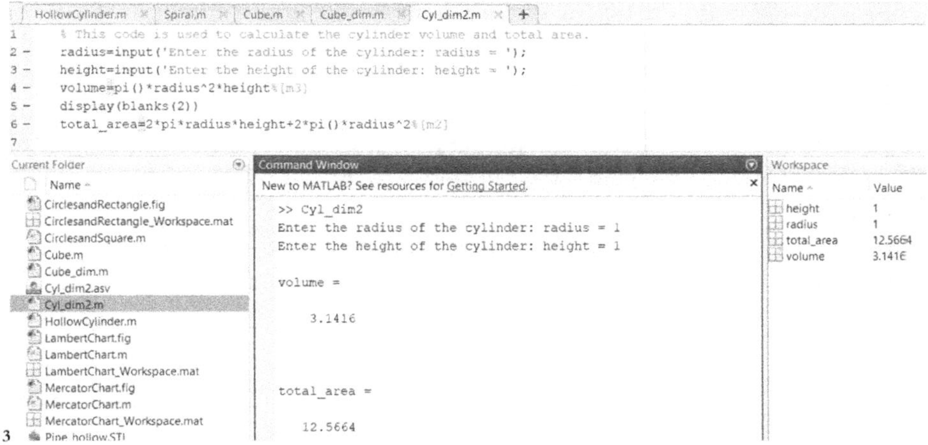

FIGURE 4.36. Displaying the results, including spaces.

fprintf and *sprintf* functions. The former outputs formatted text to a file or to the *Command Window*; the latter outputs to a text string. Both use the same formatting specifications to control how the numbers are displayed.

If text and numbers need to be output to the *Command Window*, the input arguments would start with text in single quotes that includes format operators identified by the % sign. This text is followed by the same number of numerical items to display as there are format operators within the text. Common formatting operators are %d for integer, %f for fixed-point, and %e for exponential notation display (Figure 4.37). Display of numbers can be further controlled by specifying the field width within which the number

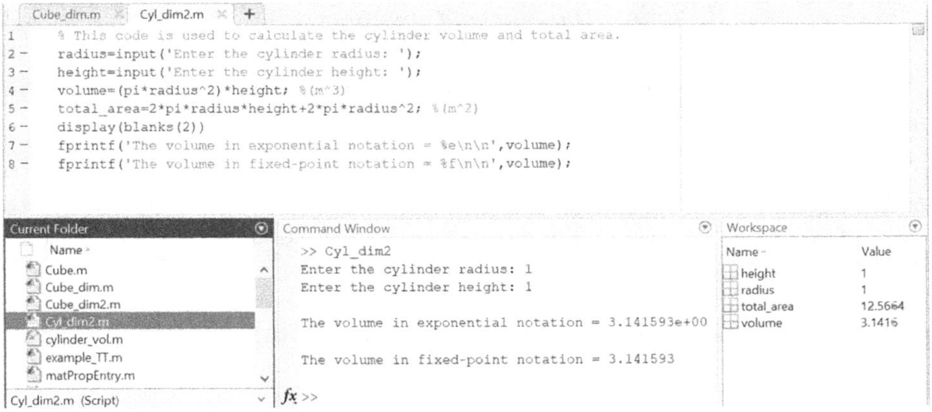

FIGURE 4.37. Specifying output format for printed numbers.

is to be printed and the precision (number of figures after the decimal), e.g., %10.2*f* will fit a number within a field of 10 spaces and will display 2 figures after the decimal. If 10 is omitted (i.e., %.2*f* or %*f*), a space is inserted by default between any preceding text and the number (Figure 4.38). The text line is often terminated by a special character \n, indicating a new line from which subsequent text display will continue. Other special characters are \t for a horizontal tab or \r for carriage return. The latter needs to be placed before \n if one is outputting to a file that must then be opened using a Windows text editor, such as Notepad.

```
test =

    10

>> test2 = 'test is %.2f W/mK\n';
>> fprintf(test2,test)
test is 10.00 W/mK
>> test2 = 'test is %10.2f W/mK\n';
>> fprintf(test2,test)
test is        10.00 W/mK
```

FIGURE 4.38. Defining format for the value embedded in the *fprintf* function.

It is possible to write the data from the *Workspace* variables into an external file by using the *save* command (Figure 4.39). The *save* command, followed by a filename only (with no extension specified) will save all the *Workspace* variables in a binary format (*.mat*) file. One can also output the specified variable data into a plain text file that can be then edited with a text editor or imported into MS Excel, for example. This is accomplished by appending the filename, variable name, and a qualifier (*–ascii*) to the *save* command. Adding the *–append* qualifier will add the variable data to the specified existing file. You can view the contents of any text file by using the *type* command; it will display them in the *Command Window* (Figure 4.39).

One can also import external data previously saved in a text file (such as, for example, a test output data set of numbers arranged in columns such as a comma separated value file (*.csv*) into the MATLAB *Workspace* by using the *load* command. In this example, such data file is created in MATLAB and saved as a text file *testf.dat* containing values from a matrix *test* (Figure 4.40). The data is imported into MATLAB by the *load* command followed by the full file name (*testf.dat*). A matrix variable *testf* is then created in the *Workspace*; the variable name is that of the file without the extension

```
Command Window

 >> a

 a =

    1     3     5     7     9
    1     5     9    13    17

 >> b

 b =

    0.5000    1.5000    2.5000    3.5000    4.5000
    0.5000    2.5000    4.5000    6.5000    8.5000

 >> save matrix.txt a -ascii
 >> type matrix.txt

    1.0000000e+00    3.0000000e+00    5.0000000e+00    7.0000000e+00    9.0000000e+00
    1.0000000e+00    5.0000000e+00    9.0000000e+00    1.3000000e+01    1.7000000e+01
 >> save matrix.txt b -ascii -append
 >> type matrix.txt

    1.0000000e+00    3.0000000e+00    5.0000000e+00    7.0000000e+00    9.0000000e+00
    1.0000000e+00    5.0000000e+00    9.0000000e+00    1.3000000e+01    1.7000000e+01
    5.0000000e-01    1.5000000e+00    2.5000000e+00    3.5000000e+00    4.5000000e+00
    5.0000000e-01    2.5000000e+00    4.5000000e+00    6.5000000e+00    8.5000000e+00
fx >>  |
```

FIGURE 4.39. Saving variable data to a new file and appending data to an existing external file.

```
>> test

test =

    1.0000    3.0000    5.0000    7.0000    9.0000
    1.0000    5.0000    9.0000   13.0000   17.0000
    0.5000    1.5000    2.5000    3.5000    4.5000
    0.5000    2.5000    4.5000    6.5000    8.5000

>> save testf.dat test -ascii
```

FIGURE 4.40. Saving data into an external file.

```
>> clear
>> load testf.dat
>> testf

testf =

    1.0000    3.0000    5.0000    7.0000    9.0000
    1.0000    5.0000    9.0000   13.0000   17.0000
    0.5000    1.5000    2.5000    3.5000    4.5000
    0.5000    2.5000    4.5000    6.5000    8.5000
```

FIGURE 4.41. Calling data from an external file.

(Figure 4.41). The variable data can also be viewed and edited by double-clicking the variable name in the *Workspace*. This opens the *Variables Editor*, displaying the data in a tabular view, like a spreadsheet (Figure 4.42).

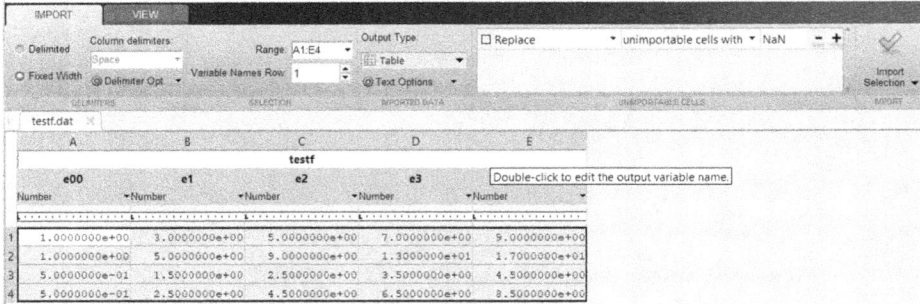

FIGURE 4.42. Viewing and editing variable data in the *Variables Editor*.

It is possible to call in an image using the *imread('image_name')* function, where *image_name* includes the file extension (e.g., **.png* or **.jpg*). This is a useful feature, making it possible to place an image over or behind a plot. In general, a wide variety of file image and data formats can be opened within MATLAB (Figure 4.43); images and figures generated within the MATLAB environment can also be saved to a broad range of the standard image types in addition to the native **.jpg* file format (Figure 4.44). Saving to the **.fig* format allows subsequent reopening and adjustment of the plots and axes parameters within the figure using the interactive tools of the *Figure* window menu.

The user may also import vectors and raster data from the Web map servers, and the files may be exported in formats such as Shapefile

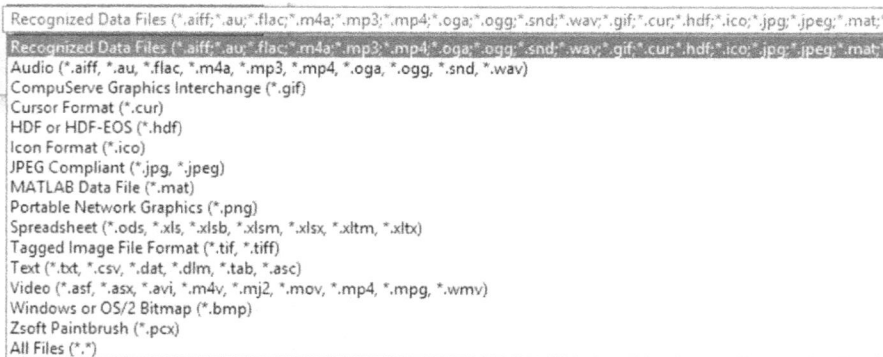

FIGURE 4.43. File formats importable into the MATLAB environment.

binary files (*shapewrite*), Keyhole Markup Language (*KML*) text files—*kmlwrite*—and *GeoTIFF*, which writes the geodata as vector coordinates and map attributes to the desired file and eventually the Web. *.*geotiff* files are like *.*tiff* files with additional attributes associated with parameters for geo referencing and projected coordinate systems. *.*kml* files are a form of *.*xml* files that store the hyperlink information and map components' relations. *.*shp* files work with vector geodata and tabular attributes. The format of the files importable into MATLAB is shown in Figure 4.43.

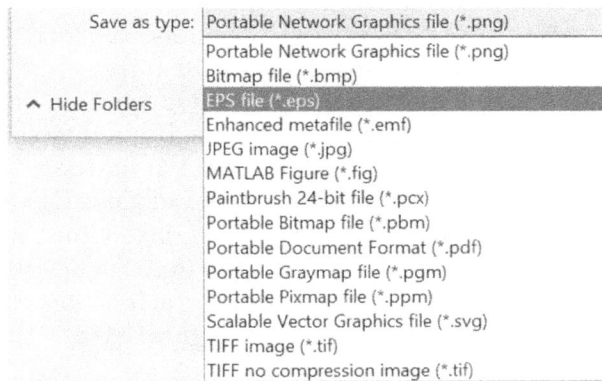

Save as type:	Portable Network Graphics file (*.png)
	Portable Network Graphics file (*.png)
	Bitmap file (*.bmp)
∧ Hide Folders	EPS file (*.eps)
	Enhanced metafile (*.emf)
	JPEG image (*.jpg)
	MATLAB Figure (*.fig)
	Paintbrush 24-bit file (*.pcx)
	Portable Bitmap file (*.pbm)
	Portable Document Format (*.pdf)
	Portable Graymap file (*.pgm)
	Portable Pixmap file (*.ppm)
	Scalable Vector Graphics file (*.svg)
	TIFF image (*.tif)
	TIFF no compression image (*.tif)

FIGURE 4.44. Image formats that may be created in the MATLAB environment.

4.8   User-Defined Functions

A good programing practice is to create functions for execution of any code that is likely to be used more than once within one program or several different programs. This improves code's reliability, as one only needs to make sure once that the function performs correctly. It also makes it easier to implement any changes as that change would only need to be done in one place (within the function). A function contains a sequence of commands, just like a script, but one can also pass input arguments to it and receive output values.

A *function* file starts with the function definition command: The keyword *function* followed by an optional list of output arguments, equal sign, function name, and an optional list of input arguments. The function name follows the same naming conventions as the MATLAB variables. The *end* command closes the function definition. The code block of the *function–end* includes the mathematical operations used to obtain the *output* or data *input/output* variables. In the example shown in

Figure 4.45a, the cylinder volume calculation that was presented earlier in Figure 4.37 is defined as the function *vol_cyl*. The function is saved as an *.m file (i.e., with the same extension as that for the MATLAB scripts); the file name should be the same as the name of the function. Using its name, the function can then be called either from a script or from the command line, as shown in Figure 4.46. If there is a mismatch between the file name and the function name, the function will be known to MATLAB by its file name, and a warning message would be given to the user about the mismatch.

In the earlier MATLAB versions, only one function definition was allowed within each *.m file. Starting from the release R2016b, the ability to have *local functions* was implemented. These functions are intended for use within the *.m file, where they are defined, and they cannot be called from the command line or from another script or function. Within the *.m file defining a function, a local function can be added at the end; it can then be called from within the main function (first one defined within the file). Within the *.m file containing a script, local function definitions are to be placed at the end of the file. An example of a local function used to calculate the cylinder volume, *vol_cyl2*, is shown in Figure 4.45b.

While it is recommended to use comments in functions, just like when writing scripts, an additional consideration for functions is that any comment lines added before the function definition within the *.m file are displayed in the *Command Window* if the user enters the *help* command followed by the function name; see Figure 4.46b. Thus, placing a brief explanation

```
1    % Calculate cylinder volume
2    % INPUT radius  height
3    % OUTPUT Volume
4    function volume=vol_cyl(radius,height)
5 -  if any(radius<0) || any(height<0)
6 -      error('Inputs cannot be negative numbers')
7 -  end
8 -  s_r=size(radius);
9 -  s_h=size(height);
10 - if any(s_r~=s_h)
11 -     error('Both input variables must be the same size')
12 - end
13   % Calculate cylinder volume
14 - volume=pi*(radius.^2).*height;
15 - end
```

```
1    % Calculate cylinder volume
2    % INPUT radius  height
3    % OUTPUT Volume
4    function volume=vol_cyl2(radius,height)
5 -  if any(radius<0) || any(height<0)
6 -      error('Inputs cannot be negative numbers')
7 -  end
8 -  s_r=size(radius);
9 -  s_h=size(height);
10 - if any(s_r~=s_h)
11 -     error('Both input variables must be the same size')
12 - end
13   % Calculate cylinder volume
14 - face_area=face_area_cyl(radius);
15 - volume=face_area.*height;
16 - end
17
18   % Local function to calculate cylinder's flat face area
19   % INPUT radius
20   % OUTPUT face_area = area of the flat face
21   function face_area=face_area_cyl(radius)
22   % Calculate cylinder face area
23 - face_area=pi*radius.^2; %(m^2)
24 - end
```

(a)   (b)

FIGURE 4.45. The *vol_cyl* function: (a) Calculating the cylinder volume, (b) Including the local function use.

about what function does and perhaps explaining, if needed, what inputs are expected and outputs are produced, will help the user seeking help for the function.

Another good programing practice when writing function code is to validate the function input argument values before continuing with calculations. In the example in Figure 4.45a, the input arrays of *radius* and *height* are checked to make sure that they contain no negative numbers and that both input arrays have the same dimensions. If a check fails, an *error* function is used to terminate the code execution and issue an error message explaining why the error occurred. Examples of errors due to mismatch of input array elements and negative radius values are shown in Figure 4.46c. Such error trapping is very helpful when debugging code. After the input validation checks, the output *volume* array is calculated using the element-by-element operations carried out with the .^ and .* operators.

```
Command Window
>> myCylVolume=vol_cyl(1,1)

myCylVolume =

    3.1416
```

```
Command Window
>> help vol_cyl
   Calculate cylinder volume
   INPUT radius  height
   OUTPUT Volume
```

(a)   (b)

```
Command Window
>> myCylVol=vol_cyl([1 2],[1])
Error using vol_cyl (line 11)
Both input variables must be the same size

>> myCylVol=vol_cyl([1 -2],[1 2])
Error using vol_cyl (line 6)
Inputs cannot be negative numbers
```

(**c**)

FIGURE 4.46. (a) Calling the *vol_cyl* function from the command line, (b) Using the *help* command with functions, (c) Invalid input examples.

Normally, each time a function is called, it creates its own local *Workspace*; any variables created locally within the function remain there and are not available in the *Workspace* of the code that called the function. Variable values are passed to the function via the input arguments and retrieved from the function via the output arguments. In some cases, however, it may be convenient to have a set of variables that are available to several functions. For example, one may have a large set of parameters that need to be used in multiple functions. In such cases, one may declare the variables using *global* command by appending their name after the command is issued. The same global statement is also made within each function that needs to access these variables. Figure 4.47 shows a script example for cylinder volume and area calculation that now uses global variables to pass the height and radius values to the functions called from within it (Figure 4.48).

```
Cyl_dim3.m    vol_cyl4.m    area_cyl4.m    +
1     % This code is used to calculate the cylinder volume and total area.
2     % Functions are used to calculate the volume & surface area
3     % Global variables are used
4  -  global radius height;
5  -  radius=input('Enter the cylinder radius: ');
6  -  height=input('Enter the cylinder height: ');
7  -  volume=vol_cyl4();
8  -  total_area=area_cyl4();
9  -  fprintf('\nFor cylinder with radius = %.2f and height = %.2f:\n',...
10        radius, height);
11 -  fprintf('Volume = %.3f and Area = %.3f\n',volume,total_area);
```

(a)

```
Command Window
>> Cyl_dim3
Enter the cylinder radius: 1
Enter the cylinder height: 1

For cylinder with radius = 1.00 and height = 1.00:
Volume = 3.142 and Area = 12.566
```

(b)

FIGURE 4.47. Script *Cyl_dim3.m*: (a) Using global variables, (b) Output.

```
Cyl_dim3.m    vol_cyl4.m    area_cyl4.m    +
1      % Calculate cylinder volume
2      % INPUT radius  height
3      % OUTPUT Volume
4      function volume=vol_cyl4()
5  -   global radius height;
6  -   if any(radius<0) || any(height<0)
7  -       error('Inputs cannot be negative numbers')
8  -   end
9  -   s_r=size(radius);
10 -   s_h=size(height);
11 -   if any(s_r~=s_h)
12 -       error('Both input variables must be the same size')
13 -   end
14     % Calculate cylinder volume
15 -   face_area=face_area_cyl(radius);
16 -   volume=face_area.*height;
17 -   end
18
19     % Local function to calculate cylinder's flat face area
20     % INPUT radius
21     % OUTPUT face_area = area of the flat face
22     function face_area=face_area_cyl(radius)
23     % Calculate cylinder face area
24 -   face_area=pi*radius.^2; %(m^2)
25 -   end
```

(a)

```
Cyl_dim3.m ✕ | vol_cyl4.m ✕ | area_cyl4.m ✕ | +
1      % Calculate cylinder total surface area
2      % INPUT radius  height
3      % OUTPUT area = Surface area
4   ⊟ function area=area_cyl4()
5 -    global radius height;
6 -    if any(radius<0) || any(height<0)
7 -        error('Inputs cannot be negative numbers')
8 -    end
9 -    s_r=size(radius);
10 -   s_h=size(height);
11 -   if any(s_r~=s_h)
12 -       error('Both input variables must be the same size')
13 -   end
14     % Calculate cylinder volume
15 -   area=2*pi*radius*height+2*pi*radius^2; %(m^2)
16 -   end
```

(b)

FIGURE 4.48. Functions called by *Cyl_dim3* script with global variable: (a) *vol_cyl4*, (b) *area_cyl4*.

4.9   Plots

Plotting data is a very useful tool for data analysis. It allows visualization of the outputs to compare them with the expected values or to obtain data trends. Data trends may be expressed in time, sample numbers, or indices in the case of frequency analysis. The data can have either an internal or external source.

Data from either source will end up stored as an array/matrix in the MATLAB *Workspace*.

To plot these data, they should represent equal numbers of columns (if they are presented by their rows headings) or equal numbers of rows (if they are presented by their column headings). For example, if Y is to be plotted against X, their lengths should be equal, assuming that X and Y are arrays of $1 \times n$, or if X and Y are arrays of vectors of $n \times 1$. They can also have the dimension $m \times n$, that will generate multiple curves.

The *figure* command opens a window within which the generated *plot*(s) or *subplot*(s) are displayed. By default, each plot appears in one figure, and the next plot replaces the previous one unless a different *figure* window has been created by issuing another *figure* command. A good practice is to define a figure with a specific ID for each plot for standardization and to have individual access to plot definitions. This is done by the *figure* (*j*) function, where *j* is the figure ID (number). To clear the MATLAB environment from all the figures, the *clf* command is used.

2D plots can be created by defining the horizontal (*x*-coordinate) and vertical (*y*-coordinate) variables (abscissa and ordinate). The *plot* function is then employed to plot the *x* versus the *y* data and represents them in the formats identified by the programmer (e.g., solid or dashed lines and circled or crossed markers) and in the selected colors. To obtain the ID of the current figure (the one to which the *plot* function output will be sent), use the *gcf* (get current figure) command.

In MATLAB, labels are the identifying features of the geometrical shapes. Descriptions of the individual curves are added via the *legend* command. Figure attributes, such as chart title and horizontal and vertical axes labels, are defined for the figures using the *title*, *xlabel*, and *ylabel* functions. The *x*- and *y*-coordinates upper and lower limits may be set by the *axis* function; the *x* limits are followed by the *y* limits. The variables to be plotted can be defined either in the form of arrays of vectors or formulae (Figure 4.49). The *x* and *y* upper and lower limits can be set independently, using the *xlim* and *ylim* functions.

```
   plot_circles.m  ×   circ_cal.m  ×   string_format.m  ×   ThermalCond.m  ×   ThermalCond2.m  ×   ThermalCond_Loop.m  ×   Thermal_Cond_Criteria.m  ×  +
1 -     close all; clear all; clc
2
3 -     for degree=1:10:360
4 -         theta=pi*degree/180;
5 -         x(degree)=sin(theta);
6 -         y1(degree)=cos(theta);
7 -     end
8
9 -     z = linspace(-1*pi,1*pi);
10 -    y2 = cos(z);
11
12 -    figure(1)
13 -        axis equal; plot(x,y1, 'k^',x,3*y1, 'k+',z,y2, 'k', 'LineWidth',1,...
14          'MarkerSize',5,'MarkerEdgeColor','k','MarkerFaceColor',[0.5,0.5,0.5])
15 -        legend('y1','3y1','y2')
16 -        title('Concentric Ellipse Plot'); xlabel('x-value'); ylabel('y-value')
17 -    figure(2)
18 -        barh(y1,'k')
19 -        title('Horizontal Bar Chart'); xlabel('x-value'); ylabel('y1-value')
20 -    figure(3)
21 -        bar(z,y2)
22 -        title('Vertical Bar Chart'); xlabel('z-value'); ylabel('y2-value')
23 -    figure(4)
24 -        histogram(y2)
25 -        title('Histogram'); xlabel('y2-value'); ylabel('Count')
26
27 -    figure(5)
28 -        axis equal;
29 -        subplot(2,2,1); plot(x,y1,'k^'); title('Ellipse Plot 1'); xlabel('x-value'); ylabel('y1-value'); legend('y1')
30 -        subplot(2,2,2); plot(x,4*y1,'k+'); title('Ellipse Plot 2'); xlabel('x-value'); ylabel('y1-value'); legend('4y1')
31 -        subplot(2,2,3); plot(z,y2,'k*'); title('Ellipse Plot 3'); xlabel('y2-value'); ylabel('y2-value'); legend('y2')
32 -        axis([-5 5 -2 2]);
33 -        subplot(2,2,4); plot(z,4*y2,'ko'); title('Ellipse Plot 4'); xlabel('y2-value'); ylabel('y2-value'); legend('4y2')
```

FIGURE 4.49. Script to create plots, bar charts, and histograms.

Different types of charts may be plotted in the **MATLAB** environment. The *scatter chart* (the plot points are connected by line of the desired styles; e.g., dotted and solid) or *separated chart*, and different markers with size and shape are available. It is possible to show single or multiple plots in one figure (Figure 4.50). Data can also be presented by means of the vertical and horizontal bar charts using the *bar* and *barh* functions (Figure 4.51) and histograms (Figure 4.52).

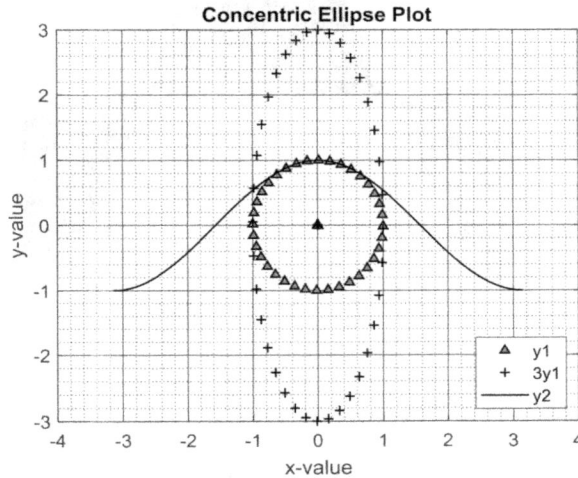

FIGURE 4.50. Multiple plots of the y-values versus the x-values in a single diagram with variables, defined in Figure 4.49.

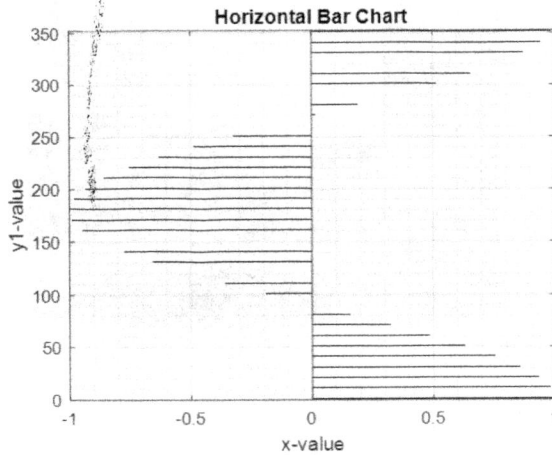

(a)

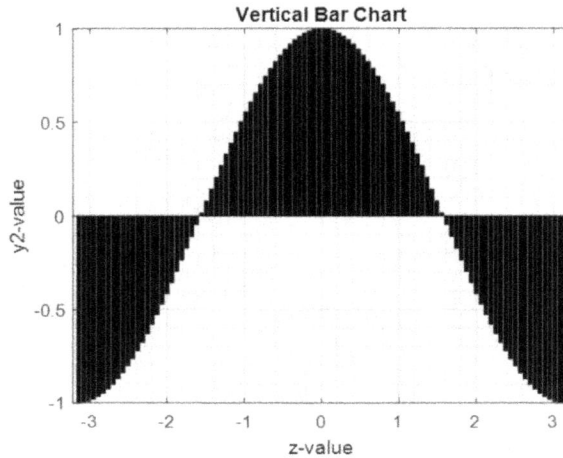

(b)

FIGURE 4.51. (a) $y1$-values versus the x-values in a horizontal bar chart, (b) $y2$-values versus the x-values in a vertical bar chart, defined in Figure 4.49.

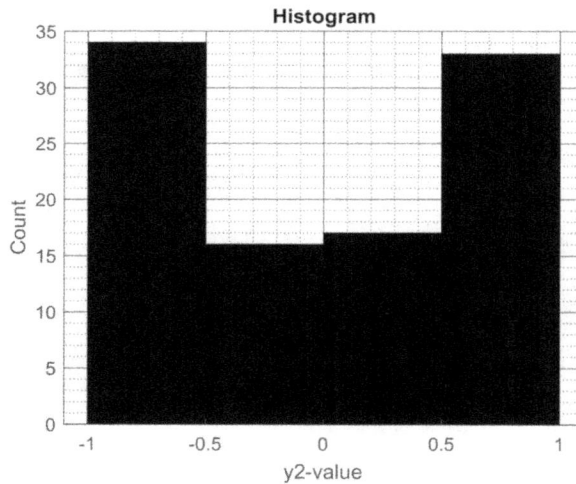

FIGURE 4.52. Histogram of the counts versus the $y2$-values, defined in Figure 4.49.

In order to show multiple plots in a single figure, the $subplot(i, j, k)$ function is used, where i and j are the grid size and k is the position of the referenced figure within the grid identified by sequential row-wise numbering starting from the top-left corner (Figure 4.53). If there are multiple curves to be displayed on the same figure, the *hold on* command may be used; this adds the output of any newly generated plot to the current figure. To release the figure, the *hold off* command is used; this means that the next figure will open on its own window. The *grid on* command adds vertical and horizontal grid lines to the current axes.

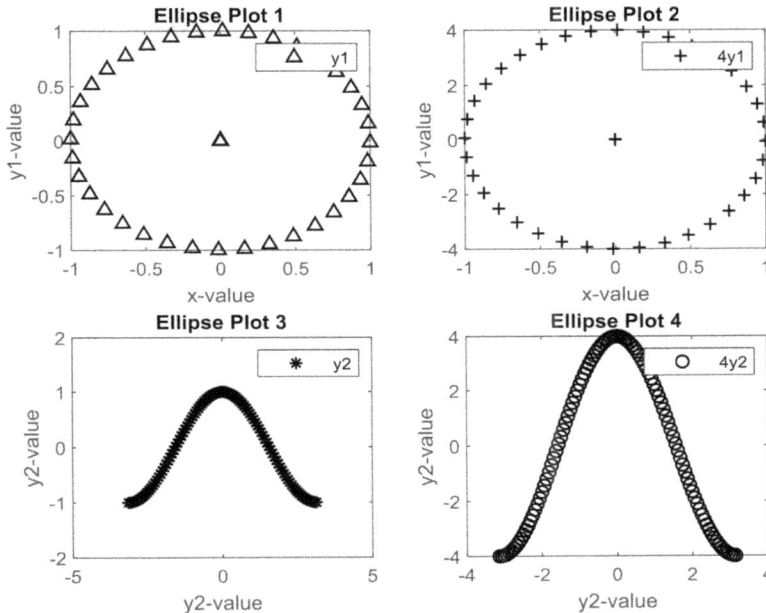

FIGURE 4.53. Four plots displayed in a single figure.

It is possible to customize the bar chart border and fill colors by accessing the *Edit > Figure Properties*, which is accessible through the *Figure* menu bar. Figure options, such as windows' appearance, position of the diagrams, units, colormap, rendering, printing, and exporting options, can be revised within each figure window (Figure 4.54). Clicking on the chart directly results in the same menu bar as that of the figure in addition to the *Debug* option that becomes accessible. All features such as plot type, line style, data source, and other chart information are available through the *Figure Properties* window. Figures can be saved or embedded later in the presentations and reports. To achieve this, on the *Figure* menu, *File > Save as* the command is activated and the figure is saved under the *.fig* format.

3D surface plots can be drawn using the *surf*(X, Y, Z) function. The function plots a surface grid with vertices above each (X, Y) point located at the height Z and with surface color determined by the value of Z. The X and Y inputs are matrices that are best generated using the *meshgrid* (X, Y) function. It takes vectors X and Y and replicates them to create matrix X, where all rows are copies of vector X, and a matrix Y, where all columns

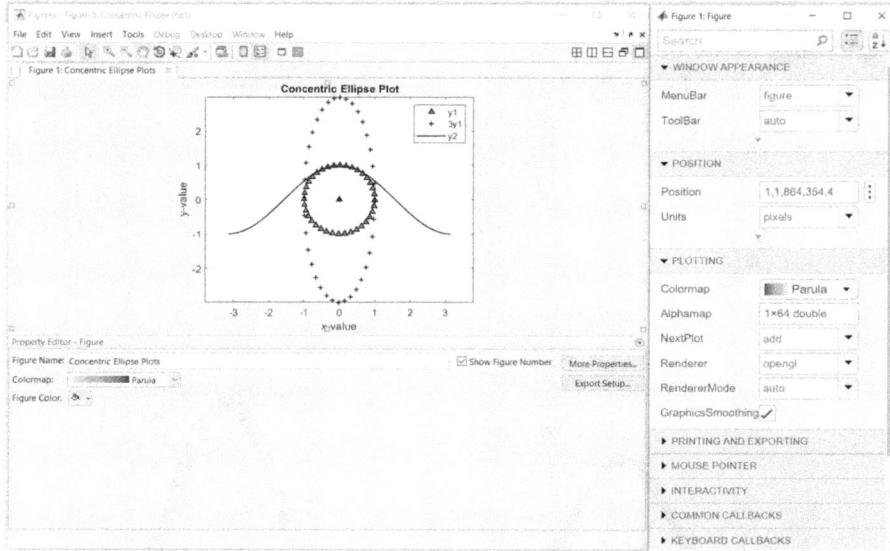

FIGURE 4.54. Figure properties options.

are copies of vector *Y*. These two matrices are then in the correct format as inputs for the *surf* function. Figure 4.55 shows an example 3D surface generated by the steps shown above the plot. It plots sum of squares of the *x*- and *y*-coordinate values.

```
>> [X,Y]=meshgrid(0:0.5:4,0:0.5:4);
>> Z=X.^2+Y.^2;
>> surf(X,Y,Z)
>> xlabel('X');ylabel('Y');zlabel('Z');
```

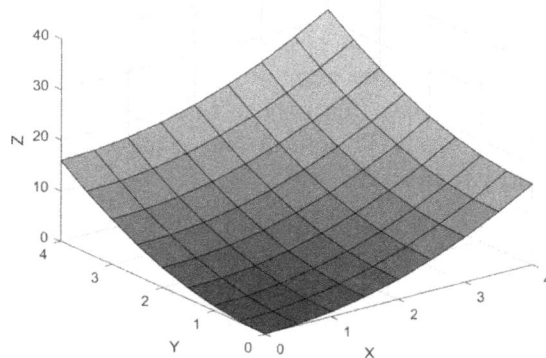

FIGURE 4.55. 3D surface plot created with the *surf* function.

4.10   Code Examples

This section includes three examples of the MATLAB code. The first example shows how code execution speed can be tested using the *tic* and *toc* commands. The second example shows the implementation of the material property entry that takes advantage of user-defined functions and shows use of the *while* loop and *switch* function. The third example shows implementation of random walk display, highlighting the use of the *while* loops and plotting.

4.10.1   Testing Code Execution Speed

The *tic* and *toc* commands are used to, respectively, start and stop a timer when executing code. The output of the *toc* command can then be assigned to a variable which would record the time in seconds elapsed from when the *tic* command was executed. Such time information is a good way to evaluate the efficiency of the code. One can try different ways to accomplish a programing task and compare the relative speed of execution between them. For example, it is known that it is more efficient to first create an array containing zeroes and then assign values to it compared to repeatedly expanding the array by adding new values to it.

In the example code in Figure 4.56, in the first *for* loop, the array *m* is expanded repeatedly over 1,000,000 loop iterations. In the alternative version, a vector array of zeroes with 1,000,000 elements is defined first using the *zeros* function and then values are assigned to this array. The

```
       tictoc.m   ×   +                      Command Window
 1 -      clear all; clc                Elapsed time is 0.076251 seconds.
 2 -      tic
 3 -      for I = 1 :1000000            tend1 =
 4 -          m(I)=2.5*I+5;
 5 -      end                              0.0764
 6 -      toc
 7 -      tend1=toc                     Elapsed time is 0.015873 seconds.
 8 -      tic
 9 -      m = zeros (1,1000000);        tend2 =
10 -      for I = 1 :1000000
11 -          m(I)=2.5*I+5;               0.0160
12 -      end
13 -      toc
14 -      tend2=toc                     timeratio =
15 -      timeratio=tend1/tend2
                                           4.7747
```

FIGURE 4.56. Code execution speed comparison using the *tic* and *toc* commands that shows benefit of predefining *zeros array*.

second code segment is completed in 0.016 seconds, which is about 4.8 times faster than the first segment.

The next code example shows efficiency improvement due to the use of element-by-element multiplication using the .* operator versus using the *for* loop. The *tic* and *toc* commands are used again to measure the elapsed time for each code segment. The execution time decreased by 6.7 times.

```matlab
timeTest.m  ×  +

1      % Code efficiency comparison for
2      % element-by-element multiplication vs. for loop
3 -    nElem=5000000;
4 -    aVec = 10*rand(nElem,1);
5 -    bVec = 10*rand(nElem,1);
6 -    abProd = zeros(nElem,1);
7      % Using the for loop
8 -    disp(['Loop product of two ' num2str(nElem)...
9          ' elem. arrays'])
10 -   tic
11 -   for i = 1:nElem
12 -       abProd(i) = aVec(i)*bVec(i);
13 -   end
14 -   Time_Loop=toc; display(Time_Loop)
15     % Using .* operator
16 -   disp(['Elem.-by-elem. prod. of two ' num2str(nElem)...
17         ' elem. arrays'])
18 -   tic
19 -   abProd = aVec .* bVec;
20 -   Time_Elem = toc; display(Time_Elem)
21
22 -   disp(['Time ratio is ' num2str(Time_Loop/Time_Elem)])


>> timeTest
Loop product of two 5000000 elem. arrays

Time_Loop =

    0.0265

Elem.-by-elem. prod. of two 5000000 elem. arrays

Time_Elem =

    0.0040

Time ratio is 6.707
```

FIGURE 4.57. Code execution speed comparison using the *tic* and *toc* commands: element-by-element operation benefit.

4.10.2 Entering Material Properties

The code in this example tells the user to enter three thermophysical properties of a material: density, thermal conductivity, and heat capacity. For each property, there is a recommended range of values that the program suggests. The thermal diffusivity is then calculated from the input properties and displayed. Additionally, the input values are summarized, and a note is made whether each is below, within, or above the recommended range.

This is accomplished by a script *matPropEntry.m* (Figure 4.58). It begins by creating a string array containing three pairs of property names and units. Note that this is an array of *strings*, and not *character* vectors. Only the former can contain text of varying length within each of its elements. Next a 3 × 2 array of lower and upper recommended property value limits is defined.

```
matPropEntry.m

1     % Ask user for thermal properties and test to make sure they are in
2     % correct range
3
4     % Array of strings containing property names and units
5     propDataStr = ["density",          "kg/m^3";...
6                    "thermal conductivity","W/mK";...
7                    "heat capacity","J/kgK"];
8     % Lower and upper limits for the input properties
9     propLim = [1500 2500; 900 3000; 100 500];
10    propVal = zeros(3,1); rangeInd = zeros(3,1); %initialize arrays
11    % Ask for property values
12    for i=1:3
13        propName=char(propDataStr(i,1));
14        propUnits=char(propDataStr(i,2));
15        propLimLow=propLim(i,1);propLimHigh=propLim(i,2);
16        [propVal(i),rangeInd(i)] = testPropLimits(propName,propUnits,propLimLow,propLimHigh);
17    end
18
19    display(blanks(2))
20    fprintf('\n ******* Input Material Properties ******\n');
21    for i=1:3
22        propName=char(propDataStr(i,1));
23        propName=[upper(propName(1)) propName(2:end)];
24        propUnits=char(propDataStr(i,2));
25        fprintf('%s = %g (%s)',propName,propVal(i),propUnits);
26        switch rangeInd(i)
27            case 1
28                fprintf('  (TOO LOW)\n')
29            case 2
30                fprintf('   (within correct range)\n')
31            case 3
32                fprintf('  (TOO HIGH)\n')
33        end
34    end
35
36    matAlpha = propVal(2)/(propVal(1)*propVal(3)); %m^2/s
37    fprintf('\n ******* Calculated Value ******\n');
38    fprintf('Thermal diffusivity = %g (m^2/s)\n',matAlpha);
```

FIGURE 4.58. Script *matPropEntry.m* that asks user to enter thermophysical properties.

Since all property-related information is stored in arrays, one can efficiently use the *for* loop to cycle through the three properties asking the user for input. For each property, one first needs to convert the strings to character vectors for property name and units. Then, the user-defined function *testPropLimits()* is called. It returns two values: the property value and an index from 1 to 3 that shows whether the returned value is below the recommended range (1), within the range (2), or above the range (3).

Next the *for* loop is used to display the entered values, with each value followed by a note saying if it is below, within, or above the recommended range. The note text is determined using the *switch* command that makes selection based on the values in the *rangeInd* array containing index values returned by the *testPropLimits* function.

The text is output to the *Command Window* using the *fprintf* command (Figure 4.59). *%s* formatting code is used to output character vectors and *%g* is used to output numbers. The latter code picks the more compact output between *%e* (exponential) and *%f* fixed-point notations. Note how the first *fprintf* function within the second *for* loop does not include the new line (\n) character at the end of the line. This allows one of the three *fprintf* functions within the switch command to append its text on the same line. After displaying the inputs, thermal diffusivity is calculated from the three properties entered and its value is displayed.

```
>> matPropEntry
Enter density value between 1500 and 2500 (kg/m^3): 1600
Enter thermal conductivity value between 900 and 3000 (W/mK): 800
WARNING: Entered thermal conductivity is less than lower limit of 900 W/mK
Enter new value (y/n)? (n=default)
Enter heat capacity value between 100 and 500 (J/kgK): 200


 ******* Input Material Properties ******
Density = 1600 (kg/m^3)  (within correct range)
Thermal conductivity = 800 (W/mK) (TOO LOW)
Heat capacity = 200 (J/kgK)  (within correct range)

 ******* Calculated Value ******
Thermal diffusivity = 0.0025 (m^2/s)
```

FIGURE 4.59. The result of running the script *matPropEntry.m*.

The *matPropEntry.m* script calls the function *testPropLimits* (Figure 4.60). It is instructive to examine in some detail its code to highlight the good programing practices employed. Before the function is defined, comment lines are added that explain the purpose of the function and describe all inputs and outputs. These comment lines will be printed if *help testPropLimits* is entered at the command line.

```
testPropLimits.m    +
1       %Check if the input is within specified limits and return the
2       %entered output value
3       %INPUTS
4       % propName = character array with material property name
5       % propUnits = character array with material property units
6       % lowerLimit = number defining property lower limit
7       % upperLimit = number defining property upper limit
8       %OUTPUT
9       % outValue = property value entered by user
10      % outIndx = indicates if the value is below lower limit (1); between limits (2);
11      % or above upper limit (3)
12      function [outValue,outIndx]=testPropLimits(propName,propUnits,lowerLimit,upperLimit)
13 -    if not(isnumeric(lowerLimit)) || not(isnumeric(upperLimit))
14 -        error('Lower and upper limits must be numbers.')
15 -    end
16 -    if not(ischar(propName)) || not(ischar(propUnits))
17 -        error('Property name and units must be character arrays.')
18 -    end
19 -    if lowerLimit >= upperLimit
20 -        error('Property lower limit must be less than upper limit.')
21 -    end
22 -    lowIndx = 1; okIndx = 2; highIndx = 3;
23 -    done = false; outIndx = 0;
24 -    while not(done)
25 -        outValue = input(['Enter ' propName ' value between ' num2str(lowerLimit)...
26              ' and ' num2str(upperLimit) ' (' propUnits '): ']);
27 -        if outValue<lowerLimit
28 -            disp(['WARNING: Entered ' propName ' is less than lower limit of '...
29                  num2str(lowerLimit) ' ' propUnits])
30 -            outIndx = lowIndx;
31 -        elseif outValue>upperLimit
32 -            disp(['WARNING: Entered ' propName ' is greater than upper limit of '...
33                  num2str(upperLimit) ' ' propUnits])
34 -            outIndx = highIndx;
35 -        else
36 -            outIndx = okIndx;
37 -        end
38 -        if outIndx ~= okIndx
39 -            yourAnswer=input(['Enter new value (y/n)? (n=default)'],'s');
40 -            % if no value is entered, default to keep same value
41 -            if isempty(yourAnswer)
42 -                yourAnswer='n';
43 -            end
44 -            if (yourAnswer(1)=='y') || (yourAnswer(1)=='Y')
45 -                done=false;
46 -            else
47 -                done=true;
48 -            end
49 -        else
50 -            done=true;
51 -        end
52 -    end %while
53 -    end %function
```

FIGURE 4.60. Function *testPropLimits* that evaluates entered values to test if they fall within specified range.

After the function is declared with the *function* statement, the inputs are tested to make sure that they are numerical values or character strings as appropriate. It is also verified that the upper limit is greater than the lower limit. If any of the inputs are found to be of incorrect type, an *error* function is used to terminate the program execution and to display a meaningful error message.

A logical variable *done* is used to control when the *while* loop is exited. It is helpful for the ease of code interpretation to use meaningful variable names: a function like *while not(done)* is then easy to understand. Within the *while* loop, the first *if–then–else* statement is used to determine whether the entered value is below, within, or above the specified value range, and appropriate *outIndx* value is assigned. Also, warnings are given to the user if the entered value is outside the range. Note also how three constants (*lowIndx*, *okIndx*, and *highIndx*) have been defined before the *while* loop entry. One could have used numbers 1, 2, and 3 instead but giving these value names makes it easier to understand the code within the first *if–then–else* statement inside the *while* loop.

If the entered value was within the recommended range, the variable *done* is assigned *true*. If the entered value was not within the range, the user is given an option to enter a different value or to keep the entered value. Depending on the answer, the variable *done* is assigned *true* and the loop terminates, or *false* and the loop continues.

The implementation of the *testPropLimits* function highlights the benefits of creating functions that are as general in their application as possible; this enhances the value of the created code. The function is applied to material properties in this case but can be applied to any other situations as well. Implementing such functions also reduces the overall complexity of the code since most of the code is now hidden within the function while the script *matPropEntry.m* that calls the function remains compact and easy to follow.

4.10.3 Random Walk Plot

This code example plots the steps of a random walk that starts at the origin and is carried out within a square enclosing box with sides of 100 units (Figure 4.61). The user enters the maximum allowed step size component in the *X* and *Y* directions and, using this value, a random integer is generated with the *randi* function at each step. This random value determines the next step location, and the new point is displayed on a plot.

The *while* loop is used to repeat the steps of the walk. The loop is repeated until either the next step goes beyond the enclosing box walls or a specified maximum number of steps is exceeded.

```
Random_Walk_Plot.m  ×  +

1     % Script displays a visual representation of a random walk
2     % Script terminates when a boundary wall is hit or the maximum number
3     % of steps is reached
4 -   clear; clc; close all;
5 -   box_size = 100; %size of the square walk enclosure box in X and Y
6 -   steps_lim = 20;  %max number of steps allowed
7 -   textOffsetX = 3; %offset in X of step number label from the step point
8 -   curr_step=[0 0];
9     % initialize walk coordinates; walk starts at origin
10 -  walk_point =zeros(steps_lim+1,2);
11 -  X=1;Y=2; % define indices for X and Y components
12    % Ask user for step size limit
13 -  max_step_size = testPropLimits('maximum step size','units',box_size*0.05,box_size);
14    % create figure to display random walk
15 -  figure(1);
16 -  plot_limit = 1.2*(box_size/2);
17 -  ylim([-plot_limit plot_limit]); xlim([-plot_limit plot_limit]);
18 -  plot((box_size/2)*[-1,1,1,-1,-1],(box_size/2)*[1,1,-1,-1,1],'Color','r','LineWidth',3)
19 -  grid on
20 -  axis padded equal
21 -  title('Random Walk Plot'); xlabel('X'); ylabel('Y');
22 -  lineID = animatedline('LineStyle',':','Color',[0.5 0.5 0.5],'LineWidth',1,'Marker','o', ...
23    'MarkerSize',5,'MarkerEdgeColor','k','MarkerFaceColor',[0.5,0.5,0.5]);
24    % Initialize while loop variables
25 -  i=1; hitWall = false; iterLim = false;
26 -  while not(hitWall)&& not(iterLim)
27 -      addpoints(lineID,walk_point(i,X),walk_point(i,Y));
28 -      drawnow
29 -      text(walk_point(i,X)+textOffsetX,walk_point(i,Y),num2str(i-1));
30 -      curr_step(X)=randi([-max_step_size max_step_size]); % step X coordinate
31 -      curr_step(Y)=randi([-max_step_size max_step_size]); % step Y coordinate
32 -      next_point=walk_point(i,:)+curr_step;
33 -      if any(abs(next_point)>= box_size/2)
34 -          hitWall = true;
35 -          hold on; plot(next_point(X),next_point(Y),'xr','MarkerSize',8);hold off
36 -          text(next_point(X)+textOffsetX,next_point(Y),num2str(i));
37 -      elseif (i-1 == steps_lim)
38 -          iterLim = true;
39 -      else
40 -          i = i + 1;
41 -          walk_point(i,:)=next_point;
42 -      end
43 -  end
44
45 -  if hitWall
46 -      disp(['Hit the wall at step ', num2str(i),'.']);
47 -      fprintf('Invalid position is (%d,%d). \n',next_point)
48 -  end
49 -  if iterLim
50 -      disp(['Reached the iteration limit of ', num2str(steps_lim),'.']);
51 -  end
52 -  fprintf('Final valid position is (%d,%d). \n',walk_point(i,:))
```

FIGURE 4.61. *Random_Walk_Plot.m script.*

The user input of the maximum step size value is obtained by employing the same *testPropLimits* function that was used in the code example described in Section 4.10.2. Recommended range limits for this input are specified to be between 5% and 100% of the box size. Employing the same function for two different applications again demonstrates the value of writing a code that is as general as possible.

A figure window is created next, and plot's X and Y axes limits are defined. A square is drawn with a thick red line to represent the enclosing box. To draw the lines joining the steps points of the walk, a function *animatedline* is used. It allows incremental addition of new points, with a line drawn to connect each new point to the previous one. The process is initialized by calling the *animatedline* function and specifying the line parameters—in this case a dotted gray line, with line width of 1, markers of small black circles filled with gray color.

The *while* loop is executed if the walk has not gone beyond the box walls and the iteration limit has not been reached. The status of these two conditions is indicated with logical variables *hitWall* and *iterLim*. These again demonstrate the benefit of using meaningful names in such cases. One can then write the *while* loop statement as repeat the code in the loop *while not(hitWall) && not(iterLim) && not(iterLIm)*.

A new point is added to the line with the *addpoints* function, which is provided with the input of the line ID and the new point coordinates; this is followed by the *drawnow* command to add the line and point to the plot.

Next, a random step is generated and evaluated with the *if–elseif–else–end* statement. It tests first if the new point is located beyond the box boundaries, setting *hitWall* to *true* if this is the case. A red *x* is then added to the plot at the location, where this step would have gone to.

Then, the *else–if* statement tests if the maximum number of steps has been reached, setting *iterLim* to *true* in that case. If neither of the above logical tests is true, the loop counter *i* is incremented and the next point is copied into the *walk_point* array that contains the coordinates of all the step points. An information summary is presented at the end of the script; the printout uses the two logical variables that controlled the loop execution to indicate the cause of the loop exit (Figure 4.62) and the plot is displayed (Figure 4.63).

The sample run results are shown for the maximum step size of 25. The results show that in this run four valid steps were made and at fifth step to (60, 4) the wall would have been hit. The last valid (within the box) point was at (39, –15).

```
Command Window

Enter maximum step size value between 5 and 100 (units): 25
Hit the wall at step 5.
Invalid position is (60,4).
Final valid position is (39,-15).
```

FIGURE 4.62. Text input/output of the *Random_Walk_Plot.m* script.

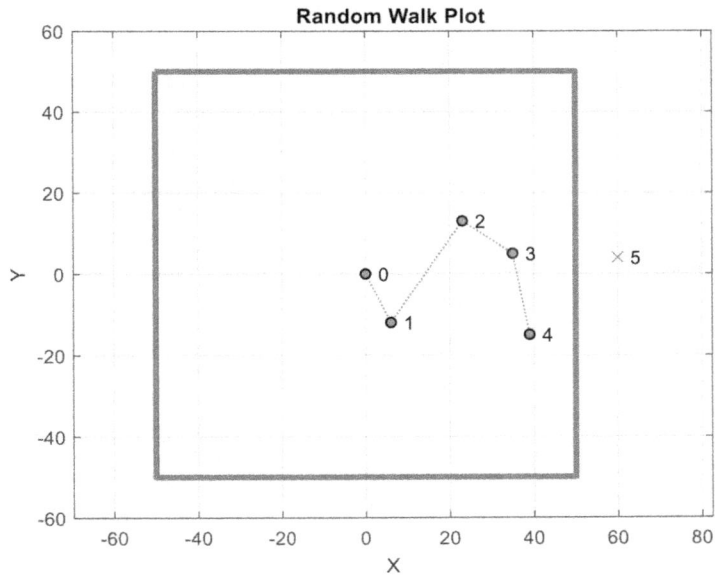

FIGURE 4.63. Random walk plot for maximum step size of 25; box boundaries indicated by red square.

End Notes

[53]   https://www.mathworks.com/products/matlab.htm
[54]   https://www.mathworks.com/products.html

CHAPTER 5

Heat Transfer Problems in MATLAB

artial Differential Equations (PDEs) can be used to describe system behavior for problems in heat transfer, structural mechanics, and numerous other fields. A general form of the heat transfer model PDE is shown in Equation (62), where T is the temperature and t is time. Parameters m, d, c, a, and f can be functions of time (t) and location within the domain, space (e.g., x, y, and z), as well as the state variable (T) or its gradients (e.g., dT/dx and dT/dt); they may also be constants. When m and d are both zero, the PDE models a stationary (*steady-state*) system. When either of m or d are nonzero, the problem represents a *transient* system. For a system of PDEs consisting of n state variables, n linearly independent equations are needed. Equation (62) can be restated as Equation (63), where its parameters are expressed as thermophysical properties.

$$m\frac{\partial^2 T}{\partial t^2} + d\frac{\partial T}{\partial t} - \nabla \cdot (c\nabla T) + aT = f \tag{62}$$

$$\rho C_p \frac{dT}{dt} - \nabla \cdot (k\nabla T) = Q + h(T_\infty - T) \tag{63}$$

The *Dirichlet* and *Neumann* boundary conditions can be applied to the edges (2D geometry) and surfaces (3D geometry). The *Dirichlet* boundary condition associates the state variable, such as temperature (T), to time (t), location within the domain $(x, y,$ and $z)$ and other parameters (e.g., h and r). These parameters can be functions of time (t) and location within the domain $(x, y,$ and $z)$ as well as the state variable (T) or its gradients; they may also be constants. *Dirichlet* boundary condition is presented by Equation (64).

$$hT = r \tag{64}$$

Neumann boundary condition associates the state variable such as temperature (T) to time (t), location within the domain (x, y, and z) and other parameters. It is described by Equation (65), where n is the unit vector, pointing outward from the edge or surface. A dot (inner) product of n with the gradient vector (∇T) returns a scalar quantity. The parameters c, q, and g can be functions of time (t) and location within the domain (x, y, and z) as well as the state variable (T) or its gradients. If any of the coefficients depend on the state variable (solution, T) or its gradient, the problem is nonlinear.

$$n \cdot (c\nabla T) + qT = g \tag{65}$$

5.1   Introduction to PDEs in MATLAB

By creating with scripts and functions in the base MATLAB installation, it is possible to solve a PDE model using the Finite Difference Method (FDM), an analysis technique that is carried out by discretizing the derivatives (the first or the second order) to linear relations and then employing elimination methods (e.g., Gauss-Sidle) to solve the problems. However, complex geometries and boundary conditions cannot be accurately addressed by this method. Instead, the FEM technique, which divides the geometry into small subdomains (elements), needs to be employed to obtain accurate solutions. Nevertheless, if the programmer is able to achieve an accurate solution using the simpler method, the benefits of faster solution times can be obtained.

There exists a specialized function within the base MATLAB installation for solving PDEs. However, this *pdepe* function only solves one-dimensional problems with one spatial variable (x) and the optional time variable (t) for *transient* analyses (parabolic relations). The *pdepe* solver transforms the partial differential equations into the Ordinary Differential Equations (ODE) using specific user-defined nodes, as shown in Equation (66), where m is the equation parameter. f and s are flux and source functions. c is a diagonal matrix with elements that are identically zero (*elliptic* functions) or positive (*parabolic* functions). f, s, and c are independent functions of the space (x) and time (t), dependent variable (u) and its special (du/dx) and temporal (du/dt) derivatives. The outputs c, f, and s correspond to the coefficients in the standard ODE equation, Equation (66), expected by the *pdepe* solver. These coefficients are coded using specific commands and functions in terms of the input variables mentioned above.

$$c\left(x,t,u,\frac{\partial u}{\partial x}\right)\frac{\partial u}{\partial t} = x^{-m}\frac{\partial}{\partial x}\left(x^{m}f\left(x,t,u,\frac{\partial u}{\partial x}\right)\right)+s\left(x,t,u,\frac{\partial u}{\partial x}\right) \quad (66)$$

Time integration is then performed using the *ode*15*s* solver. It is therefore possible to handle Jacobians (first-order partial derivatives) of vector functions.

To facilitate dealing with PDEs, MathWorks offers MATLAB, the add-on called the *PDE Toolbox*, which is part of the *Math and Optimization* product family. It provides several dedicated functions that simplify the creation of models, definition of geometry, solution, and plotting of the results. These functions can be used to solve PDEs using the finite element method (FEM). They facilitate specification of the initial and boundary conditions, as well as additional terms such as internal heat generation. The results of analyses can be displayed as temperature distributions, heat flux, and heat flow rates at the specified boundaries or domains, and other forms of temperature derivatives. Unlike the *pdepe* function of the base install, models with 2D and 3D geometry can be handled, as well.

The remaining sections in this chapter will look at how the *PDE Toolbox* facilities can be used to solve the heat transfer problems using the FEM method. First, a simpler approach using the *PDE Modeler* interface will be discussed; afterwards, a comprehensive guide will be given to using the *PDE Toolbox* function to solve these problems. As with any other FEM analysis tool, common steps need to be followed and decisions made. Following is a summary of these steps as they pertain to setting up a heat transfer FEA model with the *PDE Toolbox*:

(1) Geometry

    (a) Decide whether 2D or 3D geometry is appropriate. Consider whether the axisymmetric model can be employed that can represent a 3D geometry by a 2D model, greatly simplifying and speeding up the solution.

    (b) Decide if the geometry can be created internally within MATLAB or is to be imported from an external tool (e.g., CAD software or another FEA tool). This will depend on the model's complexity and the capabilities of the MATLAB's internal tools. Internal model creation simplifies model modifications.

(2) Boundary conditions

Decide on appropriate conditions for each boundary. Is it insulated? Is there a fixed temperature or a heat flux to be applied?

(3) Material properties

(a) Does the model consist of one material or are there multiple regions with different material properties?

(b) Find material property specifications that accurately describe the materials modeled. Are these properties constant or do they depend on location, time, or temperature? If they are not constant, then how their mathematical variations are described?

(4) Mesh

(a) What is the appropriate size of the mesh elements? For large models, small element size will lead to long solution times. One can compare solution results for progressively smaller mesh sizes to determine the appropriate level of mesh refinement. If further element size reduction does not change the results, then an appropriate element size has been found.

(b) Make sure the mesh elements are of good quality, i.e., not excessively stretched along one direction.

(5) Solution

Define initial conditions and specify solution parameters, such as the error tolerance level, which will determine when the solution is complete. Excessively small tolerance will lead to longer solution times.

(6) Post-processing of results

(a) Decide which information about the solution will need to be displayed.

(b) Generate plots, tables, or calculations to obtain the desired information.

(c) Consider extracting the solution results and processing with the third-party tools if available (e.g., Microsoft Excel).

5.2 Thermal Modeling Using the MATLAB *PDE Modeler* Application

The *PDE Modeler* application is a tool available within the *PDE Toolbox*. It presents a graphical interface that provides user with a simpler way to access the *PDE Toolbox* capabilities to solve 2D models using the FEM technique (Figure 5.1).

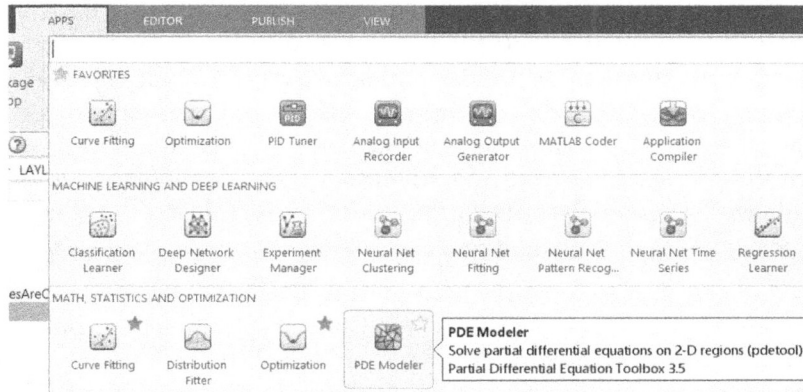

FIGURE 5.1. The *PDE Modeler* application in the *Math, Statistics,* and *Optimization* grouping.

5.2.1 The *PDE Modeler* Overview

When started, the *PDE Modeler* application presents a window with menus and a row of icons along the top edge that allow for entry of the necessary information to set up and solve a 2D FEM model (Figure 5.4). The model geometry is displayed on the plot in the middle, and the prompts are shown along the bottom strip. This application includes built-in formulae (models), with their physics already predefined for solution of thermal, structural, electromagnetic, and generic PDE problems.

A typical workflow would be to go in sequence through the menus, left to right, starting at *Options*. For the *Draw, Boundary, PDE,* and *Mesh* menus, one first needs to activate the corresponding *Mode* by clicking on the menu and selecting the first item (e.g., *Draw Mode*). Under the *Solve* and *Plot* menus, one simply executes the corresponding action. If any of the steps are skipped, the application runs with default settings (that include everything, even the geometry). One can thus start application, select *Solve* > *Solve PDE* and a solution will be presented. It is thus important to take care to define all the settings as required.

To create 2D geometries, basic 2D shapes, such as ellipses, rectangles, and polygons, can be added, transformed, and combined with Boolean operations. Note that the application can only solve 2D problems. After geometry is defined, boundary conditions are set. *Neumann* (or the default *Dirichlet*) boundary conditions can be defined for the selected boundaries, with customizable coefficients (e.g., $q = 0$ and $g = -5$). The PDE coefficients are then set using *PDE* > *PDE Specifications*. Mesh is then created by activating *Mesh* > *Mesh Mode*. The element size can be decreased by refining the mesh and various labels, and the mesh quality can be displayed.

The PDE is then solved by selecting *Solve > Solve PDE*. This command assembles the defined features (geometry shapes, equation coefficients, boundary conditions, and mesh), and solves the PDE model. The results are then plotted in the default diagram. 3D surface plots can be generated showing the variation of a dependent variable (e.g., temperature) or its derivatives (e.g., temperature gradient and heat flux) versus the position within the 2D geometry. 3D surface plots can be selected by going to *Plot > Parameters* and checking the *Height (3-D plot)* box.

One can save to a file the state of the *PDE Modeler* and restore it by opening the file later. The file is stored as the *.m* function, but it is not recommended to edit it directly. If any additional customization is required, it is suggested to either export the variables and use a script to work on them or set up and solve the problem from the beginning as a script, as described in Section 5.3.

5.2.2 Creating 2D Geometry

2D geometry is created by first activating *Draw > Draw Mode*. Shape choices are rectangle/square, ellipse/circle, or polygon. For the first two geometries, one can either create them by defining centers first or by clicking at one of the two diagonal vertices and dragging to define the shape's size. Square/circles are drawn by either using the right mouse button or by holding control key while dragging with the left mouse button. After the shape's creation, its dimensions and origin can be edited via the *Object Dialog* window that can be accessed by double-clicking on the specific shape (e.g., circle 1, *C*1) (Figure 5.2).

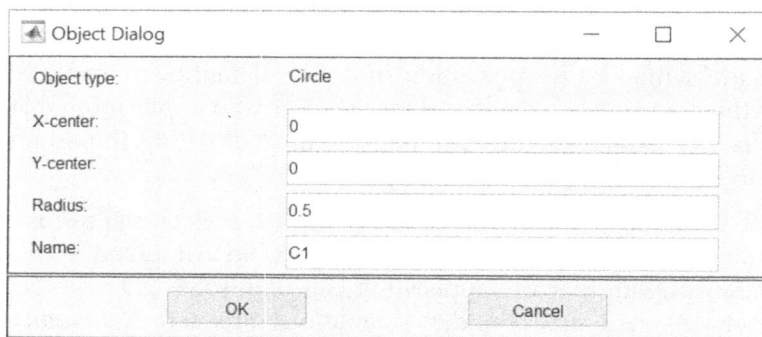

Object Dialog	— □ ✕
Object type:	Circle
X-center:	0
Y-center:	0
Radius:	0.5
Name:	C1

OK	Cancel

FIGURE 5.2. The *PDE Modeler, Object Dialog settings* for a Circle *C1, Draw Mode.*

A good practice is to define appropriate grid size so that when shapes are located or sized, the points will snap to these grid values, and accurate round numbers will result. To display and assign the grid size, select *Options* > *Grid Spacing*. The default is *Auto*; this can be unchecked, and the desired values entered (Figure 5.3). Then, select *Options* > *Snap*.

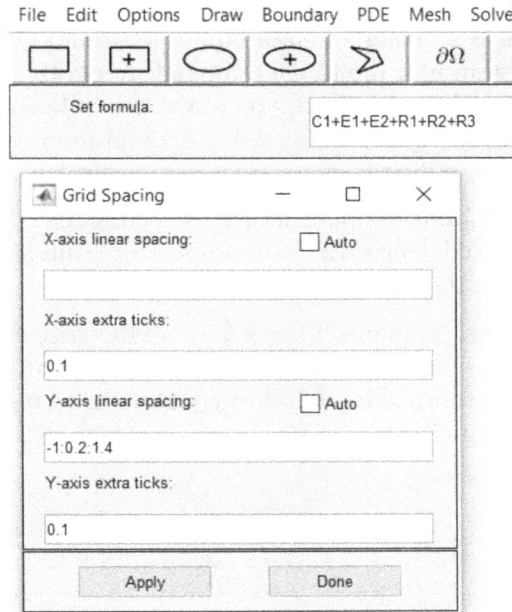

FIGURE 5.3. The *PDE Modeler, Grid Spacing settings, Options Mode.*

When each geometrical entity is created, it is automatically assigned a unique ID (*Name*); for example, the first rectangle that is generated is identified as $R1$ and the second circle is identified as $C2$. Shape IDs can be changed via the shape properties dialog box, under the *Name* field (Figure 5.2). The advantage of having shape IDs is that the geometry may be constructed by referencing these unique IDs within the *Boolean* operations (i.e., addition and subtraction). For example, to create the cross section of a hollow cylinder, given a certain thickness, the internal ($C1$) and external ($C2$), where $R_{C1} > R_{C2}$, concentric circles are defined first; the smaller circle is then subtracted from the larger one in a Boolean operation ($C1 - C2$); this operation is entered in the *Set Formula* field (Figure 5.4).

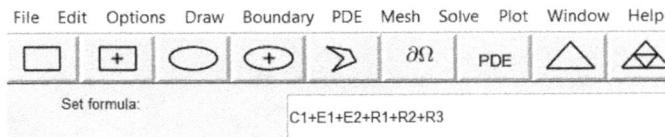

FIGURE 5.4. The *PDE Modeler,* defining *Set formula* by addition/subtraction.

5.2.3   The *PDE Modeler*: A Step-by-Step Guide

The process of setting up, solving, and analyzing a 2D PDE problem within the *PDE Modeler* application is described in the following paragraphs. The example model describes a *transient* heat transfer system, where a circular core made of aluminum is surrounded by a ring made of copper. The system is at the initial temperature of 30 °C and the environment is at 30 °C. There is also internal heat generation in the core region (100,000 W/m³). The system is modeled over an interval of 1,000 seconds and the temperature distribution at that time is displayed.

(1)   Run the *PDE Modeler* application by selecting the *APPS* tab on the menu bar and clicking on its icon (Figure 5.1); alternatively, enter *pdeModeler* at the command line.

(2)   Under the *Options* menu (Figure 5.5), set the grid, its spacing, and axis upper and lower limits. Next, set the model application type (e.g., *Heat Transfer* for a thermal model) through *Options > Application > Heat Transfer* (Figure 5.6).

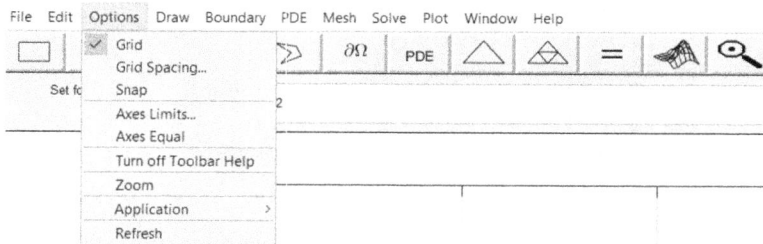

FIGURE 5.5. The *PDE Modeler*, *Options* menu items.

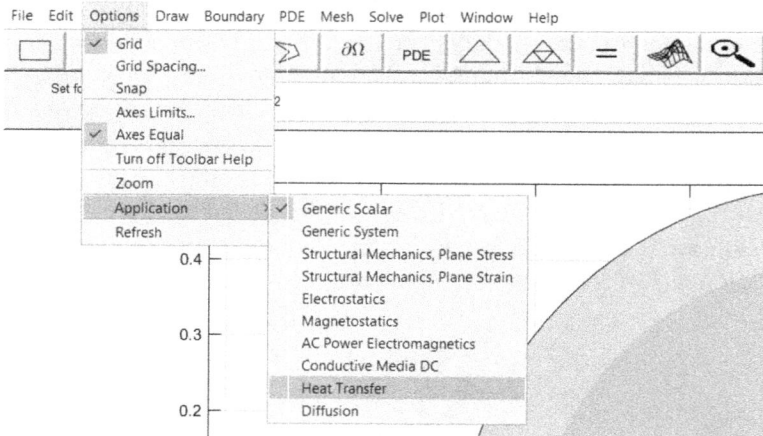

FIGURE 5.6. The *PDE Modeler*, *Options*, *Application* menu items.

(3) To create the example's 2D geometry, define two concentric circles with radii of 0.4- and 0.5-units using *Draw > Ellipse/circle (centered)*. Click at the center's location and drag with the *right* mouse button to create each circle. Two circle entities are thus created and given names *C1* and *C2* (Figure 5.7). Note that dragging with the left mouse button will allow creation of an ellipse or a circle. If you happened to have created an elliptical shape (named *E1*), it can be changed into a circle by assigning equal values to both semiaxes in the shape's properties. If you want to return to this step, *Draw > Draw Mode* can be selected again to modify the shapes or to add new ones.

Set formula is used to define how the overlapping shapes will be treated—subtracted or added. The *Set formula* field, visible in Figure 5.8, shows *C1 + C2* to indicate that the two shapes are to be combined to create two regions in the model.

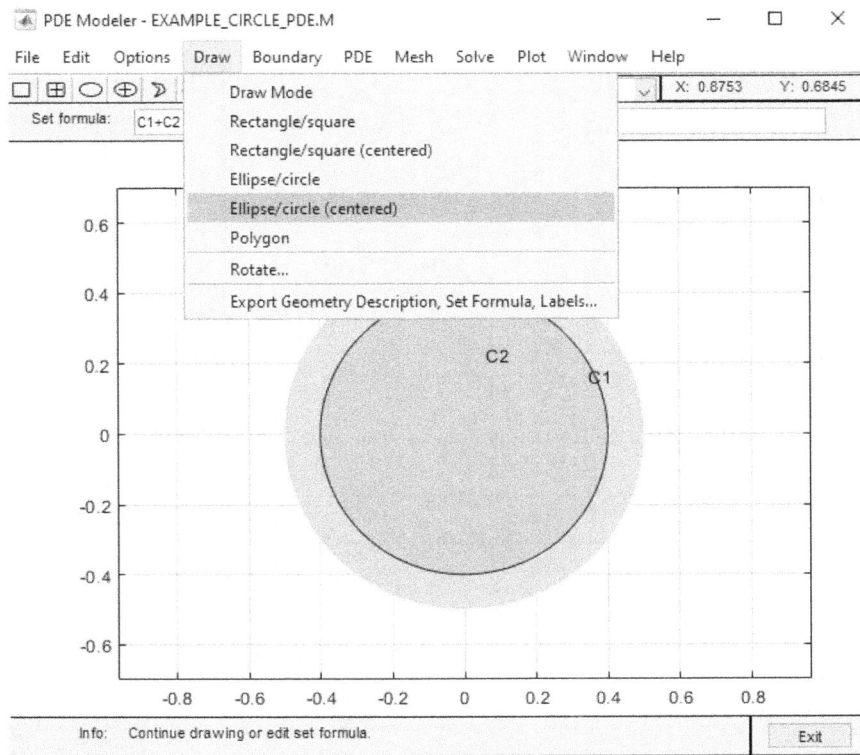

FIGURE 5.7. The *PDE Modeler, Draw* menu items.

(4) Boundary conditions on the geometrical segments are set as the next step after selecting *Boundary > Boundary Mode* (Figure 5.8). All boundaries have preassigned default insulated boundary conditions: the *Dirichlet* condition—$hT = r$ in Equation (64), where $h = 1$ and $r = 0$. If there are unneeded subdomain borders, delete them by selecting the *Remove Subdomain Border* or *Remove All Subdomain Borders*. By holding the *shift* key and clicking on the individual boundaries, multiple selections can be made. Finally, select the exterior circle boundaries (boundaries 1-5 as shown in Figure 5.8) and apply the conditions $h = 1$ and $r = 30$ as shown in Figure 5.9.

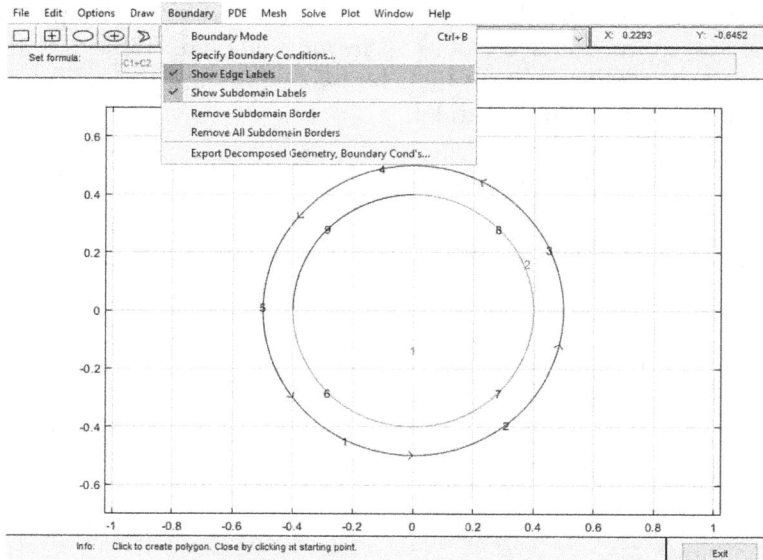

FIGURE 5.8. The *PDE Modeler*, *Boundary* menu items.

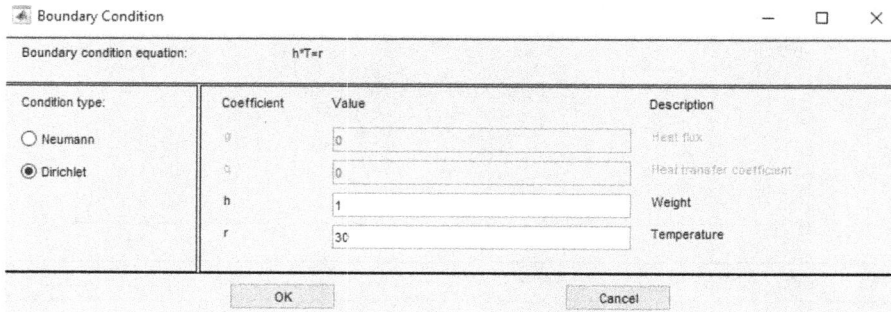

FIGURE 5.9. The *PDE Modeler*, *Boundary Condition* settings for *Dirichlet* boundary conditions.

(5) Activate *PDE > PDE Mode*. For each region, define specific PDE coefficients as needed. *PDE Specification* is selected from the *PDE* menu (Figure 5.10) to select the *Type of PDE* and to enter the PDE coefficients into the fields of the dialogue box (Figure 5.11). The coefficients must be assigned before the solution step. The coefficients do not depend on the geometry or boundary conditions, but on the physics. These coefficients can be assigned for individual regions by double-clicking each and entering the values. If regions are not selected individually, the default (initial PDE specifications) will propagate to all the domain PDEs. For a *steady-state* thermal model, an elliptic PDE should be assigned (Figure 5.11). In case of a *transient* condition, as is the case in this example, a paraboilic PDE should be selected as shown in Figure 5.12 and Figure 5.13. The coefficient values shown represent an interior region made of aluminum and an exterior region made of copper. The external temperature is assumed at 30 °C with a convection coefficient of 10 W/m²K. The internal core generates heat at 10,000 W/m³.

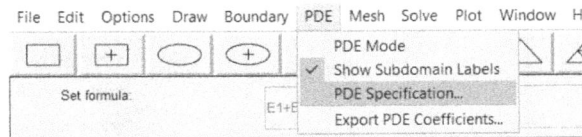

FIGURE 5.10. The *PDE Modeler*, *PDE* menu items.

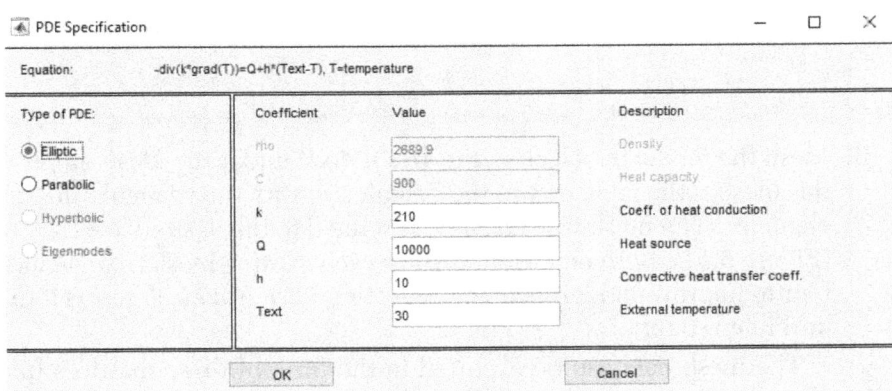

FIGURE 5.11. The *PDE Modeler*, *PDE Specification* settings for the *Elliptic* model.

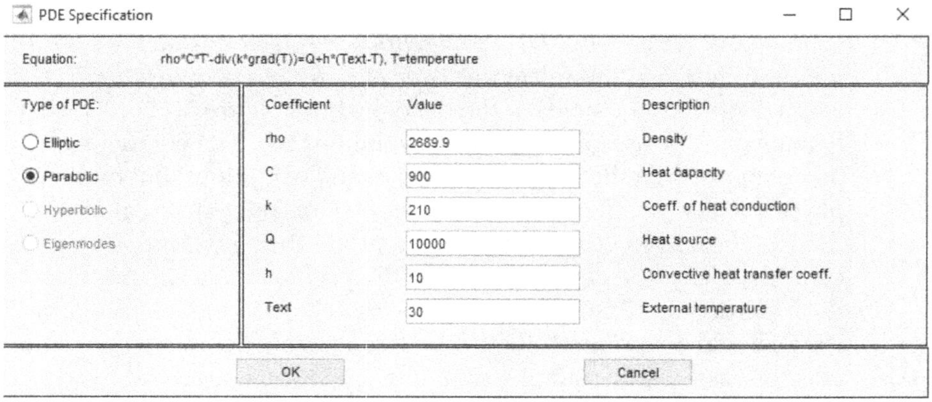

FIGURE 5.12. The *PDE Modeler, PDE Specification* settings for the *Parabolic* model (interior region, aluminum).

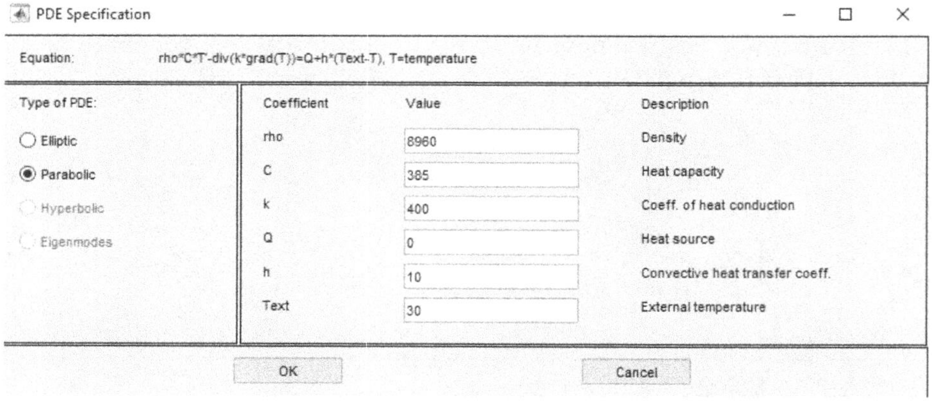

FIGURE 5.13. The *PDE Modeler, PDE Specification* settings for the *Parabolic* model (exterior region, copper).

(6) Mesh the model by selecting the *Mesh Mode* under the *Mesh* menu; this meshes the geometry to the default size with the triangular mesh elements. The mesh size can also be refined to the desired level (Figure 5.14). Refinement subdivides each existing mesh triangle into four to improve model accuracy. Selecting *Initialize Mesh* resets it to unrefined state.

The mesh data can be exported in the form of three matrices [p, e, t]. p is the matrix of nodes ($2 \times N_p$), where N_p is the number of nodes. The first and second elements in each node (matrix column) are the node coordinates (x, y). e is the matrix of edges ($7 \times N_e$), where N_e is the number of edges in the mesh. Finally, t is the matrix of triangles. For each mesh triangle, it lists indices from the p array to identify the nodes that belong to this triangle. The *PDE Modeler* uses linear elements

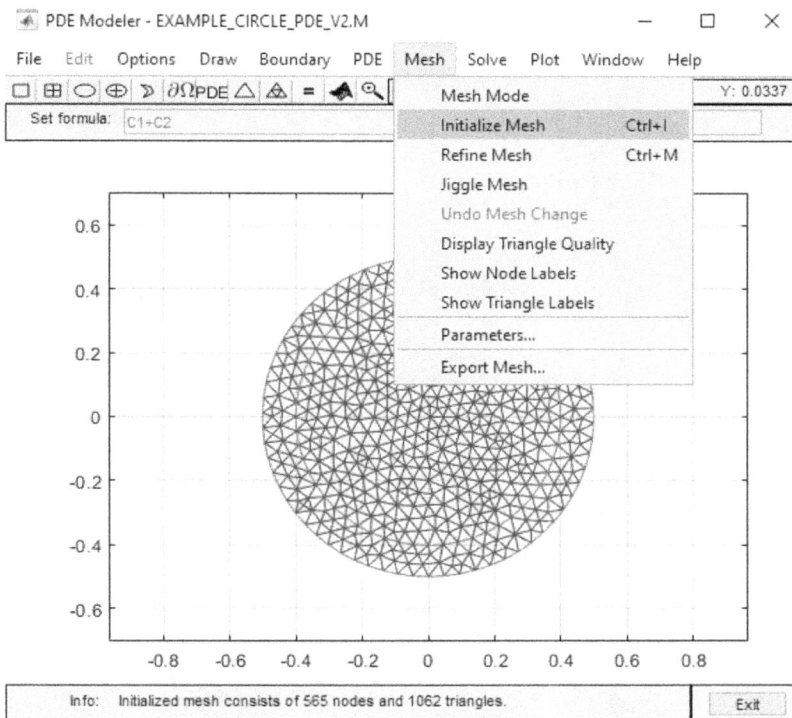

FIGURE 5.14. The *PDE Modeler*, *Mesh* menu items.

that have one node for each triangle vertex, and which result in the t matrix with dimensions $(4 \times N_t)$. When defining the mesh parameters outside of the *PDE Modeler*, one can specify either linear or quadratic elements. The latter results in the t matrix with dimensions $(7 \times N_t)$.

Node and triangle (mesh) labels as well as mesh quality can be displayed using the GUI interface. The solution accuracy level depends on the mesh selection criteria such as density and aspect ratio (quality), which can be controlled through *Mesh > Parameters* (Figure 5.15). Mesh parameters *(Maximum Edge Size* and *Mesh Growth Rate)* can be repeatedly adjusted while working on the model. The *Mesh Growth Rate* controls the rate at which the element size is allowed to increase from the narrow regions to more open areas. The higher the rate, the fewer elements will be created. After changing any of the settings, click on *Initialize Mesh* to rebuild the mesh with the new parameter values. To obtain the example mesh shown in Figure 5.14, set the maximum edge size to 0.05 and perform one mesh refinement step.

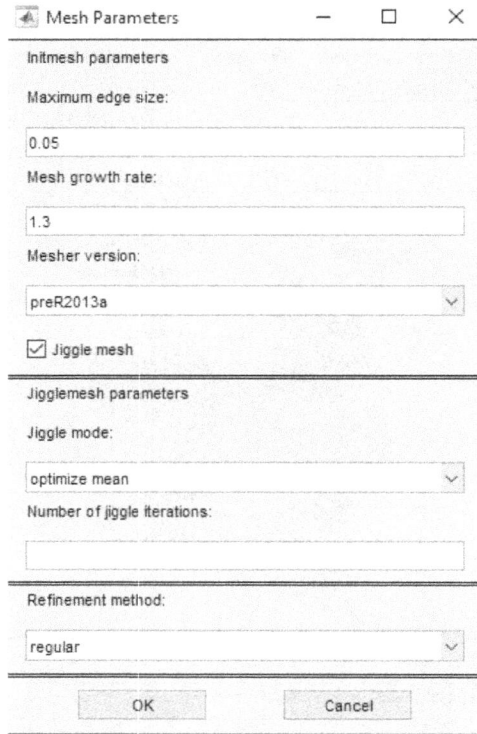

FIGURE 5.15. The *PDE Modeler, Mesh Parameters* settings.

Jiggle Mesh parameters controls the mesh optimization process carried out either automatically, if the *Jiggle Mesh* checkbox is selected, or manually by selecting the *Jiggle Mesh* item from the menu. In this process, the interior element nodes (those that are not on the region boundaries) are iteratively moved to improve the element quality (i.e., make its average value closer to one). You can observe the effect by unchecking the *Jiggle Mesh* checkbox, initializing the mesh, and then performing the *Jiggle Mesh* manually by selecting it from the menu. Select *Display Triangle Quality* from the menu after each action to observe the effect.

Choosing the appropriate mesh size requires balancing between the solution accuracy and time. Two types of the error (i.e., *cutoff* and *roundoff*) are responsible for the solution inaccuracies. These are generated either by rounding to the closest decimal or cutting off certain number of decimals. Either way, the inaccuracies that are introduced into the problem are cumulative with each iteration. This leads to the error-solution curves to take on a concave shape, with the bottom of the valley representing the optimum solution and raised sides being deviations from the optimum solution. Other considerations when setting up the mesh size in addition to the solution time and accuracy concerns are the available resources. For example, memory and hardware requirements may vary depending on the PDE type and complexity of the model.

(7) Solve the problem either by selecting the *Solve PDE* under the *Solve* menu or clicking the '=' button on the top ribbon (Figure 5.16). Under *Solve > Parameters*, one can specify time step, state variable initial guess value, and relative and absolute tolerances (Figure 5.17). The *transient* solution results are plotted automatically with the default plot settings.

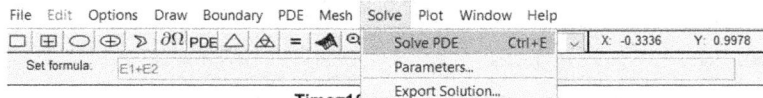

FIGURE 5.16. The *PDE Modeler, Solve* menu items.

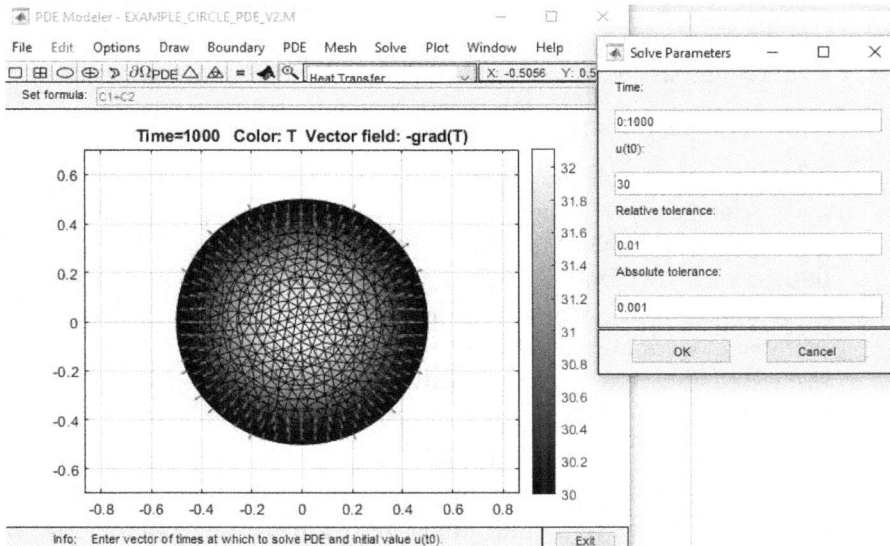

FIGURE 5.17. The *PDE Modeler, Solve Parameters* settings.

(8) Plot the solution results using the default plot settings by selecting the *Plot Solution* in the *Plot* menu (Figure 5.18). Selecting *Parameters* (Figure 5.19) allows customization of the plot display. Figure 5.17 shows the solution plot at time of 1,000 s and for initial temperature of 30 °C. The plot displays temperature contour plots with five levels and arrows showing the direction and magnitude of the temperature gradient field.

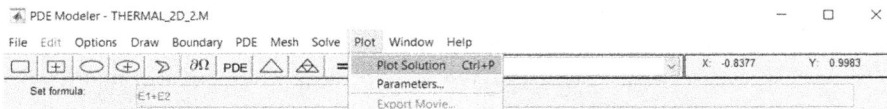

FIGURE 5.18. The *PDE Modeler, Plot* menu items.

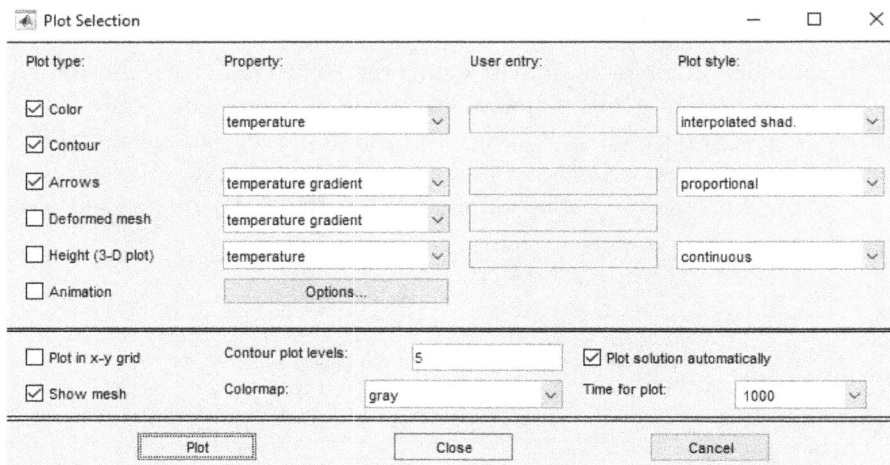

FIGURE 5.19 The *PDE Modeler, Plot Selection* settings.

After completing all the above steps, one can go back to any of them, make changes, and repeat the solution. For example, to change the geometry, select the *Draw Mode* under the *Draw* menu and make the desired changes. Output from any of the abovementioned steps (such as geometry descriptions, set formula, labels, decomposed geometry, boundary conditions, PDE coefficients, mesh, solution, and animated results) can be exported individually to the MATLAB *Workspace* or used in the MATLAB script.

When using the *PDE Modeler* application, one must remember that if the user does not define the parameters in any of the steps mentioned above, the application will complete the solution with the default settings. For example, failing to specify the geometry results in a default L-shape being used. If meshing is chosen as the first step after the L-shape geometry is created, the default boundary conditions are applied to the decomposed boundaries (the *Dirichlet* boundary conditions). If the solution is being called as the next step, the application first initializes the mesh and then runs the solution. If the user chooses to plot the results before any solution is attempted, the solution is first run using the default conditions and then the plot is generated.

This also applies when setting up the PDE coefficients. If no PDE coefficients are set, the *PDE parabolic* relation is used (provided *Heat Transfer Mode* has been selected) with the default values (Figure 5.11). The equation displayed at the top of the *PDE Specification* window shown in Figure 5.11 can be derived by simplifying Equation (62). This is done by assuming the default density ($\rho = 1$), heat capacity ($c = 1$ J/kgK), thermal conductivity ($k = 1$ W/mK), heat generation ($Q = 1$ W/m³), convection heat transfer coefficient ($h = 1$ W/m²K), and surrounding temperature ($T_\infty = 0$ K). One thus obtains Equation (67), which is what is shown in the figure. Note that in Equation (67), T' is simply a different way of writing the temperature time derivative ($T' = dT/dt$) compared to that in Equation (63). The solution to Equation (67) presenting the temperature profiles and contour plots is presented in Figure 5.20.

$$\rho C_p T' - \nabla \cdot (k\nabla T) = Q + h(T_\infty - T) \tag{67}$$

If the problem models *steady-state* conditions, the results are independent of the heat capacity and density; therefore, Equation (67) can be simplified to Equation (68) shown in Figure 5.13.

$$-\nabla \cdot (k\nabla T) = Q + h(T_\infty - T) \tag{68}$$

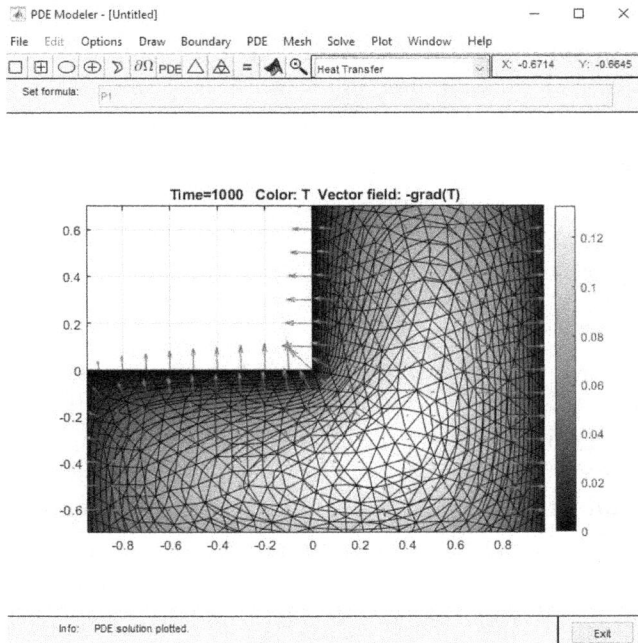

FIGURE 5.20. The *PDE Modeler*, default geometry and solution (temperature and its gradient).

5.3   Thermal Modeling Using the MATLAB Script

To solve any heat transfer model, the conservation of energy must be complied with. As discussed in Section 2.2, this law states that the total energy inputted into the system plus the energy generated within the system must be equal to the total energy outputted from the system plus the energy change within the system, Figure 2.3 and Equation (1). The result of such energy balance is one or more partial differential equations (PDE). To solve such equation in MATLAB, the equation may be discretized and solved using the finite difference approach that employs the MATLAB's base installation capabilities. The *PDE Toolbox* add-on package simplifies this task by providing functions dedicated to setting up and solving of the PDEs. While Section 5.2 described how to create a relatively simple 2D model using the GUI of the *PDE Modeler*, an application which comes with the *PDE Toolbox*, the current section outlines a more general procedure that can be employed to set up and solve a thermal model on the MATLAB platform using the *PDE Toolbox*.

5.3.1   Model Creation

Using the *PDE Toolbox* functions, to create a general PDE *scalar* (single variable-single equation) or *system* (multiple, N, variables-multiple equations) model, the *createpde*(N) function is called as shown in Equation (69). For single-equation models, $N = 1$, and brackets with N can be omitted; for multiple equations, N is equal to the number of equations.

The output of this function is a PDE model object. For example, the PDE model object has properties such as *Geometry* and *Mesh*. An object is a class or category. The object is created and behaves based on its class when the program is executed. This helps the author to write their code so that they can group the data and functions and facilitates finding the objects of the same class. In general, there are two types of programming— *procedural* and *object-oriented*. Procedural programs pass data to functions, capable of performing the intended operations on the data. Object-oriented programming condenses data and their operations in objects; these objects then interface with one another. With MATLAB, the analyst can employ both programming methods, using objects and regular functions.

To create models specifically formulated for thermal and structural applications, function calls as shown in Equation (70) are used. Two function inputs are required: *problem* type and *analysis* type. The problem type is either "thermal" or "*structural*." The analysis type describes how the system changes with time (e.g., "transient" or "*steadystate*") and whether the geometry is axisymmetric. A complete listing of available keyword strings is given in the function's help document. After creating a model of appropriate type, the model's geometry, mesh, and boundary conditions must be defined as detailed in the subsequent sections.

$$model = createpde(N) \tag{69}$$

$$thermalModel = createpde('thermal', ThermalAnalysisType)$$
$$structuralModel = createpde('structural', StructuralAnalysisType) \tag{70}$$

5.3.2 Geometry

Creating geometry that represents the model with sufficient fidelity while avoiding unneeded complexity is an important step when performing heat transfer modeling. This section explains how the geometry that can be used within the *PDE Toolbox* models can be created. The MATLAB geometry creation tools have been discussed in detail in the author's earlier publication [3]. The geometry can be created using a third-party CAD tool or generated within MATLAB. The appropriate geometry creation method depends on the model complexity. The following sections describe different approaches grouped by the number of geometry dimensions—2D or 3D.

5.3.2.1 2D Geometries

Three methods of adding 2D geometry into the model are described below.

Method 1—Build 2D geometry using the MATLAB *PDE TOOLBOX* commands

This approach uses the Constructive Solid Geometry (CSG) principles to define 2D geometry from a set of basic shapes. The geometry of each basic shape is defined by a matrix with several rows. Each row defines specific parameters for that geometry (e.g., geometry shape and related parameters such as central x- and y-coordinates and radius for a circle or number of sides and vertex coordinates for a rectangle). Table 5.1 lists shapes and settings required to define them.

To create properly formatted geometry for the *PDE Toolbox* functions, the model needs to consist of disjointed minimal regions bounded by boundary segments and border segments. This set is known as the *decomposed geometry*. In multiple domains, the geometry components are assembled from the individual basic shapes (e.g., circles and rectangles for 2D cases). This is a two-step approach, consisting of the geometry function

TABLE 5.1. The List of the basic shapes and their configurations.

Item	Row Solid Shape	1	2 (Central Data)	3 (Central Data)	4	5	6
1	Circle	1	x-coordinates	y-coordinates	Circle		
2	Polygon	2	N (the number of line segments)	The next N rows contain the x-coordinates of the starting points of the edges	The next N rows contain the y-coordinates of the starting points of the edges		
3	Rectangle	3	N = 4 (the number of line segments)	The next N rows contain the x-coordinates of the starting points of the edges	The next N rows contain the y-coordinates of the starting points of the edges		
4	Ellipse	4	Center x-coordinate	Center y-coordinate	Minor axe of the ellipse	Minor axe of the ellipse	Rotational angle of the ellipse

and the *decsg* function. The *decsg* function is used to transform the data describing the 2D geometry (as defined in Table 5.1) into a format that can be interpreted by other functions within the *PDE Toolbox*. This is described as *decomposing constructive solid 2D geometry into the minimal regions*. This method can control all the characteristics associated with the defined basic shapes; however, the user interface is not as friendly as when using the *PDE Modeler* application and the model is not viewable as it is being created. Table 5.2 provides examples for the creation of four basic 2D shapes.

TABLE 5.2. Examples of basic 2D shape creation with the *decsg* function.

Item	Solid Shape	Geometry Configuration (g)	Configuration Shape
1	Circle	decsg([1 0 0 1]')	
2	Polygon	*decsg*([2 6 0 1 2 3 2 1 1 2 2 1 0 0]')	
3	Rectangle	decsg([3 4 0 0 1 1 0 1 1 0]')	

Item	Solid Shape	Geometry Configuration (g)	Configuration Shape
4	Ellipse	$decsg([4\ 0\ 0\ 1\ 0.5\ pi/4]')$	

The example in Figure 5.21 shows how the basic shapes can be assembled into a multi-region geometry. Each basic shape is defined by a column vector the length of which varies with the shape type. To combine these shapes, a set of three matrices needs to be created, (*gd*, *sf*, *ns*), and then provided as input to the *decsg* command. These matrices are *gd* (the geometry description matrix containing the *CSG* description of the model); *sf* (the set formula); and *ns* (the name-space matrices). The *sf* character vector describes the Boolean relationships between the geometrical entities; *ns* is a character array that identifies the columns in *gd* and thus allows evaluation of the formula in *sf*. The *decsg* function then receives these three matrices as the inputs and produces the required output matrix—*dl* = *decsg*(*gd*, *sf*, *ns*).

model = *createpde*;

$C1 = [1,0,0,0.5]'$;

$T1 = [3,3,-1,1,0,0,0,2]'$;

$R1 = [3,4,-0.25,0.25,0.25,-0.25,...\ 0.75,0.75,1.25,1.25]'$;

$C1 = [C1;zeros(length(R1)-...\ length(C1),1)]$;

$T1 = [T1;zeros(length(R1)-...\ length(T1),1)]$;

$gd = [T1,C1,R1]$;

sf = *'R1-T1-C1'*;

ns = *char*(*'R1'*,*'T1'*,*'C1'*);

ns = *ns'*;

dl = *decsg*(*gd*,*sf*,*ns*);

geometryFromEdges(*model*,*dl*); *pdegplot*(*model*);

ax = *gca*; *grid on*; *grid minor*;...

ax.FontSize = 20; *xlabel*(*'x'*); *ylabel*(*'y'*);

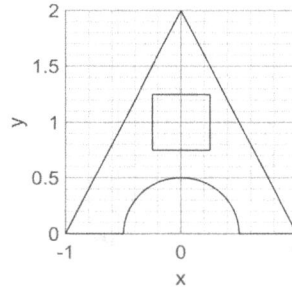

FIGURE 5.21. Basic shapes assembled into a multi-region geometry.

As one can see from Table 5.2, different geometrical shapes have definition vectors of different lengths. However, if they are to be combined into the *gd* matrix, their lengths need to be equalized. This is done by padding the shorter definition vectors with zeroes. For example, the vector length of the rectangle ($R1$) is 10 ($m = length(R1) = 10$); and that of the triangle ($T1$) is 8 ($n = length(T1) = 8$). Thus, vector $zeros(length(R1) - length(T1))$ with length $m - n = 2$ is appended to the triangle's vector. The padded triangle's vector becomes: $T1 = [T1; zeros(length(R1) - length(T1), 1)]$. One can visualize the *PDE* model geometry by executing the *pdegplot* function that takes as its input the output of the *decsg* function—e.g., *pdegplot(dl)*. The geometry can be then incorporated into the model container (*model*) by executing *geometryFromEdges(model,dl)*. Another complete example is presented in Figure 5.22 that shows the script consisting of geometry commands and their assembly and in Figure 5.23 displaying the resultant geometry.

```
Command Window
>> rect = [3 4 -1 1 1 -1 0 0 -0.5 -0.5]';
C1 = [1 1 -0.25 0.25]';
C2 = [1 -1 -0.25 0.25]';
C3 = [1 0 -0.25 0.25]';
C1 = [C1;zeros(length(rect) - length(C1),1)];
C2 = [C2;zeros(length(rect) - length(C2),1)];
C3 = [C3;zeros(length(rect) - length(C3),1)];
gd = [rect,C1,C2,C3];
ns = char('rect1','C1','C2','C3');
ns = ns';
sf = '(rect1+C1)+C2+C3';
[dl,bt] = decsg(gd,sf,ns);
pdegplot(dl,'EdgeLabels','on','FaceLabels','on')
xlim([-1.5,1.5]); ylim([-0.5 0]); ax = gca; ax.FontSize = 10;
axis equal
fx >> |
```

FIGURE 5.22. Geometry creation commands (circle and rectangle operations).

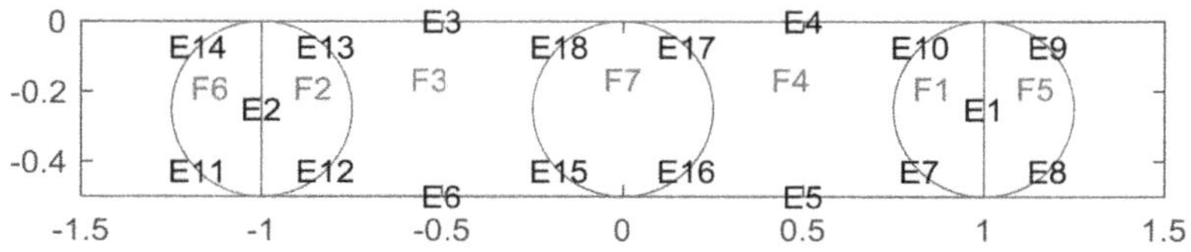

FIGURE 5.23. Geometry created using circle and rectangle operations.

Method 2—Build 2D geometry employing the *PDE Modeler* application

The *PDE Modeler* application introduced in Section 5.2 can provide a more user-friendly approach to geometry creation. Within this application, one may click and drag to generate the geometrical shapes, to view the creation progress, and to make revisions. It is also possible to confirm if the relationships between the geometry features are defined properly; for example, if the parts are connected at the desired points and lines; if the features are to be added or subtracted. Only the same basic shapes introduced in Table 5.2 can be created.

After creating the geometry in the *PDE Modeler*, as described in Section 5.2.2, one can export this geometry via the *Draw > Export Geometry Descriptions, Set Formula, Labels* menu command. This adds three matrices to the MATLAB *Workspace*, identified by default as (*gd*, *sf*, *ns*), which can then be used as input to the *decsg* function, as described under Method 1 above.

Method 3—Import geometry from a third-party CAD or FEA tool

A 2D geometry generated externally and stored in the *∗.stl* format can be imported and incorporated into the PDE model using the *importGeometry(model, geometryfile)* command, where *model* is the name of the container to which the geometry is to be added and *geometryfile* is the path to the *∗.stl* file containing the geometry.

5.3.2.2   3D Geometries

Three methods of adding 3D geometry into the model are described below.

Method 1—Build 3D geometry using the MATLAB *PDE Toolbox* commands

There are three functions in the *PDE Toolbox* that allow you to create three types of 3D geometric primitives: cuboid, cylinder, and sphere. Execution of each function can create multiple objects of the same type or a single object. One can also specify for any of the shapes that they are to be *void*. This allows creation of hollow objects by nesting the *void* shape within a solid one.

For example, to create multiple cuboids, the function $gm = multicuboid$ (W, D, H, NameN, ValueN) is employed. In this relation, W, D, and H are the width, depth, and height, respectively, which can be scalars or vectors. *Name–Value* can belong to multiple relations employed to identify if the part is offset from the default coordinate [0 0 0] by identifying the *ZOffset* value of the Nth cell and Nth cell of the cell N. The lengths of these vectors are the same as the width (W), depth (D), and height (H) vectors. Therefore, to create four *cuboids* with the same side dimensions, and the second hollow *cuboid*, which are offset from the default coordinate [0 0 0] by [0 1 3 6], the command is $gm = multicuboid(3, 1, [1 2 3 5], 'ZOffset',$ [0 1 3 6], 'Void', [false, true, false, false]).

To create four cylindrical cells (volumes) with the unit radii, heights identified by array [1 2 3 4], the third cylinder hollow, and being offset by array [0 1 3 6], the command is $gm =$ multicylinder(1, [1 2 3 4], 'ZOffset', [0 1 3 6], 'Void', [false, false, true, false]). To create one filled spherical cell with radius 5, the command is $gm = multisphere(5, 'Void', false)$.

There are, however, several limitations when using these functions. First, only objects of the same type can be combined within one geometry structure. This means, for example, that one can have multiple cuboids or multiple cylinders but not the combination of cuboids and cylinders. Second, there are some restrictions on how the geometrical entities can be combined; for example, one cannot have overlapping entities. For 3D geometries of higher complexity, it is thus recommended to use other creation methods.

Method 2—Create 3D geometry from mesh

Geometry can be created from a mesh using the *geometryFromMesh* function, as shown in Equation (71). An example using the mesh grid points generated for x-, y-, and z-coordinates is presented in Figure 5.24.

$$geometryFromMesh(model, nodes, elements \tag{71}$$

In the example, a cube with sides equal to 1 is meshed with the grid spacing of 0.1. Node coordinates (*nodes*) are created as the result. The convex hull function creates elements. Geometry (domain) is then generated based on this information (Figure 5.24b).

```
Command Window
>> [x,y,z] = meshgrid(-0.5:0.1:0.5);
x = x(:); y = y(:); z = z(:);
K = convhull(x,y,z);
nodes = [x';y';z'];
elements = K';
model = createpde();
geometryFromMesh(model,nodes,elements);
pdegplot(model,'FaceLabels','on','FaceAlpha',0.5)
>>
```

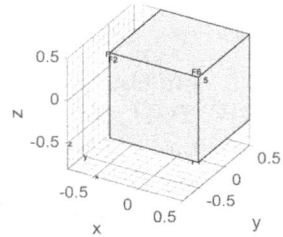

(a)   (b)

FIGURE 5.24. Use of *geometryFromMesh* function to create a convex hull element: (a) Script, (b) Element boundaries.

Method 3—Import 3D geometry from a third-party CAD or FEA tool

To create 2D and 3D geometries, the *.stl* format files can be imported using the *importGeometry(model,geometryfile)* command. For example, the sequence of commands shown in Figure 5.25a imports a CAD geometry created in SOLIDWORKS (Figure 5.25b) that was saved in the *.stl* format to a file *Test_STL_Export.stl*. Figure 5.25c, shows the triangles that are used to define surfaces in the *.stl* format. Figure 5.25d, shows how the imported geometry looks within the *PDE Toolbox* model.

Note that the imported geometry consists of uninterrupted faces, with the triangular facets removed. This is done by MATLAB automatically on the *.stl* file import. It assumes that any triangles within the same plane and having nearly identical face normal orientation belong to the same face. This can cause issues if the model contains surfaces with very small differences in orientation as they may be assumed to belong to the same face. However, in general, the *.stl* file format import is the most appropriate method for adding complex 3D geometries into the *PDE Toolbox* model.

```
>> model03=createpde;
>> gm_stl=importGeometry(model03,'Test_STL_Export.stl');
>> pdegplot(model03,'CellLabels','on','FaceAlpha',0.5)
```

(a)

(b)

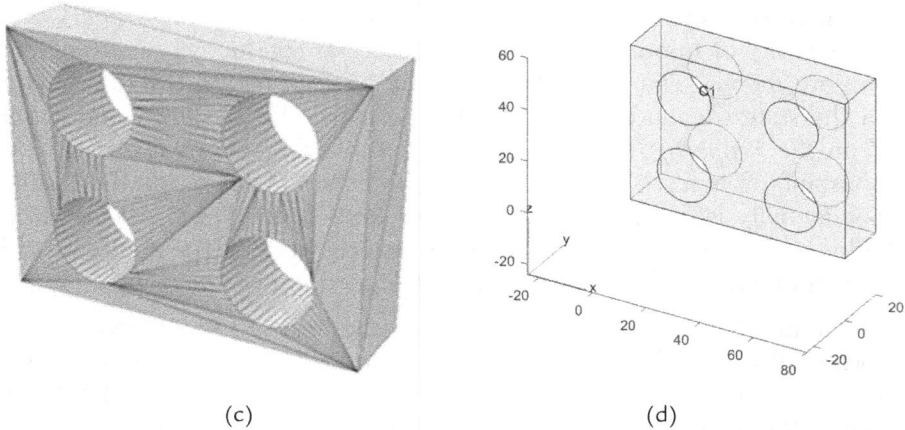

FIGURE 5.25. Importing 3D geometry from a *.stl* file: (a) Script, (b) CAD geometry, (c) *.stl* geometry, (d) Imported model.

5.3.3 Material Properties

In the most general case, material properties can vary over the model space, change over time, and be dependent on state variables. Material properties can be defined for the whole model or specifically for each region by using the *PDE Toolbox* function *thermalProperties()*. For a region-by-region definition, one needs to specify the *RegionType*, which is *Face* for 2D models and *Cell* for 3D models, followed by the *RegionID*, an integer or a vector of integers. Properties are defined by listing property *name-value pairs*.

The example in Equation (72) defines three material properties for a 3D region with ID number *i* within a model called *thermalModel*. The function output may be assigned to a variable that will be defined as a material properties object (*tp_i* in the example). Inputs *k*, *rho*, and *c_p* define, respectively, thermal conductivity, density, and heat capacity at constant pressure.

$$tp\_i = thermalProperties(thermalModel, \text{‘}ThermalConductivity\text{’}, \dots$$
$$k, \text{‘}MassDensity\text{’}, rho, \text{‘}SpecificHeat\text{”}, c\_p, \text{‘}Cell\text{’}, i) \qquad (72)$$

If the material properties are space- or time-dependent, they need to be defined using the *function handles*. For example, for metals at low temperatures, thermal conductivity may be represented by *Item* 1, in the *Property Relation* column, in Table 5.3, where *k*0 is a constant. The larger the temperature is, the larger the thermal conductivity is. In metals, as temperature increases, since the atomic mean free path is limited, thermal conductivity decreases. In metal alloys (e.g., stainless steel), since the density of impurities is large, the thermal conductivity is lower than in pure metals. The thermal conductivity (k) can be a quadratic function of the state (dependent) variable temperature (T), as shown under *Item* 2, in the *Property Relation* column, in Table 5.3, where a, b, and c are constants. When setting this property in the MATLAB script, the thermal conductivity is defined as shown under *Item* 2, the *MATLAB Definition* column, in Table 5.3, where u represents the state (dependent) variable (i.e., temperature for a heat transfer PDE model).

In another example, assume the specific heat (c_p) is a function of the location (x-, y-, and z- coordinates). *Item* 3 in Table 5.3, where a, b, and c are constants, shows how this relation can be described in the MATLAB PDE script. If the internal heat source is a function of the location, temperature, and time, a relationship can be used such as that in *Item* 4 in Table 5.3, where a and b are constants, t is time, and T is temperature.

TABLE 5.3. The property relations and their MATLAB representations.

Item	Property Relation	MATLAB Definition
1	$k = k_0 T$	k = @($\sim$, state)k_0*state.u
2	$k = aT^2 + bT + c$	k = @($\sim$, state)$a*$state.u^2 + $b*$state, $u + c$
3	$cp = ax + by + cz$	c_p = @(location, state)($a*$location.x + $b*$location.y + $c*$location.z)
4	$q = t(ax + b)T^{0.5}$	q = @(location, state)state.$t*$($a*$location.x + b) $*$ (state.u)^0.5

It is also possible to create a function in which any of the material properties vary with the location, state variable (e.g., temperature), and time. This customized function (*cust_fun*) then can be employed when defining the property—*cust_fun* = *function*(*location, state*), which can be defined as *pt* = @*cust_fun* in the MATLAB script, where *pt* can be either property (e.g., thermal conductivity) or process parameters (e.g., internal heat source and heat flux). @*cust_fun* can replace any of the input variables used to define the associated boundary and initial conditions.

The terms location.*x*, *location.y*, and *location.z* are the x-, y-, and z-coordinates of the query point(s). For boundary conditions, they are replaced with *location.nx*, *location.ny*, *location.nz*, and *location.nr*, representing the normal vectors (x, y, z, and r-components) at the query points. For *transient* and non-linear problems, the terms *state.u*, *state.ux*, *state.uy*, *state.*

uz, and state.*ur* represent the resultant *x*-, *y*-, *z*-, and *r*-components of the state variable (e.g., temperature) at the query points. *state.time* represents time at the query points. In these cases, the thermal properties (except for thermal conductivity) form a row vector in which the number of elements is equal to the number of the query points, length(*location.x*). For thermal conductivity, a matrix is formed in which the number of columns is the same as the number of query points, *length(location.x)*, and the number of rows, assuming that *Ndim* is the number of model dimensions (e.g., 1D, 2D, and 3D), is 1, *Ndim*, *Ndim**(*Ndim* + 1)/2, or *Ndim* *Ndim.

5.3.4 Analysis Type

When defining the analysis, one must define its type, such as *steady-state* or *transient*, by calling the *createpde* function. The thermal analysis type can be (a) *steadystate*, (b) *transient*, (c) *steadystate-axisymmetric*, and (d) *transient-axisymmetric*. The *steadystate* is the default when the *createpde* function is used, and the analysis type is not specified—*thermalModel = createpde('thermal')*. Types listed under options (c) and (d) above create 2D thermal models that take advantage of the axial symmetry of the 3D models.

For example, the statement in Equation (73) creates a *transient* heat transfer model *thermalModelT*. To solve an axisymmetric *transient* problem instead, replace keyword *transient* with *transient-axisymmetric*.

$$thermalModelT = createpde('thermal', 'transient') \tag{73}$$

The *thermalModelT* is a thermal PDE model object that contains all the information about the model, such as the geometry, mesh, thermophysical properties, boundary conditions, initial conditions, and number of equations.

5.3.5 Heat Generation

This section describes how to incorporate spatial and temporal heat generation into the heat transfer PDE model. Internal heat generation can be defined by using the *PDE Toolbox* function shown in Equation (74) in which *q_int* is the internal heat generation (W/m³) applied to the region (*Face* for the 2D and *Cell* for the 3D models) identified by *i*.

$$q\_internal = internalHeatSource(thermalModelT, q\_int, 'Face', i) \tag{74}$$

When solving PDEs in MATLAB, it is a good practice to make sure that temperature- or time-dependent outputs do not return valid numbers if any of the *state.time* or *state.u* variables are not-a-number (*NaN*). This means that for any *NaN* state variable, a *NaN* value should be returned. The concept of a *NaN* was introduced earlier in Section 4.5.

An example for a custom-made heat source including the testing for a not-a-number (*NaN*) input is presented in Figure 5.26. The purpose of including such capability is to both ensure that the heat generation value does exist on the selected nodes but also, that they can be defined; for example, they do not produce floating point overflow.

This function, *myHeatSource*, is used in the function *internalHeatSource* through a *function handle—@myHeatSource*. The function *Q* is a local function that can be appended at the end of a script as shown in Figure 5.26; this function then can be called by executing *internalHeatSource(thermalModelT, @myHeatSource)*. Note that, to work correctly with *transient* problems, this function must include a test for a *NaN* input and, if the result is true, return a *NaN* output.

```
function Q = myHeatSource(location,state)
  Q = zeros(1,numel(location.x));
if (isnan(state.time))
  Q(1,:) = NaN;
  return
end
if state.time < 300
  Q(1,:) = 100*state.time;
end
end
```

FIGURE 5.26. Example of function that defines heat source and correctly handles a *NaN* input.

5.3.6 Boundary and Initial Conditions

This section shows how to integrate spatial and temporal boundary and initial conditions into the model. To identify the boundary conditions (e.g., heat flux and temperature), ambient temperature (T_a) and convection heat transfer coefficient (h_c) are to be either directly or separately inputted in the associated equation (i.e., defined as a fixed value or spatial/temporal function) using the *PDE Toolbox* function *thermalBC*. The boundary condition can be identified in the form of temperature (T)—Equation (75), heat flux (q_f)—Equation (76), radiation (q_r)—Equation (77), or convection (q_c)—Equation (78), where T_i, q_i, e_i, T_{ai}, h_{ci} are the temperature, heat flux, emissivity, ambient temperature, and heat transfer convection coefficient at the region type (*Edge* for the 2D and *Face* for the 3D models) referenced by the identifier i. These parameters can be expressed as functions of space or time (e.g., heat flux (q_i_y) in Equation (79) and convection heat transfer coefficient (h_ci_y) in Equation (80).

$$T = thermalBC(thermalModelT\,,'Edge',i\,,'Temperature',T\_i) \tag{75}$$

$$q\_f = thermalBC(thermalModelT\,,'Edge',i,'HeatFlux',q\_i\_y) \tag{76}$$

$$q\_r = thermalBC(\,thermalModelT\,,'\,Edge',i,... \tag{77}$$
$$'Emissivity',e\_i,'\,AmbientTemperature',T\_ai\,)$$

$$q\_c = thermalBC(\,thermalModelT\,,'\,Edge',i,... \tag{78}$$
$$'ConvectionCoefficient\,',h\_ci,'\,AmbientTemperature',T\_ai\,)$$

$$q\_i\_y = @(region, \sim)\,q\_i*region.y \tag{79}$$

$$h\_ci\_y = @(region, \sim)\,h\_ci*region.y \tag{80}$$

Setting up appropriate initial conditions or initial guess for a thermal model is an important consideration in *transient* analyses. There are number of ways in which these conditions may be defined. One approach is to apply the initial condition or initial guess for the PDE problem to the entire geometry—T_0 in Equation (81). It is also possible to set the initial conditions for certain regions (e.g., *Vertex*, *Edge*, and *Face* for the 2D and 3D model and *Cell* for a 3-D model)— T_0_i in Equation (82). One can also have the output of one solution to be the input (or initial guess) to another problem—T_0_res in Equation (83). It is also possible to set the results of the first problem as the initial condition or initial guess for the ts_0 time index of the second problem—T_0_res0 in Equation (84), where ts_0 is a positive integer.

$$T\_0\_1 = thermalIC(thermalModelT,T\_0) \tag{81}$$

$$T\_0\_2 = thermalIC(thermalModelT,T\_0\_i,'Edge',i\,) \tag{82}$$

$$T\_0\_res = thermalIC(thermalModelT,results) \tag{83}$$

$$T\_0\_res\,0 = thermalIC\,(thermalModelT,results,ts\_0) \tag{84}$$

5.3.7 Mesh

The model geometry is meshed using the *generateMesh* function, represented by Equation (85). When the mesh is generated, it is stored in the model object (e.g., *thermalModelT*). *GeometricOrder* identifies if the elements are of *linear* or *quadratic* order. The *quadratic* order elements are the default element type choice for the MATLAB PDE problems; they produce more accurate results; however, they require more computer resources (RAM and solution time). *Hgrad* identifies the mesh growth rate (*gs_gr*). It is a number within the range $1 \leq Hgrad < 2$ with the default value of 1.5. It controls the rate at which the mesh element size is allowed to increase between narrow and open regions. *Hmax* identifies the maximum mesh edge length (*gs_max*). The smaller the *Hmax* is, the finer the mesh is. *Hmin* identifies the minimum mesh edge length (*gs_min*). *Hmax* and *Hmin* are the upper and lower limits, respectively, of the mesh size. If they are not specified explicitly, *Hmax's* and *Hmin's* default values are estimated by the generateMesh function. Mesh can be displayed with the *pdeplot(mesh)* command.

$$mesh = generateMesh(thermalModelT, 'GeometricOrder',$$
$$'quadratic/linear', ...'Hgrad', gs\_gr, 'Hmax', gs\_max, \quad (85)$$
$$'H\min', gs\_min)$$

5.3.8 Solver Options

Several settings can be adjusted from their default values before the solution is run. The defaults determined by the MATLAB *PDE Toolbox* applications in general provide suitable values, but there may be special circumstances for some problems, where changes from the defaults will need to be made. PDE solver options that can be adjusted are the absolute tolerance, relative tolerance, residual tolerance, maximum iterations, minimum step, residual norm, and report statistics. Equation (86) shows the general format for setting these property values. This section provides a summary of these options, with the focus on tolerances, due to their importance in thermal modeling applications and particularly internal ODE solvers.

$$thermalModelT.SolverOptions.PDE\ SolverOptions\ Properties =$$
$$Property\ Value \quad (86)$$

Absolute tolerance identifies how accurate the final solution is expected to be when the solver terminates the iterations. It specifies the solution error estimate, which is the acceptable threshold below which the value of the solution error is considered insignificant—$y = f(x) = 0$. The default value for the absolute tolerance is $1e{-}6$—Equation (87), where a is the absolute tolerance (e.g., $2e{-}6$).

$$thermalModelT.SolverOptions.AbsoluteTolerance = a \quad (87)$$

Relative tolerance identifies how accurate the solution is relative to the exact solution; in other words, it specifies the number of correct digits for the solution components. For the components, which are smaller than the threshold, it is overridden by the absolute tolerance. The default value for the relative tolerance is $1e–3$ with an acceptable accuracy of about 0.1% — Equation (88), where b is the relative tolerance (e.g., $2e–3$).

$$thermalModelT.SolverOptions.RelativeTolerance = b \qquad (88)$$

Residual tolerance identifies the accuracy of the iterative solutions. Residues are the solution deviation (*zero-sum* problems) from zero. The solution iterations continue until the set tolerance is met when performing the analyses. The default value for the residual tolerance is $1e − 4$, as shown in Equation (89), where c is the residual tolerance (e.g., $2e–4$).

$$thermalModelT.SolverOptions.ResidualTolerance = c \qquad (89)$$

Maximal number of Gauss-Newton iterations identifies the highest number of solutions that the nonlinear solver is allowed to iterate. If when this limit is reached the solution still has not reached the relative (or absolute) tolerance, the solution has not converged. In such cases, either the tolerance is too tight and should be relaxed or the maximal number of iterations should be increased; it is also possible that something is not right with the model setup or the geometry. The default value for this property is 25. See Equation (90), where d is the maximal number of Gauss-Newton iterations for internal non-linear solver (e.g., 20).

$$thermalModelT.SolverOptions.MaxIterations = d \qquad (90)$$

Minimum step identifies the damping factor when searching for solutions. It determines the smallest distance that the solution will jump to on the next iteration. The default value for the minimum damping step is $1.5259e − 07$. See Equation (91), where e is the minimum step size (e.g., $1.7e–7$).

$$thermalModelT.SolverOptions.MinStep = e \qquad (91)$$

Solution statistics, such as the number of successful steps (excluding the comment lines), failed attempts, function evaluations, partial derivatives, Lower Upper (LU) decompositions, and solutions of linear systems can be displayed by activating the *ReportStatistics* feature within the *SolverOptions* by setting it equal to *on*; the default value is *off*—Equation (92).

$$thermalModelT.SolverOptions.ReportStatistics = `on' \qquad (92)$$

5.3.9 Solution and Postprocessing

This section describes how to obtain the spatial and temporal solutions and their derivatives (e.g., spatial gradients and heat fluxes) for the heat transfer problems at any given coordinate and time. *Transient* and *steady-state* thermal *results* are obtained by calling the *solve* function—Equation (93). A *transient* thermal result object (*results*) contains the temperature and gradient values in a form convenient for plotting and post-processing.

$$results = solve(thermalModelT, tlist) \tag{93}$$

Transient thermal *results* can be expressed in a variety of forms. They include the temperature (T) in Equation (94); temperature gradients along the x-, y- and z-coordinates, given as Tx, Ty, and Tz, respectively, in Equations (95) to (97); and the solution times, T_{st} in Equation (98).

$$T = results.Temperature \tag{94}$$

$$Tx = results.XGradients \tag{95}$$

$$Ty = results.YGradients \tag{96}$$

$$Tz = results.ZGradients \tag{97}$$

$$T_{st} = results.SolutionTimes \tag{98}$$

Data derived from the solution output, such as the heat flux, heat rate, temperature gradient and temperature interpolation, can be evaluated using the *evaluateHeatFlux*, *evaluateHeatRate*, *eavaluateTemperatureGradient*, and *interpolateTemperature* functions at points that are not necessarily the same as the grid points. Heat flux can be evaluated for the selected nodes, at specific coordinates, and time. It is possible to add arrows to show a derivative of the dependent variable gradients (e.g., direction of the heat flow).

The *interpolateTemperature* function can be applied to results to obtain the temperature values at the selected coordinates. It evaluates the temperature for the nodal grid points, queried points, or any desired spatial locations. For the nodal locations, the points are specified by the coordinates, or query points; see Equation (99), where t is time.

The minimum requirement to define the heat flux function is to identify the output variable name (*results*). The query points are converted to column vectors (e.g., $x(:)$, $y(:)$) before the evaluation can take place. n determines the time interval at which the data is stored. The smaller the n, the more frequently the data is stored. For example, for the following case study, given the total time of 5.7 h, there are 204 data points, given the time step (100 s). If $n = 1$, there are 204 evaluations made at every time interval, resulting in 204 evaluations and the selected query points (x, y, z).

If $n = 10$, there will be 21 evaluations made at every time interval and the selected query points (x, y, z), Equation (99), where t is time.

$$[T\_x, T\_y, T\_z] = \text{interpolateTemperature}(results, x, y, z, 1:n:length(t)) \quad (99)$$

The *evaluateHeatFlux* function can be applied to results for the heat flux values to be found at the selected nodes. It evaluates heat fluxes for the nodal grid points, queried points, or any desired spatial locations. Heat flux (T_x, y, z) is measured in energy per unit time (W). For the nodal locations, the points are specified by the coordinates, or query points, as shown in Equation (100), where t is time.

$$[q\_x, q\_y, q\_z] = \text{evaluateHeatFlux}(results, x, y, z, 1:n:length(t)) \quad (100)$$

The *evaluateHeatRate* function evaluates the integrated heat flow rates normal to specified *region type* or boundaries identified by i, j, and k variables in Equation (101). It can be applied to *Edge* for the 2D or *Face* for the 3D models. Heat rate (Qn) is measured in energy per unit time (W) and flows in the direction normal to the boundary (region). Positive values represent the heat flowing out of the domain, and negative values represent the heat flowing into the domain.

$$Qn = \text{evaluateHeatRate}(results, 'Edge', [i, j, k]) \quad (101)$$

The *evaluateTemperatureGradient* function can be applied to results for the temperature gradient values, Equation (102), to be found at the selected nodes. It evaluates temperature gradients for the nodal grid points, queried points, or any desired spatial locations. For the nodal locations, the points are specified by the coordinates, or query points, as given in Equation (103), where t is time. To present the vector field plot of the temperature gradient calculations, the *FlowData* command can be employed as shown in Equation (104).

$$[dT/dx, dT/dy, dT/dz] = [Tx, Ty, Tz] \quad (102)$$

$$[Tx, Ty, Tz] = \text{evaluateTemperatureGradient}(results, x, y, z, 1:n:length(t)) \quad (103)$$

$$fPlot = \text{pdeplot}(thermalModel, 'FlowData', [Tx \ Ty \ Tz]) \quad (104)$$

It is possible to create a mesh grid and identify the points at which the thermal data is to be displayed. The spacing between the grid points along the x- and y-coordinates does not need to be the same and can be defined independently. Thermal data (e.g., temperature) then can be evaluated at the newly defined points even if it may differ from the grid-size data, as shown in Figure 5.27. Note that sets of $x1$, $x2$, $y1$, $y2$, $z1$, $z2$, and $t1$, $t2$,

dt, and n are the initial (subscript 1) and final (subscript 2), x-, y-, and z-coordinates, time, time step, and number of divisions, respectively.

$t = t1{:}dt{:}t2;$
$w = linspace(x1,x2,n);$
$h = linspace(y1,y2,n);$
$l = linspace(z1,z2,n);$
$[Xw,Yh,Zl] = meshgrid(w,h,l);$
$T\_xyz = interpolateTemperature(results,Xw,Yh,Zl,1{:}n{:}length(t));$
$[Tx,Ty,Tz] = evaluateTemperatureGradient(results,Xw,Yh,Zl,1{:}n{:}length(t));$
$[qx,qy,qz] = evaluateHeatFlux(results,Xw,Yh,Zl,1{:}n{:}length(t));$
$Qn = evaluateHeatRate(results,'Edge',[i,j]);$

FIGURE 5.27. Use of the object functions to analyze the thermal model results.

After the solution is run, the results and mesh can be further processed in the MATLAB environment by exporting the data to the MATLAB *Workspace*. The following steps may be taken to further investigate the solution process: (a) the PDE may be modified, and the solution rerun to study the data sensitivity to the PDE coefficients; (b) the specific material or nodal properties can be displayed; and (c) the mesh parameters can be reset, and the solution rerun to study the solution sensitivity to the mesh criteria.

5.3.10   Verifying the Model Inputs

This section describes how the model's input conditions (e.g., heat fluxes and boundary conditions) can be verified. It also shows how to check the values of the boundary (e.g., edge, face, or cell) settings that have been applied. After the solution is performed, to confirm what types of boundary conditions have been applied to the geometry boundaries, the function given in Equation (105) is used. In this equation, the boundary condition (bc) is a vector whose length equals the number of boundary condition queries made. The *Edge* and *Face* options are applied to 2D and 3D models, respectively. $i, j,$ and k are the *Face* or *Edge* IDs.

To confirm the initial conditions and their settings for specified geometry elements (*Edge* for the 2D and *Face* for the 3D models), the function given in Equation (106) is used. In this equation, the initial condition (ic) is a vector whose length equals the numbers of initial condition queries made. *Vertex*, *Edge*, and *Face* are employed for the 2D and 3D models, and *Cell* is used for the 3D models. $i, j,$ and k are the *Face* or *Edge* IDs.

$$bc = findThermalBC(thermalModelT.BoundaryConditions,'Edge',[i,j,k]);$$
$$(105)$$

$$ic = findThermalIC(thermalModelT.InitialConditions,'Face',[i,j]) \quad (106)$$

5.4   Summary of the Steps to Create a Thermal Model in MATLAB

(1) Create a PDE model object.

(2) Specify if the problem is *transient* or *steady-state*.

(3) Create or import the model's geometry.

(4) Define mesh parameters and mesh the model.

(5) Specify the material thermophysical properties for each domain. This includes thermal conductivity, heat capacity, and density. These properties can be space- or temperature-dependent.

(6) Define the internal heat generation terms for each domain, as applicable.

(7) Define the boundary conditions for edges or faces, as applicable:
   (a) For the convection boundary conditions, the ambient temperature and convection heat transfer coefficient are specified.
   (b) For the radiation boundary conditions, the ambient temperature, emissivity, and Stefan-Boltzmann constant are specified.

(8) Define the initial conditions (for *transient* problems).

(9) Solve the problem.

(10) Postprocess the solution results by presenting them (or derived quantities) using plots, such as scatter diagrams, histograms, bar charts, and contour plots.

(11) Save the results as plots (figures), the *Workspace* data for further postprocessing, or images that can be embedded into a report.

CHAPTER 6

The MATLAB Heat Transfer Problem Case Studies

6.1 Case Study 1—Axisymmetric Pipe: Single-Domain, Steady-State Thermal Model

6.1.1 Setup

This case study applies the PDE modeling approach presented in Section 5.3 to predict the steady-state temperature distribution for a copper pipe (Figure 6.1). The geometry dimensions represent a pipe with one-inch nominal diameter—a type of pipe used in plumbing applications. The case considered is that of hot water at a constant temperature of 80 °C flowing through the pipe. The environment surrounding the pipe is at 25 °C. The parameter of primary interest in this study is the rate of heat loss from this pipe to the environment, expressed in terms of power per unit length of the pipe. This would allow one to estimate the cumulative heat loss for any length of hot water piping.

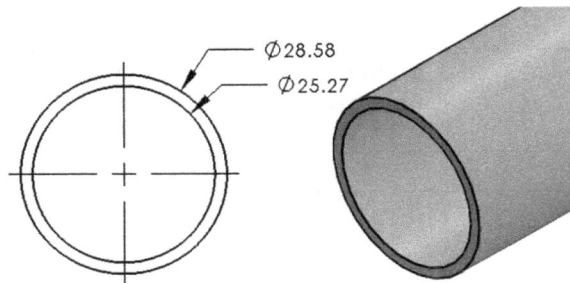

FIGURE 6.1. Axisymmetric copper pipe geometry (dimensions in mm).

Since the model's geometry has axial symmetry (a cylinder) and the model's boundary conditions are similarly symmetrical, it can be accurately represented by a 2D axisymmetric model (Figure 6.2). This model geometry forms a rectangle. In the figure, the pipe's axis of symmetry is aligned with the z-coordinate and is located at $r = 0$. The upper ($E2$) and lower ($E4$) edges correspond to the pipe's ends; these edges are insulated ($q = 0$). The left edge ($E1$) is the pipe's interior surface. A constant temperature (80 °C) boundary condition is applied to this edge; it represents a case of fast-flowing hot water. The right edge ($E3$) is the exterior surface; a convective heat flux ($h_c = 10W/m^2K$) is applied to this edge. The thermal model data are presented in Table 6.1. Copper (Cu), medium- or high-density cross-linked polyethylene (PEX), and fiberglass (FG) are used in subsequent studies, as well as the temperature settings.

TABLE 6.1. Thermal model parameters for axisymmetric pipe models [55,56].

Thermal Model	Type	Grid Size (mm)	Thickness (mm)
Cu	Axisymmetric-Steady-State	0.17	1.65
PEX		0.32	3.18
FG		1.44	14.35
Pipe (25.4 mm Nominal Diameter)	**Internal Diameter (mm)**	**External Diameter (mm)**	**Length (mm)**
Cu	25.27	28.57	10
PEX	22.22	28.57	10
FG	28.57	53.97	10
Thermophysical Properties	**Thermal Conductivity (W/mK)**	**Density (kg/m³)**	**Heat Capacity (J/kgK)**
Cu	400	8,960	385
PEX	0.41	935	2,100
FG	0.04	150	700
Initial Temperature (°C)	**Ambient Temperature (°C)**	**Water Temperature (°C)**	**Heat Generation (W/m³)**
25	25	80	0

Geometry with edge and face labels displayed

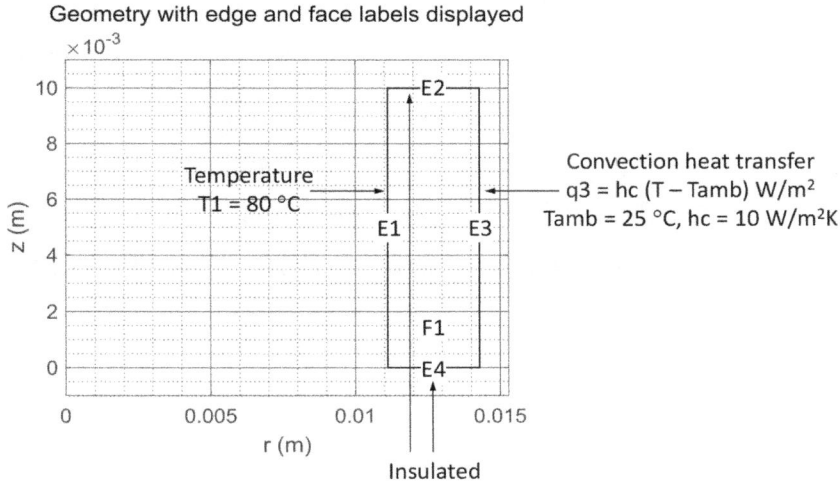

FIGURE 6.2. Axisymmetric pipe geometry showing the edge and face IDs.

The pipe's steady-state thermal model, named *thermalModelS*, is created by the function in Equation (107). This name will be referenced by subsequent functions that define the model's parameters.

$$thermalModelS = createpde(\text{‘thermal’}, steadystate - axisymmetric\text{’}) \quad (107)$$

When specifying the boundary conditions, it is a good practice to activate the display of the geometry labels (*Edge, Face, Node, Element,* and *Cell*) on the plot produced with the *pdegplot* function (as was done in Figure 6.2). This facilitates assigning the boundary conditions. The *Vertex* (2D and 3D), *Edge* (2D and 3D), *Face* (2D and 3D), and *Cell* (3D) IDs are particularly important to assign the heat generation, as well as the initial and boundary conditions (physics), to the applicable regions. Edge IDs are usually created in the order by which the points are introduced into the model; for example, in the geometry presented in Figure 6.2, the bottom-left node (intersection of *Edges E*1 and *E*4) is first created followed by the top-left node, in a clockwise order. The edge numbers can be seen to also follow the same order, increasing clockwise starting from the interior edge (*E*1).

Material properties are defined for each region by the function in Equation (108) and Figure 6.3. Variables identified by $k1$, $rho1$, and $cp1$ are the thermal conductivity, density, and heat capacity at constant pressure, respectively; the last two properties are only applicable to transient models. For the steady-state analysis presented here, the heat capacity and density are not needed. For the region (*F*1) presented in Figure 6.2, the function shown in Equation (109) defines the material properties (Figure 6.4). Material property assignments can be confirmed by entering on the command line the name of the container (*tp*1) to which the above function output was assigned.

$$tp1 = thermalProperties(thermalModelS, 'ThermalConductivity',$$
$$k1, '...MassDensity', rho1, 'SpecificHeat', cp1, 'Face', 1); \quad (108)$$

```
Command Window
>> tp1 = thermalProperties(thermalModelS,'ThermalConductivity',k1,...
    'MassDensity',rho1,'SpecificHeat',cp1,'Face',1)
Warning: Steady-state model does not use MassDensity and SpecificHeat.
> In pde.ThermalModel/thermalProperties (line 107)

tp1 =

  ThermalMaterialAssignment with properties:

            RegionType: 'face'
              RegionID: 1
    ThermalConductivity: 400
            MassDensity: 8960
           SpecificHeat: 385
```

FIGURE 6.3. Material properties assignment for a transient model.

$$tp1 = thermalProperties(thermalModelS, 'ThermalConductivity', \quad (109)$$
$$k1, 'Face', 1);$$

```
Command Window
>> tp1 = thermalProperties(thermalModelS,'ThermalConductivity',k1,'Face',1)

tp1 =

  ThermalMaterialAssignment with properties:

            RegionType: 'face'
              RegionID: 1
    ThermalConductivity: 400
            MassDensity: []
           SpecificHeat: []
```

FIGURE 6.4. Material properties assignment for a steady-state model.

While there is no internal heat generation in this model, if present, it could be applied to the interior *Face* 1 (*F*1) using the function in Equation (110). In this function input, *q*1 represents the internal heat generation value, which is set to zero here (*q*1 = 0). Heat source assignments can be confirmed by entering *thermalModelS.HeatSources* on the command line; alternatively, one can enter the name of container (*q_internal*) to which the output of the above function was assigned (Figure 6.5).

$$q\_internal = internalHeatSource(thermalModelS, q1, 'Face', 1) \qquad (110)$$

```
Command Window

>> q_internal = internalHeatSource(thermalModelS,q1,'Face',1)

q_internal =

  HeatSourceAssignment with properties:

    RegionType: 'face'
      RegionID: 1
    HeatSource: 0
```

FIGURE 6.5. Heat source assignments.

To generate the mesh, the function in Equation (111) is executed; it uses the simplified form of the *generateMesh* function shown in Equation (85). The main additional specifications given by the extra parameters in the function shown in Equation (85) are the element edge size limits. The *GeometricOrder* remains at the default setting (*quadratic*) and *Hgrad* (growth rate) at the default value of 1.5. The mesh properties can be retrieved by entering *thermalModelS.Mesh* on the command line (Figure 6.8). Triangular elements are employed for this 2D model (Figure 6.6). One can zoom in on the mesh to identify the associated elements and nodes (Figure 6.7).

$$mesh1 = generateMesh(thermalModelS, 'H\max', grid\_size) \qquad (111)$$

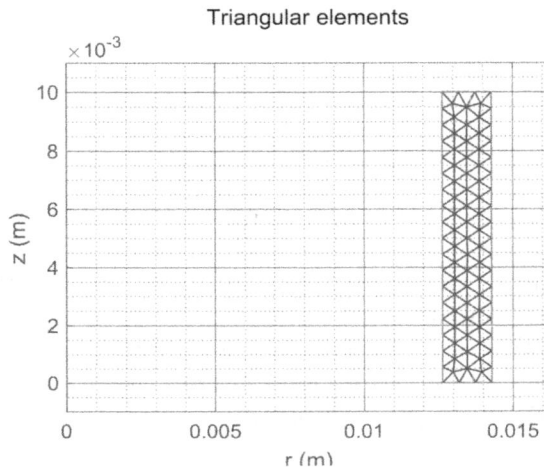

FIGURE 6.6. Triangular mesh for axisymmetric pipe mesh.

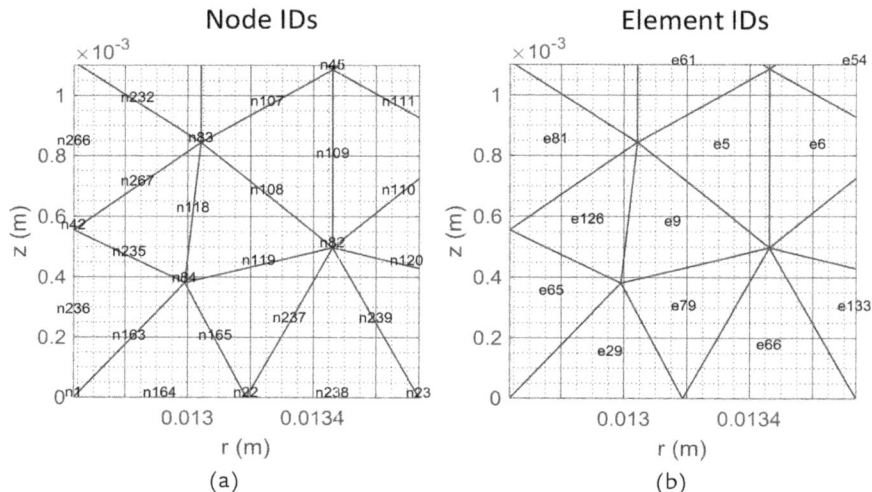

FIGURE 6.7. Triangular elements for the 2D pipe: (a) Node IDs on, (b) Element IDs on.

```
Command Window
>> thermalModelS.Mesh

ans =

    FEMesh with properties:

                  Nodes: [2×335 double]
               Elements: [6×146 double]
         MaxElementSize: 5.5033e-04
         MinElementSize: 2.7517e-04
          MeshGradation: 1.5000
         GeometricOrder: 'quadratic'
```

FIGURE 6.8. Mesh statistics.

The functions listed in the following equations define the model's boundary conditions and initiate the solution. A constant temperature $T\_water = 80\,°C$ is applied to the interior edge ($E1$) in Equation (112). The pipe ends (*Edges E2* and *E4*) are insulated. However, to set up the problem in the most general way and show the capability of non-constant and non-zero heat flux setting, the heat flux here is defined as a function *top_BC_HF*, which can be, if required, defined to be dependent on time and position, as shown in Equation (113). However, in this case, a zero heat flux is set by $qs = 0$. Heat flux on the top and bottom boundaries is set by the function in Equation (114).

The external pipe surface (*Edge* 3, *E3*) is exposed to convection heat transfer. As was the case for the heat flux on the pipe ends, it is defined here in the most general way using the function *outerCC_V*, Equation (115), that would allow it to be defined dependent on position and time. Again, in this case, it is set to a constant value $h_c = 10$ W/m²K. The convective boundary condition is then applied by the function in Equation (116), where *T_ambient* = 25 °C. Steady-state thermal results are obtained by calling the *solve* function (*results*), as shown in Equation (117).

$$Tw = thermalBC(thermalModelS,'Edge',1,'Temperature',T\_water) \quad (112)$$

$$top\_BC\_HF = @(region,\sim)qs \quad (113)$$

$$heat\_flux = thermalBC(thermalModelS,'Edge',[2\ 4], \quad (114)$$
$$'HeatFlux',top\_BC\_HF)$$

$$outerCC\_V = @(region,\sim)hc \quad (115)$$

$$conv\_heat = thermalBC(thermalModelS,'Edge',3,...$$
$$'ConvectionCoefficient',outerCC\_V,'AmbientTemperature', \quad (116)$$
$$T\_ambient)$$

$$resultS = solve(thermalModelS) \quad (117)$$

6.1.2 Results for Copper Pipe

Temperature distribution obtained by the solution is displayed as a function of the *r*- and *z*-coordinates using contour plots in Figure 6.9. It shows an extremely small temperature difference between the interior

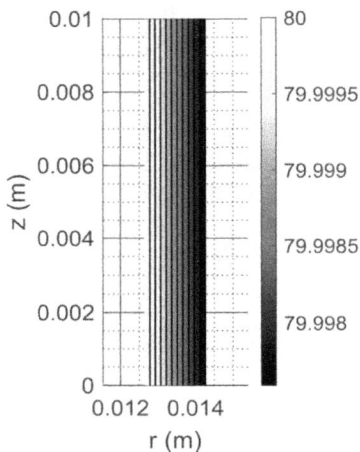

FIGURE 6.9. Axisymmetric pipe temperature contours.

surface at 80 °C and the exterior at 79.998 °C. This is due to the very high conductivity of copper. A plot of the temperature gradient in Figure 6.10 shows that the gradient is negative and decreases in magnitude from 1.55 °C/m at the interior surface to 1.38 °C/m at the exterior surface. As expected, there is no variation in the gradient along the z-coordinate. A negative gradient means that the heat is flowing from the interior to the exterior of the pipe.

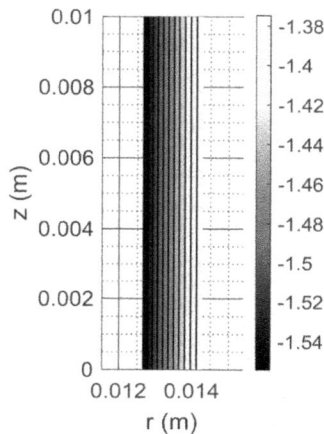

FIGURE 6.10. Axisymmetric pipe temperature gradient contours.

The following plots extract the temperature, gradient, and heat flux data for specific nodes of the model. These nodes are selected within the script by specifying the desired range of coordinates and the required data is plotted. The plots are shown, first, of the spatial location of the selected nodes, and second for the values of interest at these nodes.

Figure 6.11a shows the location of nodes selected along the midplane of the pipe, along the radial direction. The temperature profile for the same midplane nodes is presented in Figure 6.11b. It confirms that there is only a very small decrease in temperature along the radial direction. The midplane radial temperature gradient and radial heat flux profiles are presented in Figure 6.12.

From the pipe's geometry, we can calculate that the pipe's external surface area is 898 mm². The heat flux at the pipe's exterior surface is 550 W/m² (Figure 6.12). Therefore, the heat flux per unit length of the pipe is equal to 49.36 W/m $\left(Heat\ flux \times \dfrac{Pipe\ area}{Pipe\ length} = 49.36 \right)$.

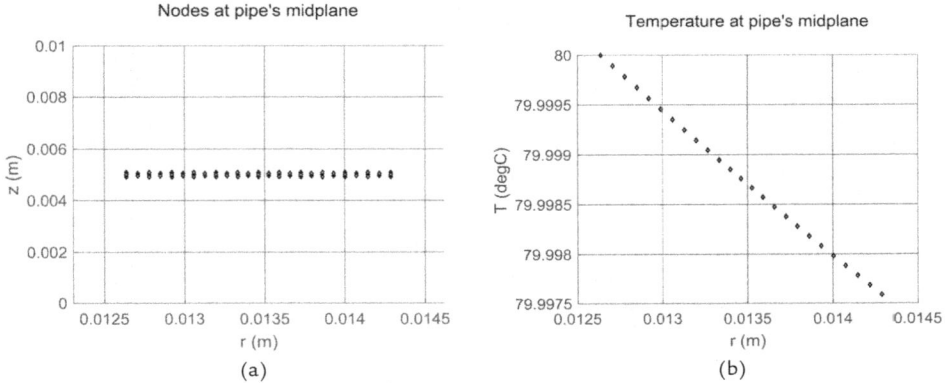

FIGURE 6.11. Axisymmetric copper pipe (midplane): (a) Nodes, (b) Radial temperature profile.

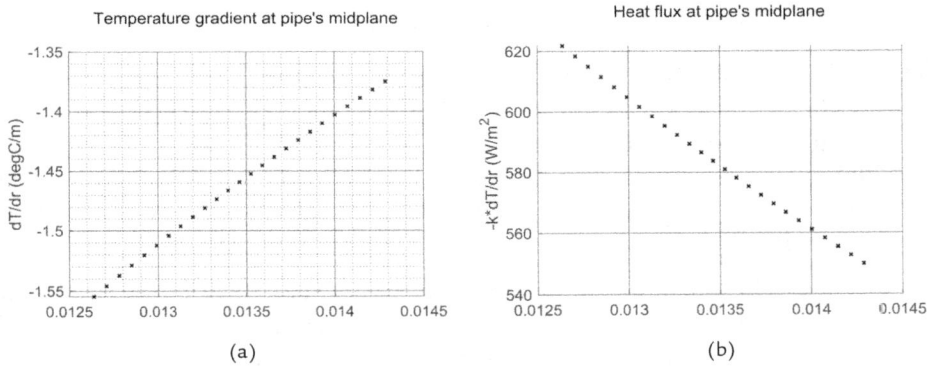

FIGURE 6.12. Axisymmetric copper pipe midplane radial profiles: (a) Temperature gradient, (b) Heat flux.

The solution statistics (e.g., the number of iterations, residual error, step size, and solver method, Jacobian option) are displayed by activating the *ReportStatistics* feature within the *SolverOptions* (Figure 6.13). The steady-state model configuration can be displayed for reference at any time by entering the model object name (e.g., *thermalModelS*) on the command line. The output includes data inputted into the model such as the analysis type, geometry, material properties, boundary and initial conditions, internal heat sources, mesh, and solver options (Figure 6.14).

```
Command Window
>> resultS = solve(thermalModelS); % identify PDESolverOptions Properties
Iteration       Residual      Step size   Jacobian: Full
   0            4.4409e-15
```

FIGURE 6.13. Solution statistics.

```
Command Window

  >> thermalModelS.SolverOptions.ReportStatistics = 'on'

  thermalModelS =

    ThermalModel with properties:

                  AnalysisType: "steadystate-axisymmetric"
                      Geometry: [1×1 AnalyticGeometry]
            MaterialProperties: [1×1 MaterialAssignmentRecords]
                   HeatSources: [1×1 HeatSourceAssignmentRecords]
       StefanBoltzmannConstant: []
            BoundaryConditions: [1×1 ThermalBCRecords]
             InitialConditions: []
                          Mesh: [1×1 FEMesh]
                 SolverOptions: [1×1 pde.PDESolverOptions]
```

FIGURE 6.14. Thermal model configurations.

6.1.3   Results Comparison for Copper and PEX Pipes

The analysis performed for the pipe made of copper, which is a highly conductive material, is repeated for a less thermally conductive material (PEX). The geometry of the PEX pipe is based on a nominally one-inch nominal diameter pipe used in plumbing applications. Its wall thickness is greater than that of the copper pipe (3.175 mm versus the 1.651 mm).

The following figures show comparisons between the solution results for the copper and PEX pipes. First, the temperature variation along the radial direction at the midplane is much greater for the PEX pipe, as expected, due to its much lower conductivity. For the PEX pipe, the temperature varies from 80 °C at the inside surface to 76 °C at the exterior (Figure 6.15a) compared with the nearly constant 80 °C for copper.

The greater variation in the radial temperature profile for the PEX pipe also results in a much higher temperature gradient; this value ranges from −1,426 to −1,109 °C/m for PEX, which is nearly 1,000 times higher than that for copper (Figure 6.15b). The gradient magnitude decreases along the radial direction. The heat flux decreases from 655 W/m² at the interior surface to 510 W/m² at the exterior for the PEX pipe (Figure 6.16). Note that while the quantity of the heat escaping from the pipe is the same, the heat flux, which is the heat per unit surface area, decreases as the surface area of the cylindrical pipe increases with radius. The heat flux at the PEX pipe's exterior surface is 510 W/m² compared to 550 W/m² for the copper pipe. The heat loss per unit length of the PEX pipe can be thus calculated to equal about 45.8 W/m compared to 49.4 W/m for copper (note that the

PEX pipe's surface area is about 898 mm^2). The result is interesting, as it points out that though the PEX is much less conductive than copper, the effect on the heat loss is a reduction of only about 7%. This can be explained by considering that the heat loss is due to convection at the exterior surface, and that is a function of the difference between the surface temperature and the environment. This difference is 55 °C for copper and 51 °C for PEX, and consequently, the heat loss difference is of a similar magnitude.

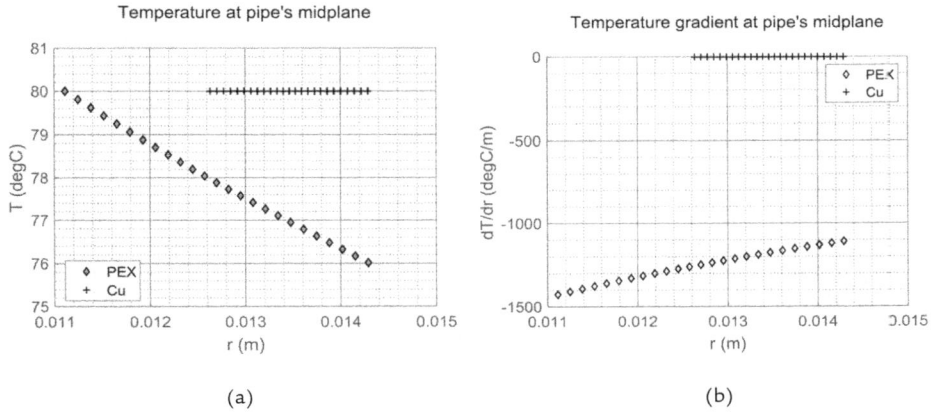

(a)

(b)

FIGURE 6.15. Axisymmetric pipe midplane results radial profiles comparisons: (a) Temperature, (b) Temperature gradient.

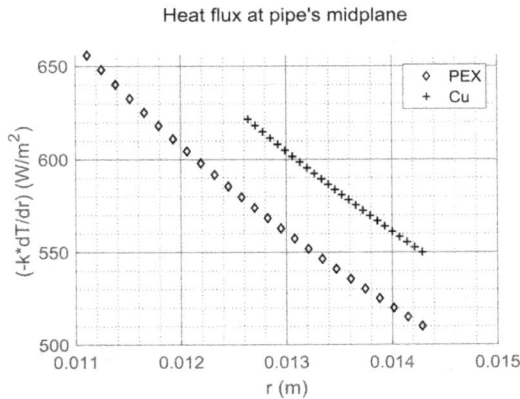

FIGURE 6.16. Axisymmetric pipe midplane rdaial heat flux comparison.

6.2   Case Study 2—Axisymmetric Pipe: Multi-Domain, Steady-State Thermal Model

6.2.1   Setup

This case study models a nominally one-inch nominal diameter copper pipe covered by half-inch (12.7 mm) layer of the fiber glass (FG) insulation. The objective is to investigate the effect of the insulating material on the exterior temperature and on the heat loss per unit length of the pipe. As in the previous case study, water at 80 °C is flowing through the pipe and the environment is at 25 °C. The pipe is transferring heat to the environment by convection. Since the insulated pipe geometry and the boundary conditions are axisymmetric, the pipe can be modeled using a 2D axisymmetric model. However, in this case study, there are two different materials (copper and fiberglass), and therefore it requires a multi-domain model. The length of the modeled pipe is 10 mm (Figure 6.17). The external diameter of the pipe, including the insulating layer, is 54 mm. The geometry dimensions and the material thermophysical properties are listed in Table 6.1.

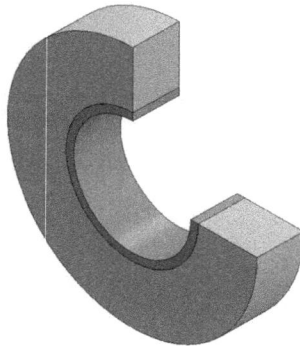

FIGURE 6.17. Axisymmetric insulated pipe geometry with quarter section removed to show the interior structure.

As in the previous, single-domain case study, the model object is first created and then the material properties, as well as the boundary and initial conditions, are applied. The main difference between multi-domain modeling and single-domain modeling is the need to define distinct material properties for each domain. For a 2D model, thermophysical properties are associated with surfaces, each one representing a specific domain. Each surface is identified by the *Face* feature and its identity (ID). For comparison, for a 3D model, thermophysical properties are assigned to volumes associated with the corresponding domains. Each volume is identified by the *Cell* feature and its ID. The initial and boundary conditions are applied to the *Vertex*, *Edge*, and *Face* for the 2D and 3D models and *Cell* for the 3D models. The heat generation terms are applied to the *Face* for the 2D geometries and *Cell* for the 3D cases.

Edge and *Face* IDs along with the boundary conditions are presented in Figure 6.18. The edge and face ID's display is activated in this figure by turning on the *FaceLabels* and *EdgeLabels* options in the *pdegplot* function. *Edge* 3 (*E3*) is at the interface of the two domains (*Face* 1 and *Face* 2); no boundary conditions are assigned to this interface. The upper and lower boundaries, corresponding to the two ends of the pipe/insulation, are insulated ($q4 = q5 = q6 = q7 = 0$). The internal edge (*Edge* 1, *E1*) is exposed to hot water at 80 °C, and therefore, a constant temperature boundary condition is applied. The external edge (*Edge* 2, *E2*) is exposed to the ambient at constant temperature (25 °C) and convection heat transfer is applied, with the convection heat transfer coefficient set to 10 W/m²K.

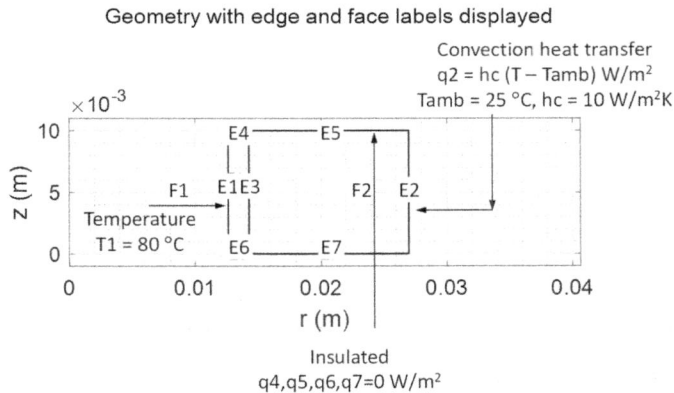

FIGURE 6.18. Axisymmetric 2D geometry of the insulated pipe, including the edge and face IDs.

As the model used in this study is the same axisymmetric steady-state thermal model type as the one in the Case Study 1, the same function is used to create it as the one shown in Equation (107). Similarly, for the material properties, the function shown in Equation (109) can be used to define the copper domain properties for *Face* 1; the same function, but with the insulating material (FG) properties, needs to be applied for *Face* 2. The mesh is generated, as for Case Study 1, with the function in Equation (111).

Geometry meshed using 2D triangular elements is presented in Figure 6.19. There were 636 quadratic elements and 1,341 nodes generated (Figure 6.20). Note that when performing a multidomain analysis using the MATLAB *PDE Toolbox*, the domains cannot have different grid sizes.

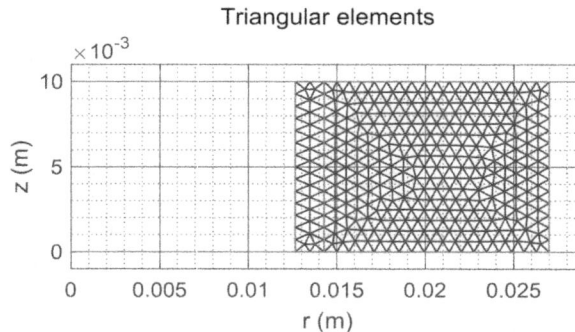

FIGURE 6.19. Triangular mesh for the 2D axisymmetric model of the insulated pipe.

```
Command Window
 >> thermalModelS.Mesh

 ans =

    FEMesh with properties:

              Nodes: [2×1341 double]
           Elements: [6×636 double]
     MaxElementSize: 7.1755e-04
     MinElementSize: 3.5877e-04
      MeshGradation: 1.5000
     GeometricOrder: 'quadratic'
```

FIGURE 6.20. Mesh statistics for the 2D axisymmetric model of the insulated pipe.

6.2.2 Results

The solution results are presented in Figure 6.21 as temperature contour plots. They are obtained by plotting the temperature data $T(:, end)$ versus the x- and y-coordinates ($XYData$). In addition, the heat flow data $[qx, qy]$, where $[qx, qy] = evaluateHeatFlux(results)$, is displayed on the same plot by arrows along the left boundary. It shows that the heat is flowing left-to-right in the radial direction. The temperature gradient contours are presented in Figure 6.22. To generate this plot, the data $Tx(:,end)$ are plotted versus the xy-plane.

Figure 6.23a shows the selected nodes at the pipe's midplane for which the temperature profile versus the radial distance is plotted in Figure 6.23b. The temperature is seen to be a nearly constant 80 °C through the copper pipe wall. The temperature decreases steeply through the insulation, going down to about 35 °C at the exterior. These results show the effect of the very large difference in the thermal conductivity between copper and fiberglass: copper is about 10,000 times more conductive.

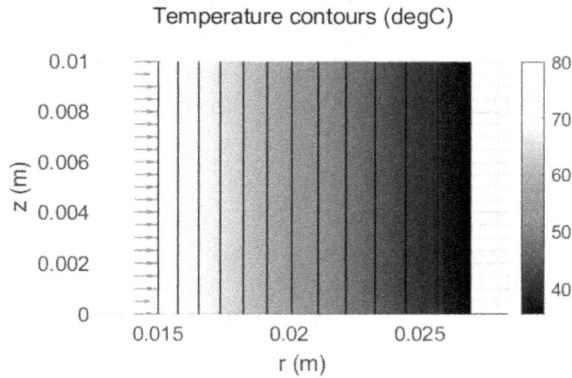

FIGURE 6.21. Axisymmetric insulated copper pipe: temperature contours.

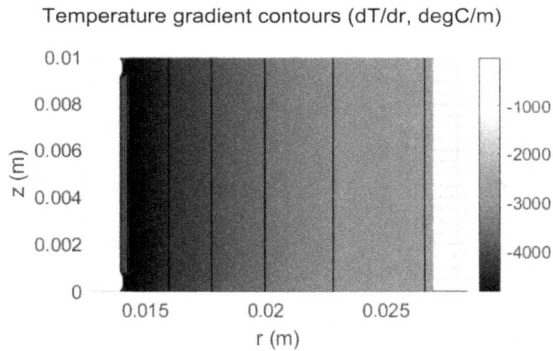

FIGURE 6.22. Axisymmetric insulated copper pipe: temperature gradient contours.

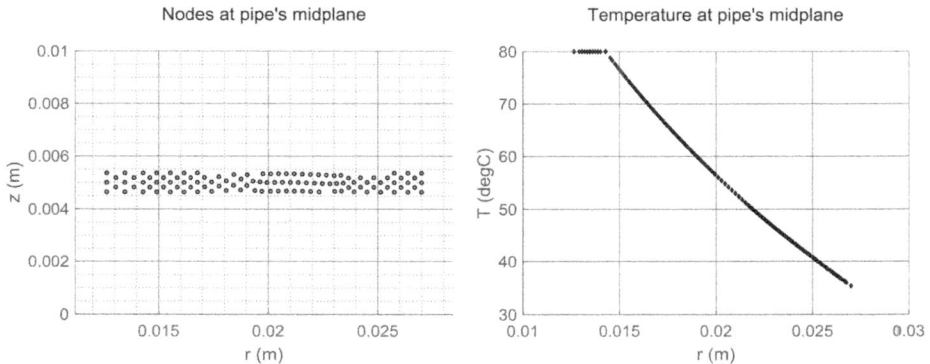

(a)   (b)

FIGURE 6.23. Axisymmetric pipe with insulation (midplane): (a) Nodes, (b) Radial temperature profile.

For the same midplane nodes, Figure 6.24a displays the temperature gradient and Figure 6.24b shows the heat flux along the radial direction. The temperature gradient is nearly zero within copper, corresponding to a nearly constant temperature distribution within the pipe's wall. The gradient drops quickly to a negative value, indicating a steep decline in the temperature within the insulation. The gradient gradually decreases in magnitude, corresponding to a less steep temperature profile slope towards the outer boundary.

Correspondingly, the heat flux within the insulation starts at about 193 W/m² and decreases to about 104 W/m² at the exterior surface. This decrease relates to the increasing cylindrical surface area as the radial distance increases. Using the heat flux at the exterior surface reported above, the pipe exterior surface area (1,695 mm²), and length (10 mm), one can calculate the heat loss per unit length of the pipe is equal to 17.64 W/m.

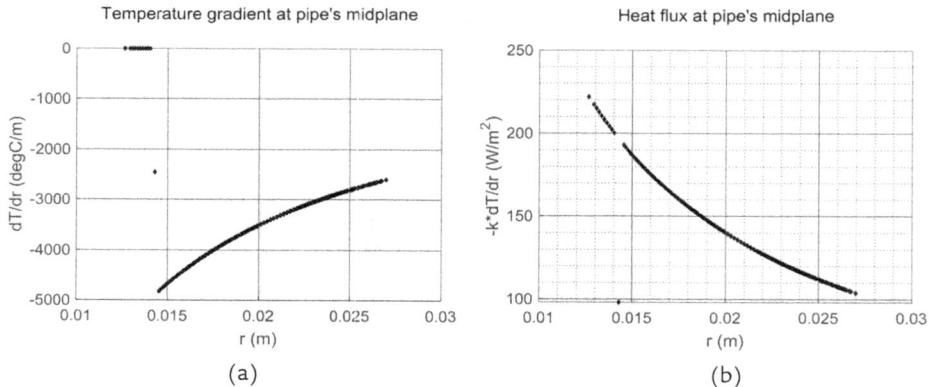

FIGURE 6.24. Axisymmetric pipe with the insulation midplane radial profiles: (a) Temperature gradient, (b) Heat flux.

6.2.3 Validation by an Analytical Model

To validate the results presented in Figure 6.23, the heat diffusion Equation (45), presented in Section 2.4, can be employed. This equation can be simplified to Equation (118), if there are no angular or axial temperature variations. There is also no heat generated inside the cylinder and the process is steady-state. In this relation, k is the thermal conductivity of the insulation and only the radial temperature variation is present. The analytical solution to such a problem is presented by Equation (119). To obtain the $C1$ and $C2$ constants, the pipe interior and exterior surface temperatures can be employed. In this case, the water temperature (80 °C) is assumed to be applicable for the pipe insulation interior surface temperature since the copper thermal conductivity is very large.

$$\frac{1}{r}\frac{\partial}{\partial r}\left(k_r r \frac{\partial T}{\partial r}\right) = 0 \tag{118}$$

$$T = \ln(r)C1 + C2 \tag{119}$$

The pipe insulation exterior surface temperature should be calculated. This can be achieved by assuming a constant heat flow from the interior to the exterior surface of the FG insulation. The FG insulation exterior surface is transferring heat to the ambient by means of convection (h_c = 10 W/m²K). The thermal resistor analogy, where the heat flow is considered the equivalent to the electric current, can be adopted in this case. The total thermal resistance of a hollow cylinder (R_{total}) is given by Equation (120). The internal and external radii of the hollow cylinder are r_1 and r_2, respectively. Since the radial heat flow remains constant—Equation (121)—it is possible to present the heat flow relation by Equation (122). By rearranging this equation, T_2, which is the temperature at the pipe insulation's exterior, is obtained. Note that k is the thermal conductivity of the FG, L is pipe length, T_∞ is ambient temperature (25 °C), and T_1 is pipe insulation's interior surface temperature (80 °C). After carrying out the derivation, Equation (124) gives the radial temperature profile (T).

$$R_{total} = \frac{\ln(r_2/r_1)}{2\pi kL} + \frac{1}{2\pi r_2 L h_c} \tag{120}$$

$$Q = \frac{(T_\infty - T_1)}{R_{th\,cond}} \tag{121}$$

$$Q = \frac{(T_\infty - T_1)}{R_{total}} = \frac{(T_\infty - T_2)}{\frac{1}{2\pi r_2 L h_c}} = \frac{(T_2 - T_1)}{\frac{\ln(r_2/r_1)}{2\pi kL}} \tag{122}$$

$$T_2 = \frac{(T_1 + CT_\infty)}{1 + C}, \text{ where } C = \frac{r_2 h_c}{k}\ln\left(\frac{r_2}{r_1}\right) \tag{123}$$

$$T = \frac{(T_2 - T_1)}{\ln(r_2/r_1)}\ln(r_1/r_2) + T_2 \tag{124}$$

Figure 6.25 compares the plot of the radial temperature variation derived via the analytical solution with the numerical results obtained via the PDE solution. The analytical solution plot exactly overlaps the numerical (PDE) results, thus validating the PDE solution.

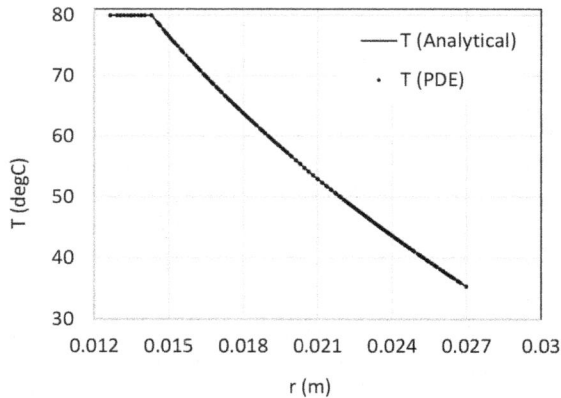

FIGURE 6.25. Radial temperature profiles for analytical and PDE solutions for the axisymmetric pipe with insulation.

6.2.4 Heat Loss Comparison

In this section, the results of Case Studies 1 and 2 are compared. These results show how the temperatures and heat loss depend on the pipe's material (Cu versus PEX) and on the addition of insulation. Figure 6.26a shows the radial temperature profiles. The temperatures in the copper pipe wall are nearly identical (overlaid) for the bare and insulated cases; the temperature at the exterior of the PEX pipe is only about 4 °C lower that of the interior. As expected, insulation is shown to produce much more significant effect, decreasing the exterior temperature to about 35 °C. The heat flux along the radial direction for the three cases is compared in Figure 6.26b. The exterior surface heat flux decreases from 550 W/m^2 for copper, to 510 W/m^2 for PEX, and down to 104 W/m^2 for the FG-insulated copper pipe.

The heat flux at the exterior surface can be used to calculate the total heat loss for the modeled surface area; this is then divided by the modeled pipe length (10 mm) to obtain the heat loss per unit pipe length. It is instructive, then, to see how this specific heat loss translates into an approximation of a real-world case, where a pipe of; for example, 25-m is used to transport the hot water. Multiplying the specific heat loss by the pipe length gives the total heat loss rate in Watts for the 25-m pipe segment. Figure 6.27a shows the heat rate per unit length of the pipe calculated for the three cases. The results vary from a maximum of 49.4 W for uninsulated copper pipe to minimum of 16.6 W for the insulated one, giving about a three-fold reduction in heat loss.

The above information can be further used to estimate the annual cost of the lost heat, assuming the water is heated by a gas water heater (efficiency 100%). The following additional information is used: 37 MJ/m³ is obtained by burning of the natural gas; natural gas cost is 13.3 ¢/m³ (the Canadian dollar value was taken from a gas bill in Ontario, Canada, on July 1, 2021). The calculation results show the annual heat loss cost (in Canadian dollars) decreasing from about $140 to $47, for savings of about $93 per year. These results indicate why it may be worthwhile to add hot water pipe insulation.

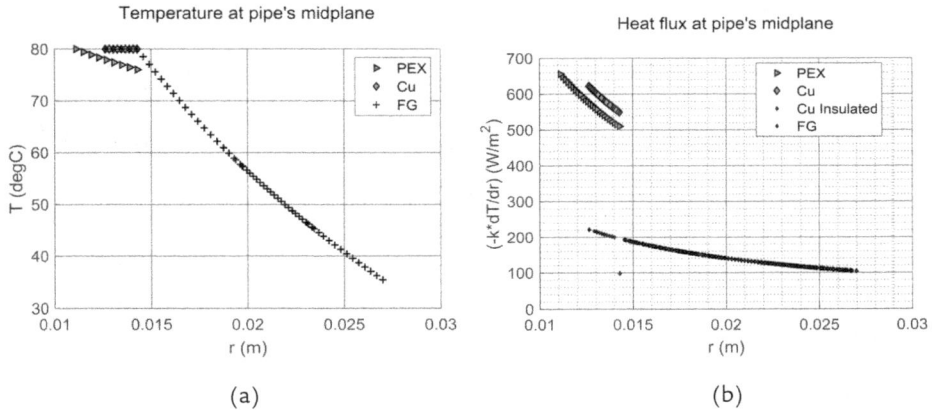

FIGURE 6.26. Axisymmetric pipe model comparison radial profiles comparisons (midplane): (a) Temperature, (b) Heat flux.

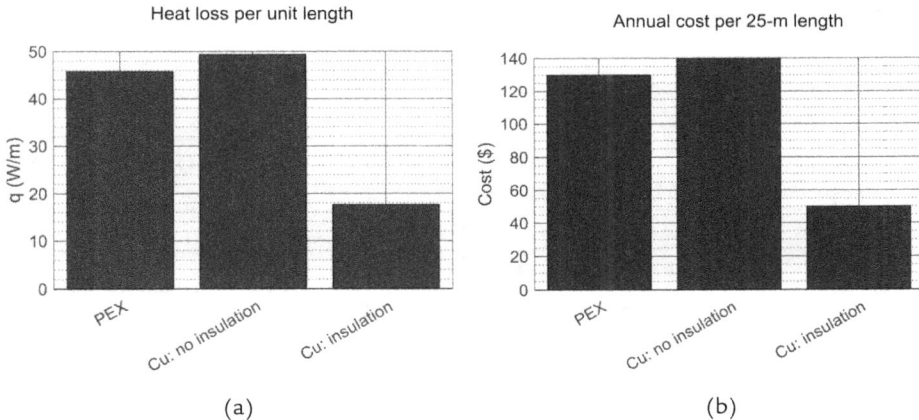

FIGURE 6.27. Axisymmetric pipe model savings comparison: (a) Heat loss per unit length of the pipe, (b) Annual cost.

6.3   Case Study 3—Axisymmetric Pipe: Multi-Domain, Transient Thermal Model

6.3.1   Setup

In this case study, a transient model is investigated. The model consists of a copper pipe with aluminum radial fins. Figure 6.28 shows the 3-D model of the finned pipe to clarify the model's structure. Each radial fin is 1-mm thick and 19-mm deep, with a 66.5-mm external diameter. The thermal model data are presented in Table 6.2. Similar to the previous case studies, water at a constant temperature of 80 °C flows through the pipe. The exposed surfaces of the pipe and fins transfer heat to the environment by means of convection. The ambient temperature is 25 °C and convection heat transfer coefficient is 10 W/m²K. The two surfaces at the pipe's ends are insulated ($q = 0$). Figure 6.29 presents the boundary conditions applied to the geometry of the axisymmetrical pipe (Figure 6.29a) and the geometry, including the face and edge labels (Figure 6.29b). A triangular mesh is shown in Figure 6.30.

TABLE 6.2. Thermal model setup [57,58].

Thermal Model	Type	Grid Size (mm)	Thickness (mm)
Cu	Axisymmetric-Transient	0.25	1.65
Al		0.25	19.0
Pipe (25.4 mm Nominal Diameter)	Internal Diameter (mm)	External Diameter (mm)	Length (mm)
Cu	25.3	28.6	25
Al	28.6	66.5	25
Thermophysical Properties	Thermal Conductivity (W/mK)	Density (kg/m³)	Heat Capacity (J/kgK)
Cu	400	8,960	385
Al	238	2,700	900
Initial Temperature (°C)	Ambient Temperature (°C)	Water Temperature (°C)	Heat Generation (W/m³)
25	25	80	0

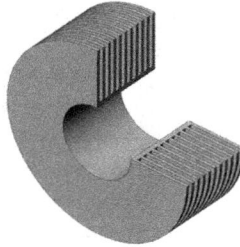

FIGURE 6.28. Axisymmetric pipe with radial fins, with a quarter section removed to show the interior structure.

Geometry with edge and face labels displayed

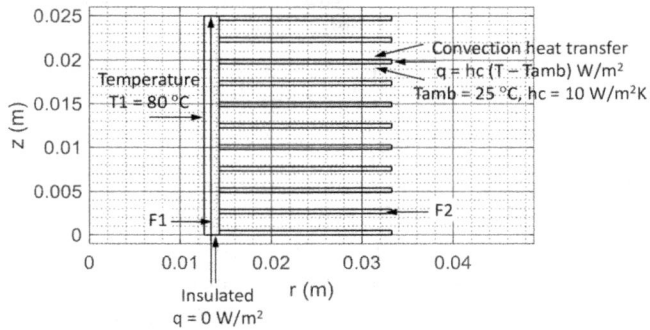

(a)

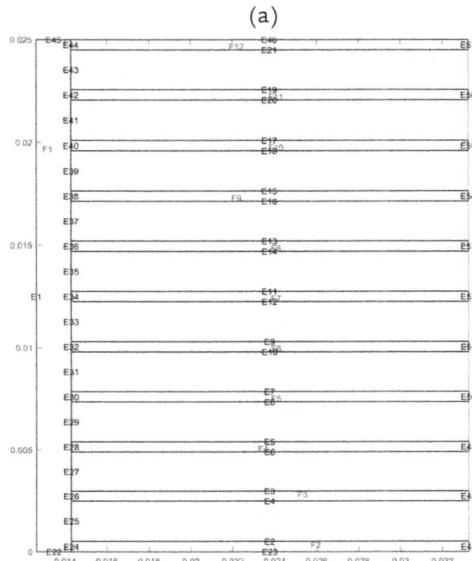

(b)

FIGURE 6.29. Axisymmetric pipe with radial fins: (a) Geometry, (b) Edge and face identities.

Triangular elements

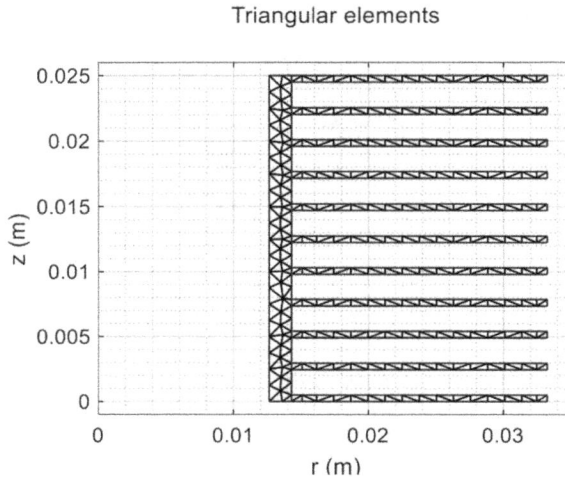

FIGURE 6.30. Triangular mesh for an axisymmetric pipe with radial fins.

The transient thermal model named *thermalModelT* is created by the function in Equation (125). Thermal properties are defined separately for copper (*Face 1, F1*) and aluminum (*Face 2, F2*) domains—Equations (126) and (127). There is no internal heat generation in this model. As a placeholder, a heat generation of $q2 = 0$ W/m³ is set by the function in Equation (128).

$$thermalModelT = createpde('thermal','transient') \tag{125}$$

$$tp1 = thermalProperties(thermalModelT,'ThermalConductivity',... \\ k1,'MassDensity',rho1,'SpecificHeat',cp1,'Face',1); \tag{126}$$

$$tp2 = thermalProperties(thermalModelT,'ThermalConductivity',... \\ k2,'MassDensity',rho2,'SpecificHeat',cp2,'Face',[2:12]); \tag{127}$$

$$int\,ernal\_heat = int\,ernalHeatSource(thermalModelT,q2,'Face',2); \tag{128}$$

The exterior of the copper pipe and all the exterior fin surfaces are assigned boundary conditions of convection heat transfer with $h_c = 10$ W/m²k and *T_ambient* = 25 °C. This is done in the most general way by creating a function *outerCC_V* by using Equation (129) and making it return a constant value of *hc* independent of position and time. The *BC* assignment is made with the function in Equation (130).

$$outerCC\_V = @(region,\sim)hc \tag{129}$$

$$conv\_heat = thermalBC(thermalModelT,'Edge',[1:57],... \\ 'ConvectionCoefficient',outerCC\_V,'AmbientTemperature',T\_ambient) \tag{130}$$

The two edges corresponding to the ends of the pipe ($E45$ and $E22$) are insulated; they are exposed to zero heat flux. Also insulated are the adjacent fin edges on both ends ($E46$ and $E23$) to represent the modeled pipe segment being stacked indefinitely. This is implemented in a general way by defining a function top_BC_HF in Equation (131) and making it return a constant value of $qs = 0$ W/m^2. These BCs are assigned by the function in Equation (132).

The interior edge ($E1$) is assigned a *BC* of constant temperature of $T\_water = 80$ °C to represent the hot water flowing in the pipe interior, as defined by the function in Equation (133). An initial temperature (T_01) of 25 °C is assigned to the entire domain. The first assignment is made by the function in Equation (134) and the second assignment by the function in Equation (135).

For a transient thermal analysis, it is necessary to select the solution time step, which determines the solution output time interval. The model thermal response is monitored for the period of 6 min. Given the selected time step (*time_step*) of 0.25 s, this results in 1,440 saved solution steps (*tlist = tinitial : time_step: tfinal*). The solution is executed by the function in Equation (136). *tinitial* and *tfinal* are the initial (0.1 s) and final times (360 s), respectively.

$$top\_BC\_HF = @(region,\sim)qs \tag{131}$$

$$heat\_flux = thermalBC(thermalModelT,'Edge',[22\ 23\ 45\ 46],' \\ HeatFlux',top\_BC\_HF) \tag{132}$$

$$Tw = thermalBC(thermalModelT,'Edge',1,'Temperature',T\_water) \tag{133}$$

$$T\_i1 = thermalIC(thermalModelT,T\_01,'Face',1) \tag{134}$$

$$T\_i2 = thermalIC(thermalModelT,T\_01,'Face',2) \tag{135}$$

$$resultT = solve(thermalModelT,tlist) \tag{136}$$

Solution statistics (e.g., number of successful steps, failed attempts, and function evaluations) are displayed when the *ReportStatistics* option is turned on under the *SolverOptions*, as shown in Figure 6.31.

```
Command Window

>> thermalModelT.SolverOptions.ReportStatistics = 'on'
resultT = solve(thermalModelT,tlist) % identify PDESolverOptions Properties

thermalModelT =

  ThermalModel with properties:

              AnalysisType: "transient-axisymmetric"
                  Geometry: [1×1 AnalyticGeometry]
        MaterialProperties: [1×1 MaterialAssignmentRecords]
               HeatSources: [1×1 HeatSourceAssignmentRecords]
   StefanBoltzmannConstant: []
        BoundaryConditions: [1×1 ThermalBCRecords]
         InitialConditions: [1×1 ThermalICRecords]
                      Mesh: [1×1 FEMesh]
             SolverOptions: [1×1 pde.PDESolverOptions]

131 successful steps
0 failed attempts
230 function evaluations
1 partial derivatives
29 LU decompositions
229 solutions of linear systems

resultT =

  TransientThermalResults with properties:

      Temperature: [11853×1440 double]
    SolutionTimes: [1×1440 double]
       RGradients: [11853×1440 double]
       ZGradients: [11853×1440 double]
             Mesh: [1×1 FEMesh]
```

FIGURE 6.31. Thermal model solution statistics.

6.3.2 Results

The temperature contours for the domains are presented along the z- and r-coordinates using the contour plots (Figure 6.32) and Equation (137). The temperature variation is represented by the *hot colormap* setting. Since T is the transient temperature data, it is a function of the z- and r-coordinates (*XYData*) and time (t). The number of levels of contour plots can be specified by the above function (e.g., 10 in this case). Also note that the thermal data for the last time step (6 min) are selected for display.

$$pdeplot(thermalModelT,'XYData',T(:,end),$$
$$'colormap','hot','Contour','on','Levels',10); \quad (137)$$

For the total time (6 min) and time step (0.25 s), 1,440 steps are required for the solution to be completed. This is confirmed when the length of the time vector is called $(length(t) = 1440)$. Note that the total time is formed into grids (vector) based on the selected time step, and therefore the data at the selected time grids become available when being enquired. The rest of the data are interpolated when the node evaluation queries are made.

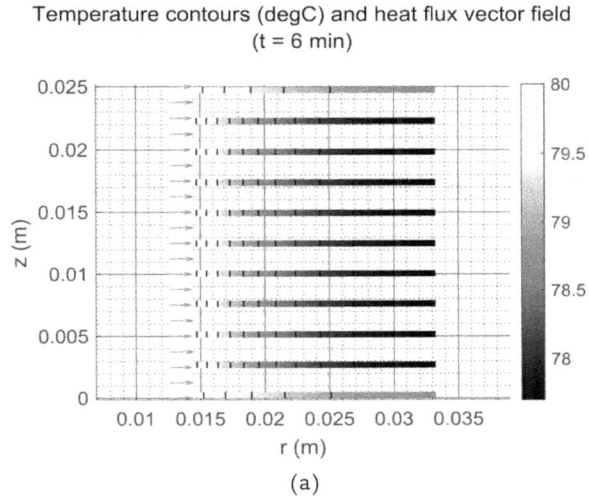

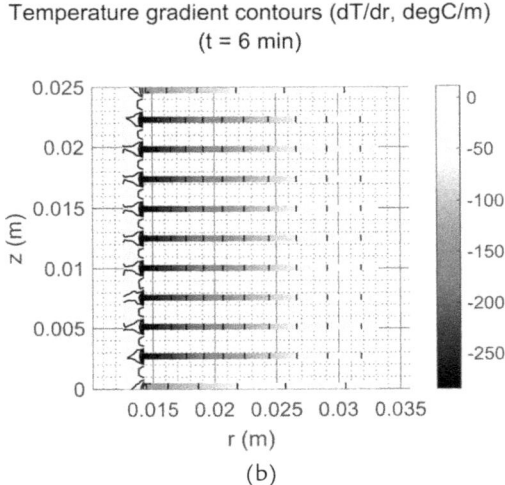

FIGURE 6.32. Axisymmetric pipe with radial fins: (a) Temperature contours and heat flux vector fields, (b) Temperature gradient contours.

The author recommends treating the steps of the PDE solutions as vector variables and assigning variable names to them. This approach can be even applied to the plots, facilitating calling the results, making it possible to define their properties (e.g., line style) in groups and treat them as variables.

To confirm the type of the boundary conditions applied to the *Edges 1* (*E*1), and *2* (*E*2), the function in Equation (138) can be used. This function returns a vector *bc* of length 2, which is equal to the number of boundary condition queries made: '*Edge*'[1, 2]. The *Edge* feature is used since the problem is a 2D model (Figure 6.33). If there are more than one boundary following the same condition (e.g., *Edge 4*, *5*, and *7* in addition to *Edge 6* in Figure 6.33), all the related edges are shown at the time of enquiry. Note that the ambient temperature 25 °C is returned for *Edge 2*.

$$bc = findThermalBC(thermalModelT.BoundaryConditions,'Edge',[1,2]) \quad (138)$$

```
Command Window

   >> bc = findThermalBC(thermalModelT.BoundaryConditions,'Edge',[1 2])

   bc =

     1×2 ThermalBC array with properties:

       RegionType
       RegionID
       Temperature
       HeatFlux
       ConvectionCoefficient
       Emissivity
       AmbientTemperature
       Vectorized

   >> bc(1)

   ans =

     ThermalBC with properties:

                  RegionType: 'Edge'
                    RegionID: 1
                 Temperature: 80
                    HeatFlux: []
       ConvectionCoefficient: []
                  Emissivity: []
          AmbientTemperature: []
                  Vectorized: 'off'

   >> bc(2)

   ans =

     ThermalBC with properties:

                  RegionType: 'Edge'
                    RegionID: [1×57 double]
                 Temperature: []
                    HeatFlux: []
       ConvectionCoefficient: 10
                  Emissivity: []
          AmbientTemperature: 25
                  Vectorized: 'off'
```

FIGURE 6.33. Query to determine the boundary conditions on *Edges 1* and *2* (*E1* and *E2*).

To confirm the initial conditions applied to *Face 1* (*F1*), and *Face 2* (*F2*), the function in Equation (139) is employed. This function returns vector *ic* of length 2, which is equal to the number of the initial condition queries made: '*Face*' [1, 2]. The Face feature is used since the problem is a 2D model (Figure 6.34). For a 3D model, to enquire about the initial conditions, a *Cell* feature would be referenced instead.

$$ic \quad findThermalIC(thermalModelT.InitialConditions, 'Face', [1, 2]) \quad (139)$$

```
Command Window

>> ic = findThermalIC(thermalModelT.InitialConditions,'Face',[1 2])

ic =

  1×2 GeometricThermalICs array with properties:

    RegionType
    RegionID
    InitialTemperature

>> ic(1)

ans =

  GeometricThermalICs with properties:

            RegionType: 'face'
              RegionID: [1 2 3 4 5 6 7 8 9 10 11 12]
    InitialTemperature: 25

>> ic(2)

ans =

  GeometricThermalICs with properties:

            RegionType: 'face'
              RegionID: [1 2 3 4 5 6 7 8 9 10 11 12]
    InitialTemperature: 25
```

FIGURE 6.34. Query to determine the initial conditions on *Faces 1* and *2* (*F1* and *F2*).

Being able to query the nodes at specific coordinates when analyzing the PDE results is very useful. To demonstrate the approach required, some examples are presented here. To find the point indices associated with the coordinates, the function sequence in Equation (140) is employed. In these functions, the thermal model mesh node coordinates are first identified (point), which creates a $(2 \times n)$ data array. The array has two columns identifying the 2D coordinate data, the first column being the *r*-coordinate and the second column being the z-coordinate ($r00$, $z00$). Therefore, to obtain the r- and z-coordinates, the data for each column can be called and set into a separate data vector—($r00$, $z00$). The $r00$ array is sorted and assigned to *radius_sorted* array. This facilitates the creation of connected data points on plots. If the points are not sorted, the connecting lines will interconnect with the points scattered through the plot area.

The function sequence in Equation (141) first selects the radial nodes at the top surface (among all nodes) (*nodesTop_1_h*), as shown in Figure 6.35a. Note that the insulated boundary conditions were applied to these nodes, where $z = length\_P$ and *length_P* is the pipe length (z-coordinate). The array has two columns, identifying the r- and z-coordinate data ($x1, y1$).

The function sequence in Equation (142) is used to query the midplane data ($z = 0.5*length\_P$). Note that the data have a margin length (*eps* is a very small value approaching zero, $eps = 2.2204e{-}16$), which accounts for the small variations from the exact coordinate values; $0.5*length\_P\text{-}eps <= z00 <= 0.5*length\_P + eps$, as shown in Figure 6.35a.

Similar to the previous cases in which the y-data were queried for a given x-value (the horizontal query points in Figure 6.35a), it is possible for the x-data to be queried for certain y-values (the vertical query points in Figure 6.35b) and Equations (143). In these sets of relations, the ($x3, y3$) coordinate represent the midplane nodes at the interior, pipe-fin interface, and exterior surfaces. Note that *radius_O*, *radius_P*, and *radius_E* are the pipe interior, pipe-fin interface, and fin exterior radii, respectively. *grid_size* is the grid size that is uniformly selected for this analysis (0.25-mm).

$$point = thermalModelT.Mesh.Nodes$$
$$nodesIndex = find(point)$$
$$xy00 = point;\ r00 = xy00(1,:);\ z00 = xy00(2,:)$$
$$radius\_sorted = sort(r00);$$
$$index\_sort = find(radius\_sorted);$$
$$length\_sorted = z00(index\_sort);$$

$$(140)$$

$$nodesTop\_1\_h = find(and(length\_P - 0.5*grid\_size <= z00,...$$
$$z00 <= length\_P + 0.5*grid\_size))$$

$$(141)$$

$$xy1 = point(:,nodesTop\_1\_h);\ x1 = xy1(1,:);\ y1 = xy1(2,:)$$

$$nodesCenter\_h = find(and(0.5 * length\_P - 0.5 * grid\_size <= z00,...$$
$$z00 <= 0.5 * length\_P + 0.5 * grid\_size)) \tag{142}$$
$$xy2 = po\operatorname{int}(:, nodesCenter\_h); x2 = xy2(1,:); y2 = xy2(2,:)$$

$$nodesExterior\_v\_O = find(and(offset - eps <= r00, r00$$
$$<= offset + eps))$$
$$nodesExterior\_v\_P = find(and(radius\_P - eps <= r00, r00$$
$$<= radius\_P + eps))$$
$$nodesExterior\_v\_E = find(and(radius\_E - eps <= r00, r00$$
$$<= radius\_E + eps))$$
$$xy3\_O = po\operatorname{int}(:, nodesExterior\_v\_O); x3\_O = xy3\_O(1,:);$$
$$y3\_O = xy3\_O(2,:) \tag{143}$$
$$xy3\_P = po\operatorname{int}(:, nodesExterior\_v\_P); x3\_P = xy3\_P(1,:);$$
$$y3\_P = xy3\_P(2,:)$$
$$xy3\_E = po\operatorname{int}(:, nodesExterior\_v\_E); x3\_E = xy3\_E(1,:);$$
$$y3\_E = xy3\_E(2,:)$$

Horizontal and vertical query points along the x- and y-coordinates are presented in Figure 6.35a. Note that the radial (horizontal) query nodes $(x1, y1)$ are located at the top $(y1 = 25$ mm$)$ and midplane $(y1 = 12.5$ mm$)$, as shown in Figure 6.35a. The axial (vertical) query nodes $(x3, y3)$ are located at the interior $(x3\_O = 12.6$ mm$)$ interface $(x3\_P = 14.3$ mm$)$ and exterior $(x3\_E = 33.2$ m$)$ surfaces of the pipe, pipe-fin, and fin domains. Equation (144) presents the function that plots the queried data shown in Figure 6.35b. In this relation, the $x3\_P$ array is plotted against the $y3\_P$. Since there are large number of data points at the interface of the two parts, it is possible to plot every other point, which is implemented by the *MarkerIndices* feature.

$$plot(x3\_P, y3\_P, 'd', 'Mar\ker\,Size', 3, 'Mar\ker\,FaceColor',...$$
$$[1\ 1\ 1], 'Mar\ker\,Indices', 1:2:length(x3\_P)) \tag{144}$$

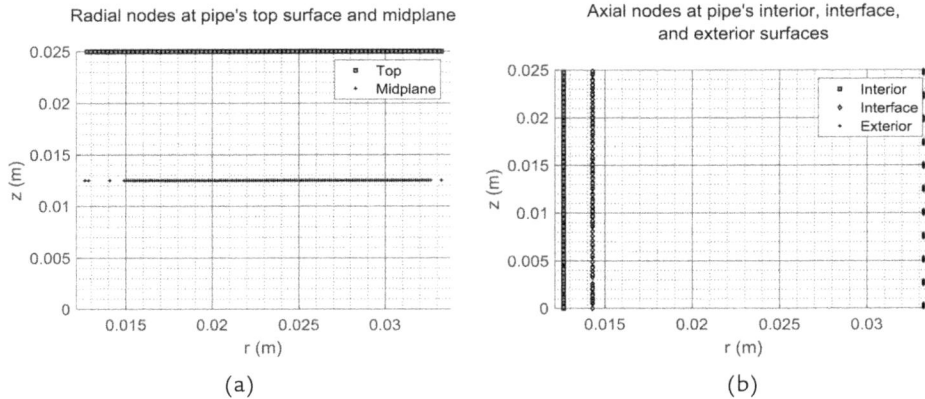

Radial nodes at pipe's top surface and midplane

Axial nodes at pipe's interior, interface, and exterior surfaces

(a)

(b)

FIGURE 6.35. Query points to determine the nodes: (a) Radial ndoes, (b) Axial nodes.

Figure 6.36 presents the temperature profiles versus the radius (Figure 6.36a) and length (Figure 6.36b), starting from the pipe's axisymmetric axis ($x1 = 0$, $y1 = 0$) to the end of the fin ($x1 = 33.2$ mm, $y1 = 25$ mm). Note that in these diagrams, $x1$ represents the r-coordinate. Two plots are shown in this diagram, belonging to the upper edge, representing the insulated top surface ($y1 = 25$ mm), and the midplane ($y1 = 12.5$ mm). As expected, the temperature of the top surface is higher than the middle because it is insulated. Figure 6.44 presents the temperature gradient ($Tx1 = dT/dr$) and heat flux ($-k*Tx1 = -k*dT/dr$) for the data points presented in Figure 6.35a. It is paramount to identify the regions correctly. This is done by enquiring about the associated nodes at given distances from the center, as shown in Equation (145). The conductivity for each region is then multiplied by the corresponding temperature gradient to obtain the heat flux of that region.

$$copper\_x1\_P = find(x1 <= radius\_P + 0.5*eps)$$
$$aluminum\_x1\_E = find(and(radius\_P - 0.5*eps <= x1,x1 \qquad (145)$$
$$<= radius\_E + 0.5*eps))$$

Figure 6.37 represents the temperature gradient (dT/dr in Figure 6.37a) and heat flux ($-k*A*dT/dr$ in Figure 6.37b) along the radial direction. At the pipe-fluid interface (the pipe's interior boundary), the temperature gradient has the largest value, and it is reduced to zero as it approaches the fin's exterior surface. The temperature gradient on the top surface has a smaller magnitude than that at the midplane. The top surface is insulated and the midplane transfers heat by convection to the surrounding surfaces. Equation (145) is used to identify the pipe and fin regions. This information

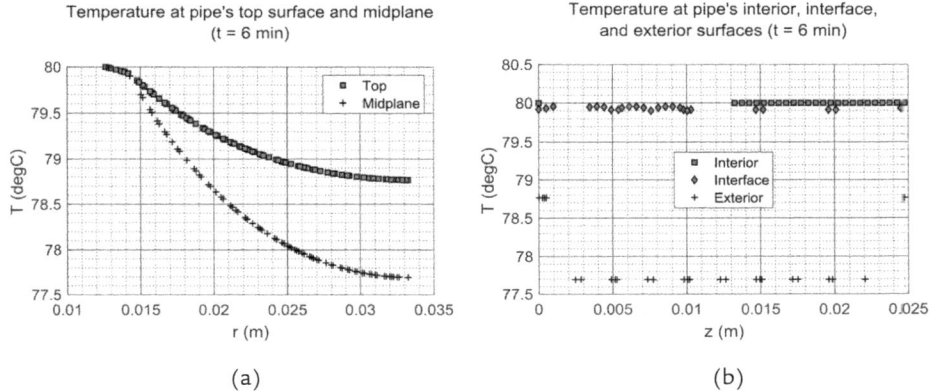

FIGURE 6.36. Temperature profiles at selected planes: (a) Radial temperature, (b) Axial temperature.

is then employed to determine which thermal conductivity to use when calculating the heat flux.

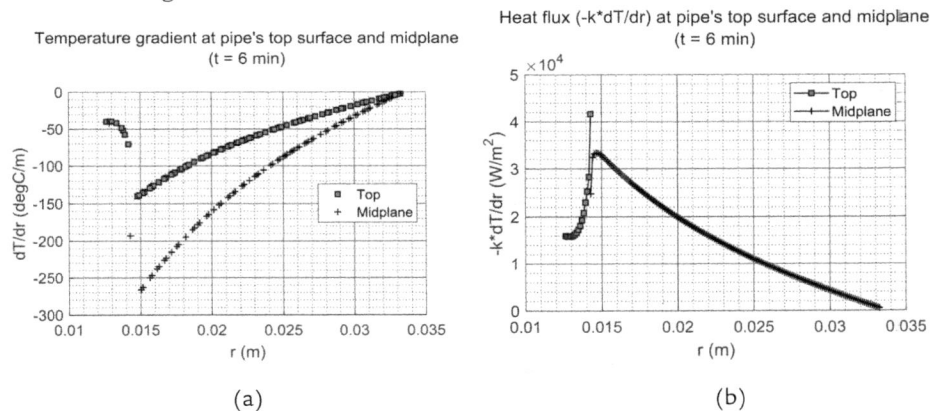

FIGURE 6.37. Radial profiles at the top surface and midplane: (a) Temperature gradient, (b) Heat flux.

The heat rate $(-k*A*dT/dr)$ is presented in Figure 6.38a in Watts at the time of 6 min and per the edge length in W/m versus the time in Figure 6.38b for the selected edges. These edges are either in direct contact with the surrounding environment (*Edges 11, 12, 52,* and *35*) or at the pipe-fin interface (*Edge 34*). *Edge 35* (on the copper pipe exterior surface) shows the largest heat loss rate by convection and that is because of the high temperature difference between this edge and the surrounding environment (Figure 6.39). To calculate the temperature, temperature gradient, and heat rate for the selected coordinates as well as the heat rate (per unit length of the edge), the functions in Equation (146) are employed. The nodes created at the mesh grids, given their spacing ($dx = 0.5051$ mm,

$dy = 0.3030$ mm), are obtained by calling the consecutive nodes along the x- and y-coordinates, (Xw, Yh) $[Xw(1, 2), Yh(2, 1)] = [0.2525, 0.1443]$ mm.

Transient heat rate at pipe's top surface and midplane
(t = 6 min)

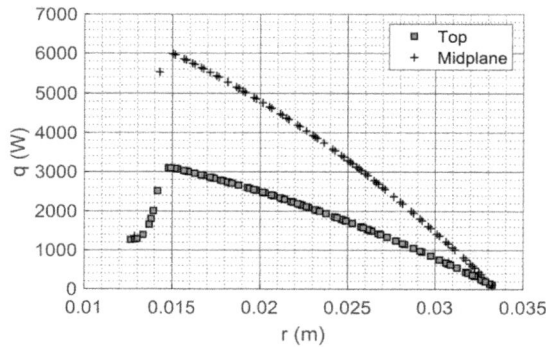

(a)

Trasient heat rate per edge length at pipe's edges

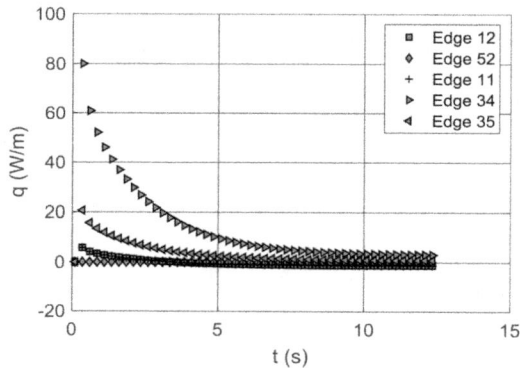

(b)

FIGURE 6.38. Heat rate: (a) Radial, top surface and midplane, (b) Transient, selected edges, per unit length of the edge.

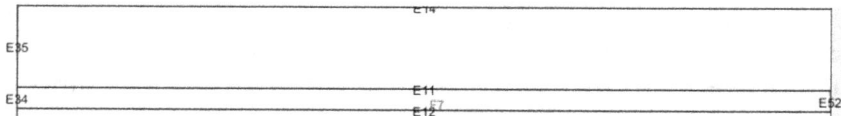

FIGURE 6.39. Edges whose heat rates are calculated.

$$w = linspace(0, length\_P, 100);$$

$$h = linspace(0, radius\_P, 100);$$

$$[Xw, Yh] = meshgrid(w, h);$$

$$T\_xy = interpolateTemperature(resultT, Xw, Yh, 1 : length(t) \qquad (146)$$

$$[Tx, Ty] = evaluateTemperatureGradient(resultT, Xw, Yh, 1 : length(t))$$

$$[qx, qy] = evaluateHeatFlux(resultT, Xw, Yh, 1 : length(t))$$

$$Qn = evaluateHeatRate(resultT, 'Edge', [11\ 12])$$

Further data points can be extracted from any of the horizontal or vertical sets of data explained above. For example, to find the midplane points at the interior, pipe-fin interface, and exterior surfaces—($x4$, $y4$), Equation (147) can be employed, which is a subset of Equations (143) (Figure 6.38a).

$$nodesInterior\_2\_O = find(and(-eps + radius\_O <= x2, x2$$
$$<= + eps + radius\_O))$$

$$nodesInterior\_2\_P = find(and(-eps + radius\_P <= x2, x2$$
$$<= + eps + radius\_P))$$

$$nodesInterior\_2\_E = find(and(-eps + radius\_E <= x2, x2$$
$$<= + eps + radius\_E)) \qquad (147)$$

$$xy4\_O = xy2(:, nodesInterior\_2\_O);\ x4\_O = xy4\_O(1,:);$$
$$y4\_O = xy4\_O(2,:)$$

$$xy4\_P = xy2(:, nodesInterior\_2\_P);\ x4\_P = xy4\_P(1,:);$$
$$y4\_P = xy4\_P(2,:)$$

$$xy4\_E = xy2(:, nodesInterior\_2\_E);\ x4\_E = xy4\_E(1,:);$$
$$y4\_E = xy4\_E(2,:)$$

Midplane query points at the interior ($x4\_O$, $y4\_O$) = (12.6, 12.5) mm interface ($x4\_P$, $y4\_P$) = (14.3, 12.5) mm, and exterior surfaces ($x4\_E$, $y4\_E$) = (33.2, 12.5) mm are the combination of the two said query points, as shown in Figure 6.40a. Temperatures at these query points are plotted along the radial direction in Figure 6.40b. Transient temperature profiles at the midplane pipe-fin interface and exterior surfaces are presented in Figure 6.41.

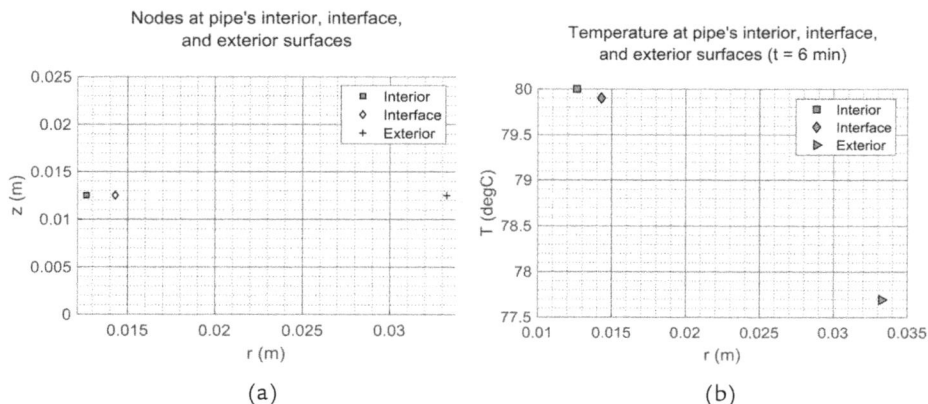

FIGURE 6.40. Data at the midplane-interior, interface, and exterior surfaces:
(a) Query points (b) Temperature at the last time step.

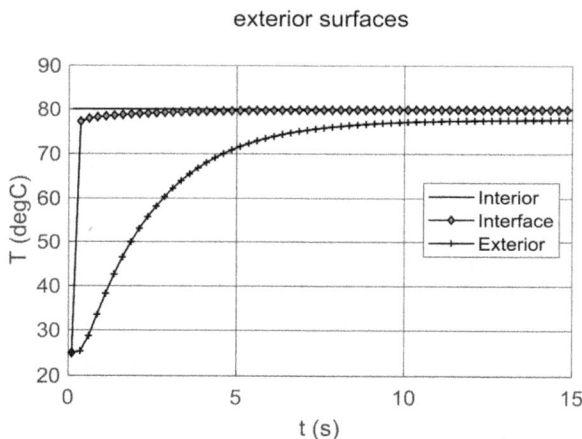

FIGURE 6.41. Transient temperature profiles at the midplane pipe-fin interface and exterior surfaces.

The heat flux field vectors at the pipe's interface and exterior surfaces at the last time step are presented in Figure 6.42. Note that, as mentioned earlier, the heat flow rate returns a real number or, for time-dependent results, a vector of real numbers. This number (or vector) represents the integrated heat flow rate and is normal to the boundary. It is positive if the heat flows out of the domain and is negative if the heat flows into the domain. Note that the arrow direction and length at each data point (x- and y-coordinates) represent the associated values for the temperature gradient ($Tx = dT/dr$ and $Ty = dT/dz$) or temperature (T).

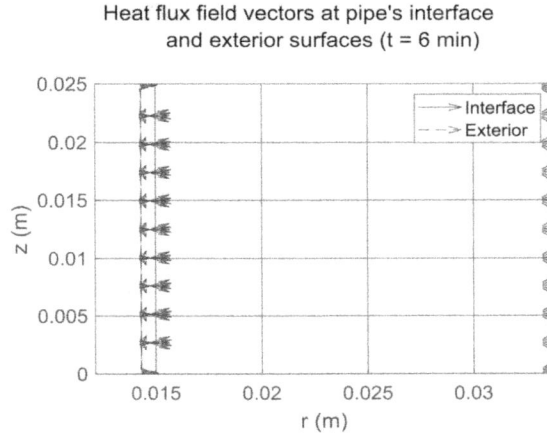

FIGURE 6.42. Heat flux field vectors at the pipe's interface and exterior surfaces at the last time step.

The model configuration data are shown in Figure 6.43 using the *ReportStatistics* function for the *thermalModelT* model. Entering this model object name (*thermalModelT*) in the *Command Window* returns the model object type (*thermal*), analysis type (*transient*), and information about the solver options, heat source, and mesh (Figure 6.44). Appending any of the properties presented in Figure 6.44 after the model object name (*thermalModelT*) returns the properties' related information. For example, inputting *thermalModelT.Geometry* returns the information in Figure 6.45, which represents the number of faces, edges, and vertices. Inputting *thermalModelT.SolverOptions* returns the information in Figure 6.46, which includes the absolute, relative, and residual tolerances, maximum iterations, and minimum time steps. Inputting *thermalModelT.Mesh* returns the mesh data, such as the number of nodes, elements maximum and minimum element size, gradation, and geometric order (e.g., quadratic), as given in Figure 6.47.

To obtain the node coordinates, execute the *size* function with *thermalModelT.Mesh.Nodes* as its input. This will return [2 1 1 8 5 3], with the latter number equal to the number of nodes. The two indices of the first component (1,:) and (2,:) are related to the r- and z-coordinates, respectively. To identify the extent of the r- and z-coordinates (the domain boundaries), the *max* and *min* functions are employed, as shown in Equations (148) and (149).

```
Command Window
>> thermalModelT.SolverOptions.ReportStatistics = 'on'

thermalModelT =

  ThermalModel with properties:

                 AnalysisType: "transient-axisymmetric"
                     Geometry: [1×1 AnalyticGeometry]
           MaterialProperties: [1×1 MaterialAssignmentRecords]
                  HeatSources: []
        StefanBoltzmannConstant: []
           BoundaryConditions: [1×1 ThermalBCRecords]
            InitialConditions: [1×1 ThermalICRecords]
                         Mesh: [1×1 FEMesh]
                SolverOptions: [1×1 pde.PDESolverOptions]
```

FIGURE 6.43. Thermal model query to determine model configurations.

```
Command Window
>> thermalModelT

thermalModelT =

  ThermalModel with properties:

                 AnalysisType: "transient-axisymmetric"
                     Geometry: [1×1 AnalyticGeometry]
           MaterialProperties: [1×1 MaterialAssignmentRecords]
                  HeatSources: []
        StefanBoltzmannConstant: []
           BoundaryConditions: [1×1 ThermalBCRecords]
            InitialConditions: [1×1 ThermalICRecords]
                         Mesh: [1×1 FEMesh]
                SolverOptions: [1×1 pde.PDESolverOptions]
```

FIGURE 6.44. Thermal model properties records.

```
Command Window
>> thermalModelT.Geometry

ans =

  AnalyticGeometry with properties:

          NumCells: 0
          NumFaces: 12
          NumEdges: 57
       NumVertices: 46
          Vertices: [46×2 double]
```

FIGURE 6.45. Thermal model geometry records.

```
Command Window
  >> thermalModelT.SolverOptions

  ans =

    PDESolverOptions with properties:

       ReportStatistics: 'on'

     ODE solver
       AbsoluteTolerance: 1.0000e-06
       RelativeTolerance: 1.0000e-03

     Nonlinear solver
       ResidualTolerance: 1.0000e-04
          MaxIterations: 25
                MinStep: 1.5259e-05
            ResidualNorm: Inf

     Lanczos solver
               MaxShift: 100
              BlockSize: []
```

FIGURE 6.46. Thermal model solver records.

```
Command Window
  >> thermalModelT.Mesh

  ans =

    FEMesh with properties:

                Nodes: [2×11853 double]
             Elements: [6×4982 double]
       MaxElementSize: 2.5000e-04
       MinElementSize: 1.2500e-04
         MeshGradation: 1.5000
        GeometricOrder: 'quadratic'
```

FIGURE 6.47. Thermal model mesh records.

$$min(thermalModelT.Mesh.Nodes(1,:)) = 0.0126$$
$$max(thermalModelT.Mesh.Nodes(1,:)) = 0.0332 \quad (148)$$
$$min(thermalModelT.Mesh.Nodes(2,:)) = 0$$
$$max(thermalModelT.Mesh.Nodes(2,:)) = 0.025 \quad (149)$$

6.4 Case Study 4—Non-Axisymmetric Pipe: Transient Thermal Model with Spatial and Temporal Boundary Conditions

6.4.1 Setup

This case study investigates transient heat transfer in a non-axisymmetric pipe. The pipe is made of PEX plastic exposed to a moving heat source, such as may be encountered in laser welding.

This model takes advantage of the MATLAB's import geometry feature. While it is possible to create simple 3D and 2D geometries in the MATLAB environment, it is more practical to create complex geometries in a dedicated CAD tool and import them into the MATLAB environment.

First, the PDE thermal model (*thermalModelT*) was created using the model-creation function for transient analysis in Equation (125). The 3D geometry (*geom*) was created in SOLIDWORKS CAD software and exported in **.stl* format (*Pipe_hollow_shortened.stl*). It was then imported into MATLAB using the *importGeometry* function, shown in Equation (150). One could also add the path information to the file specification when importing (e.g., *../geometrics/Pipe_hollow_shortened.stl*). The function in Equation (151) displays the model geometry after the import. The function makes the model semi-transparent (*'FaceAlpha'* = 0.5) and activates *Edge*, *Face*, and *Cell* labels, *EdgeLabels/FaceLabels/CellLabels*. Figure 6.48 shows the resulting model with the *Cell* ID (labelled *C1*) and *Face* IDs (labelled *1-26*) shown.

$$geom = importGeometry(thermalModelT,'Pipe\_hollow\_shortened.stl') \quad (150)$$

$$geom = importGeometry(thermalModelT,'Pipe\_hollow\_shortened.stl') \quad (151)$$

Examining the imported geometry displayed in Figure 6.48, one can observe that the dimensions shown are in m. However, the modeled pipe should have the equivalent dimensions in mm. One could return to the CAD tool and try to fix this issue, but there is another approach available within MATLAB. It makes it possible for the geometry to be scaled as required. The scaling may be carried out independently with respect to each of x-, y-, and z-coordinates by employing the function $scale(geometry,[x\ y\ z])$, where geometry is the geometry name (*geom* in this case) and $[x\ y\ z]$ is the scale vector that defines scaling for each coordinate. For this example, the scale factors $[x\ y\ z] = [1\ 1\ 1]/1000$ have been selected, decreasing the size of the entire geometry along all coordinates by a factor of 1,000, as shown in Figure 6.49.

Geometry with face and cell labels displayed

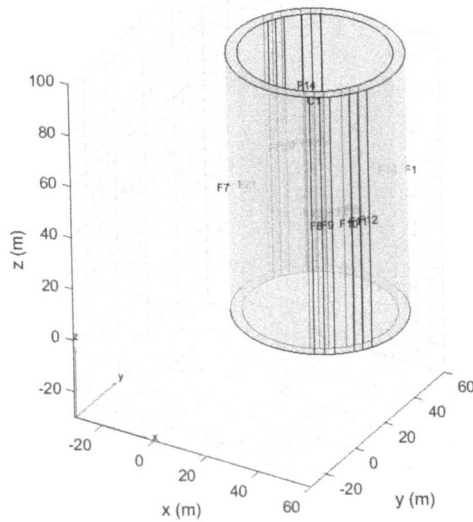

FIGURE 6.48. 3D pipe geometry after *.stl* file import.

Geometry with face and cell labels displayed

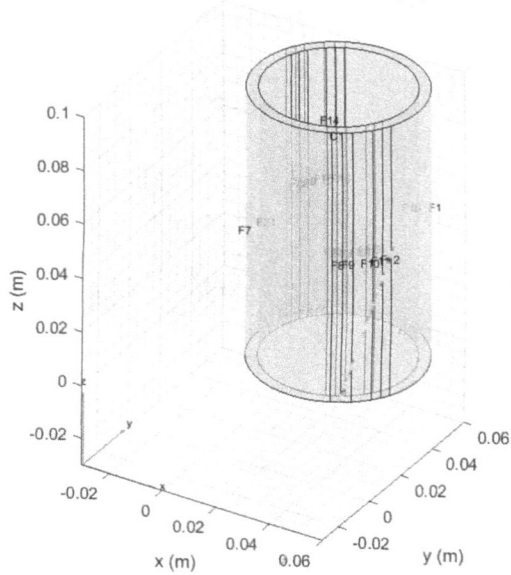

FIGURE 6.49. 3D pipe geometry after scaling by [1 1 1]/1,000.

The *pdeplot3D(mesh)* function displays the meshed view of the geometry. For this analysis, the mesh grid size is set to 1 mm. It may be useful to display the 3D geometry from different points of view. To achieve this, the view function is used—*view(az, el)*. The inputs to this function are two rotation angles in degrees for the line of sight: azimuth (*az*) and elevation (*el*). The azimuth corresponds to the rotation about the *z*-axis, with value measured from the negative *y*-axis, and with increasing values leading to counterclockwise rotation of the line of sight relative to the axes. The second input angle is measured between the horizontal *xy*-plane and the line of sight and ranges from -90 to 90 degrees.

One can also obtain the 2D view of the *xy*-plane by the *view(2)* function, equivalent to the *view(0, 90)* function. The default 3D view is obtained by the *view(3)* function, equivalent to the *view(−37.5, 30)* function. Figure 6.50 presents the result of the *view(0, 0)* function, which shows the front view of the pipe (2D view of the *xz*-plane). Figure 6.51 shows the result of the *view(2)* and *view(0, 90)* functions and the result of the *view(3)* function.

Tetrahedral mesh

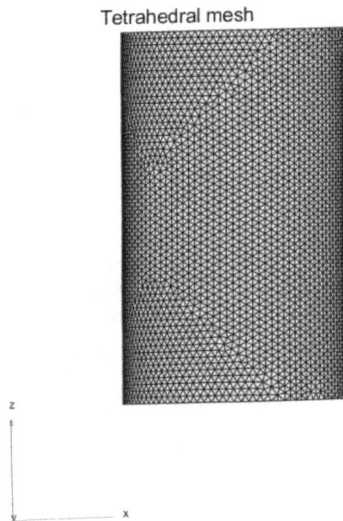

FIGURE 6.50. 3D mesh, front view (0,0).

Tetrahedral mesh   Tetrahedral mesh

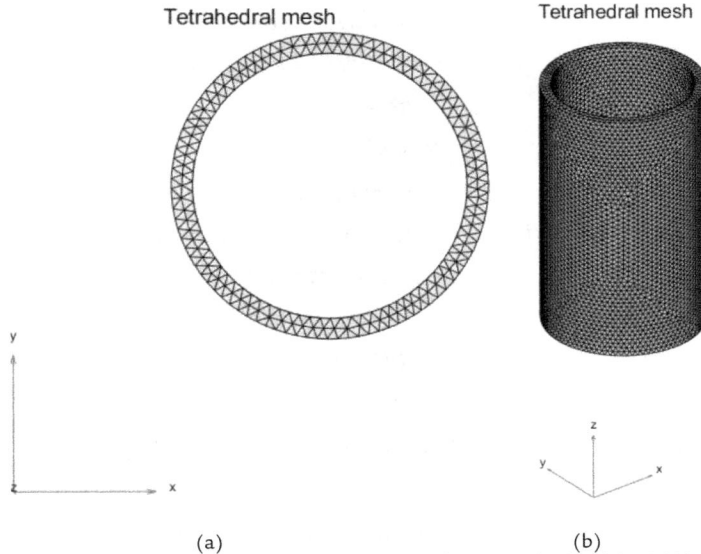

(a)   (b)

FIGURE 6.51. 3D mesh: (a) Top view (0,90), (b) Isometric view (-37.5,30).

The functions that define model configuration, including material properties, are listed in Figure 6.52. The material properties are defined using the same methodology presented in the previous case studies—Equation (125) for the transient models. Equation (128) is employed to assign internal heat generation term. The boundary conditions are presented in Figure 6.53. The thermal model conditions are set by employing Equations (129) to Equation (133) for the boundary conditions and Equation (134) for the initial condition.

For this 3D model, the *Face* geometry region type is used to assign boundary conditions. Heat source (internal heat generation), if present, would be applied to *Cell* for the 3D model. There are a total of 26 faces identified in this geometry, with *Face 13* and *Face 14* being the upper and

%Pipe material properties(PEX)

$k1 = 0.41; \%thermal conductivity(W/mK)$

$rho1 = 935; \%density(kg/m^3)$

$cp1 = 2100; \%heat capacity(J/kgK)$

$tp1 = thermalProperties(thermalModelT, 'ThermalConductivity', k1, ...$

$'MassDensity', rho1, 'SpecificHeat', cp1, 'Cell', 1);$

FIGURE 6.52. Material properties in the MATLAB script for the non-axisymmetric transient pipe model.

%Boundary condition : convection

$num\_faces = thermalModelT.Geometry.NumFaces;$

$hc = 10;$

$outerCC\_V = hc;$

thermalBC(thermalModelT,' Face',1 : num_faces,...

'ConvectionCoefficient',' outerCC_V,' AmbientTemperature',

T_ambient,'Vectorized',' on');

%Cell condition : heat source

$q1 = 0; \%W / m^3$

$ih1 = \text{int } ernalHeatSource(thermalModelT, q1,' Cell',1);$

%Boundary condition : radiation

$thermalModelT.StefanBoltzmannConstant = 5.670373E - 8;$

$emis = 0.91;$

thermalBC(thermalModelT,' Face',[1 : 12], Emissivity', emis,...

'AmbientTemperature', T_ambient,'Vectorized',' on');

%Boundary condition : heatflux

thermalBC(thermalModelT,' Face',1,' HeatFlux',@q_func,

'Vectorized',' on');

%Boundary condition : temperature (optional − not included

in this analysis)

%thermalBC(thermalModelT,' Face',[12],'Temperature',@Twater_func);

% Identify boundary conditions

bca = findThermalBC(thermalModelT.BoundaryConditions,' Face',[1 : 26]);

FIGURE 6.53 Boundary conditions MATLAB script for the non-axisymmetric transient pipe model.

lower end surfaces of the cylinder, respectively. The exterior surfaces (*Face 1* to *Face 12*) are exposed to the ambient air (25 °C), and transferring heat by convection, having a convection heat transfer coefficient of 10 W/m²K.

The upper and lower end surfaces (*Face 13* and *Face 14*) are exposed to zero heat flux. *Face 1* to *Face 6* are exposed to radiation heat transfer with the ambient. *Face 1* is additionally exposed to a spatial and temporal heat flux (*q_func*) that models the moving heat source on the exterior surface. This term is defined in the form of a function, appended to the end of the complete script (Figure 6.54). The function represents a heat source with 1 W power and dimensions of 4×4 mm² scanning the pipe at 0.5 mm/s.

function q_out = q_func(region, state)

global Velocity q_heat hc T_ambient x_c spot_w y_max grid_size

t = state.time;

*z_c = Velocity * t;*

x = region.x;

y = region.y;

z = region.z;

*q_out = hc * (T_ambient - state.u);*

if isnan(t)

　　q_out = nan(1, numel(x));

end

*indx = (abs(z - z_c) < spot_w * 0.5) & (abs(x - x_c) < spot_w * 0.5) ...*

*& (y > (y_max - 0.5 * grid_size));*

xxx = x(indx); yyy = y(indx); zzz = z(indx);

$q_o$ut(1, indx) = $q_h$eat / (spot_w ^ 2);

q_o

end

FIGURE 6.54. Moving heat source boundary condition
MATLAB script for the non-axisymmetric transient pipe model.

To work correctly, the function needs to have two inputs: the first is the point locations (*region*) and the second is the times (*state*). These are PDE model objects passed internally to the function. Their names within the function are arbitrary. Input time values are extracted by appending *time* to the object name (e.g., *state.time*); location coordinates are extracted by appending the coordinate name (e.g., *region.x*). All other parameters are passed to the function using the global variable definitions. These are defined by listing the variable names after the keyword *global* both at the start of the script and the function (Figure 6.55).

The time step is 0.02 s and the entire heating process is 100 s. Analysis settings are presented in Figure 6.56. Note the use of the *tic* and *toc* functions to monitor the solution time, which was 5.83 hr for this case study.

% Global data
global T_low T_high t_low t_high Velocity q_heat hc T_ambient ...
x_c spot_w y_max grid_size

FIGURE 6.55. Global variables in the MATLAB script for the non-axisymmetric transient pipe model.

%Initial parameters
tfinal = 100;
time_step = 0.02;
tlist = 0 : time_step : tfinal;
t = tlist;
t_length = length(tlist);
T0_1 = 25; % deg C
IC1 = thermalIC(thermalModelT, T0_1, 'Cell', 1);
t2 = tlist / 60;

%Solve properties
thermalModelT.SolverOptions.ReportStatistics = 'on';
thermalModelT.SolverOptions.AbsoluteTolerance = 1.0000e - 05;
tic
timeStart1 = datetime('now');
hSt1 = hour(timeStart1); mSt1 = minute(timeStart1); sSt1 = round(second(timeStart1));
results = solve(thermalModelT, tlist);
timeStart2 = datetime('now');
hSt2 = hour(timeStart2); mSt2 = minute(timeStart2); sSt2 = round(second(timeStart2));
sol_time = toc;

%Model results
T1 = results.Temperature; % identify tempeature data for the selected nodes
tt = round(tfinal / 60, 1); % identify final time step in min
Tx1 = results.XGradients; % identify tempeature gradient along x - coordinate
Ty1 = results.YGradients; % identify tempeature gradient along y - coordinate
Tz1 = results.ZGradients; % identify tempeature gradient along z - coordinate
T_st1 = results.SolutionTimes; % identify solutine times

FIGURE 6.56. Solution settings in the MATLAB script for the non-axisymmetric transient pipe model.

Node distribution is displayed in Figure 6.57 to help with locating of the heat source coordinates. Nodes where the moving heat source applies its energy are shown in Figure 6.58 at the end of the heating process (100 s).

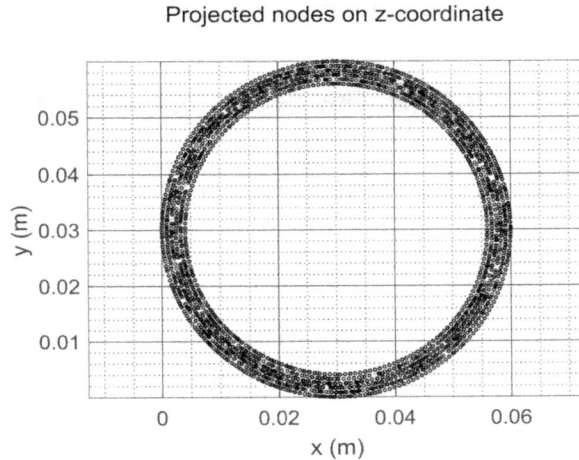

FIGURE 6.57. Projected node density on the z-coordinate.

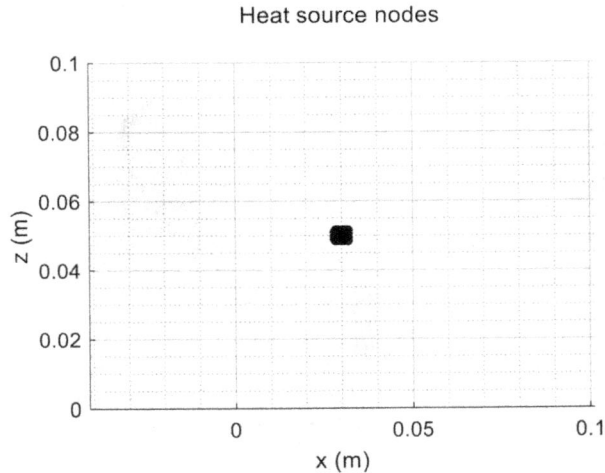

FIGURE 6.58. Heat source nodes at the end of the heating process ($t = 100$ s).

6.4.2 Results

Figure 6.59 presents the solution statistics. Figure 6.60 shows temperature contours at 50 s and 100 s. The contours show a hot spot followed by a trail of progressively cooling material, as is typically expected when modeling a moving heat source. Figure 6.61 shows the temperature gradient contours with respect to the x- and y-coordinates at the end of the process.

Following the *x*-coordinate gradient as the *x* value increases shows variation from a very high positive value, where the temperature rapidly rises as the location approaches the heat source path to the equivalent in magnitude negative values as the temperature falls on the other side of the heat source path.

```
Command Window

  results =

     TransientThermalResults with properties:

        Temperature: [121343×5001 double]
       SolutionTimes: [1×5001 double]
          XGradients: [121343×5001 double]
          YGradients: [121343×5001 double]
          ZGradients: [121343×5001 double]
                Mesh: [1×1 FEMesh]
```

FIGURE 6.59. 3D thermal transient model solution statistics for the non-axisymmetric pipe.

Temperature contours (degC)
(t = 50 s)

Temperature contours (degC)
(t = 100 s)

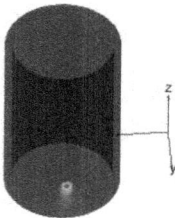
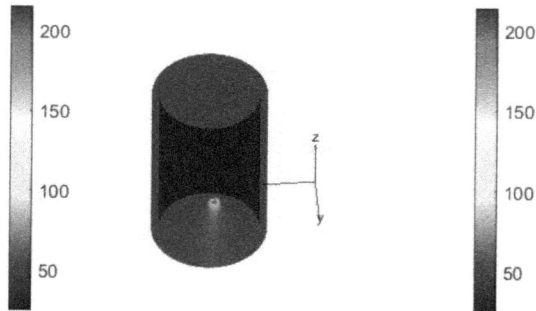

(a)   (b)
FIGURE 6.60. 3D thermal transient model temperature contours for the non-axisymmetric pipe: (a) *t* = 50 s, (b) *t* = 100 s.

Temperature gradient contours (dT/dx in degC/m)
(t = 100 s)

Temperature gradient contours (dT/dy in degC/m)
(t = 100 s)

(a)

(b)

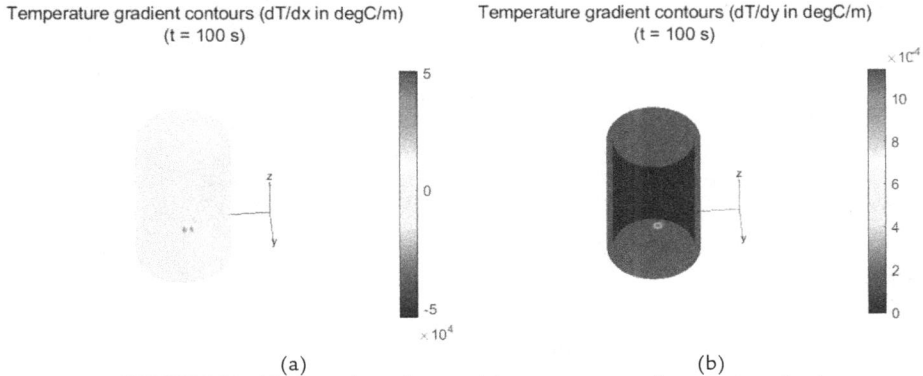

FIGURE 6.61. 3D thermal transient model temperature gradient contours for the non-axisymmetric pipe: (a) dT/dx, (b) dT/dy.

Figure 6.62 shows the transient temperatures within the pipe wall (exterior, middle, and interior). The location for which the temperatures are plotted is where the beam center is located after scanning 0.02 m ($z = 0.02$ m) of the pipe, which is 40 s after the movement start. The transverse (x) coordinate value is $x = 0.03$. This temperature information is useful in determining the appropriate process parameters. The maximum temperature of 202 °C on the exterior surface occurs just after the beam passes the location for which the temperature is plotted, at the coordinate (0.03, 0.06, 0.02) m. The maximum is reached at progressively later times at the middle, 71 °C, and the interior of the pipe wall, 52 °C. The delay is due to the time it takes for the heat to conduct from the exterior to these points.

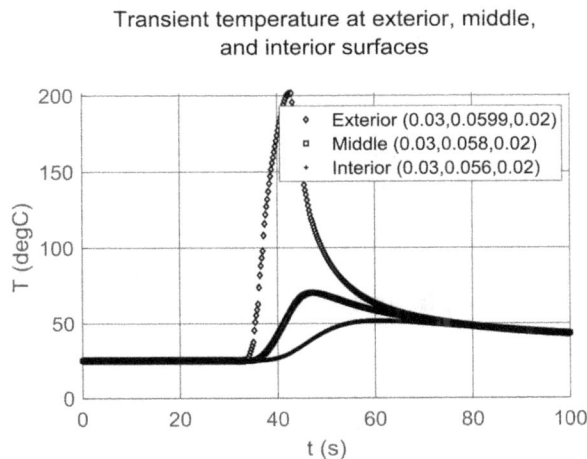

Transient temperature at exterior, middle,
and interior surfaces

FIGURE 6.62. Transient temperature profiles at the exterior, middle, and interior surfaces ($x = 0.03$ m, $z = 0.02$ m).

An optional setting has been added to the program, which implements an additional transient boundary condition applied to the pipe's interior surface (*Face 12*). This BC represents fast-flowing water, the temperature of which rises over 3 s from 35–80 °C (Figure 6.60).

```
function T_out = Twater_func(location, state)
t = state.time;
global T_low T_high t_low t_high
if t_high < t_low
    error('t_low must be less than t_high')
end
if any(isnan(t))
  T_out = NaN(size(t));
else
   T_out = zeros(size(t));
   T_out(t <= t_low) = T_low;
   T_out(t >= t_high) = T_high;
   t_indx = (t > t_low) & (t < t_high);
   if (any(t_indx))
      aa = (T_high - T_low) / (t_high - t_low);
      bb = T_low - aa * t_low;
      T_out(t_indx) = aa * t(t_indx) + bb;
   end
end
end
```

FIGURE 6.63. Changing temperature boundary condition in the MATLAB script for the non-axisymmetric transient pipe model.

6.5 Case Study 5—Combining the MATLAB Script and the *PDE Modeler* Application

There are two approaches to the development of custom scripts for FE modeling with the help of the *PDE Modeler* application: (a) Exporting the model created using the *PDE Modeler* application to the MATLAB script (**.m*) file and then revising it, and (b) Creating the geometry in the

PDE Modeler application, exporting its data, and then incorporating that geometry into a script developed using the *PDE Toolbox* commands or functions.

If the former approach is selected, the structure and functions employed in the stored code need to be understood and decoded so that correct revisions can be made. The first section below helps with this task. The section describes creation of the model and then the exporting of its PDE data at the end of each operation (e.g., geometry and mesh creation or assigning the boundary conditions and solution settings). The subsequent section begins with export of the geometry parameters from the *PDE Modeler* and continues with the definition of the model and solution.

The presented model was primarily intended as a demonstration of the alternative approaches described above. It consists of multiple overlapping simple 2D shapes. Physically, the model represents a pump. The heated fluid enters the geometry through *Edge 1* and exits at *Face 12* (outside the xy-plane). The thermal model is transient (parabolic) and requires complete set of the thermophysical properties. This case study presents an advanced usage of the MATLAB coding to generate and revise 2D thermal models. There are three types of boundary conditions applicable to this geometry: (a) Temperature, where it is assumed constant—*Dirichlet*; (b) Convection, where heat is transferred to the ambient at a constant temperature (35 °C) and convection coefficient (10 W/m^{2}K)—*Neumann*; and (c) Heat flux, which is constant and applied to the fluid at the inlet—*Neumann*.

6.5.1 The *PDE Modeler* Script

This section addresses creation of a 2D model in a script which is based on the model specifications (e.g., PDE settings, thermophysical properties, and geometry data) exported from the *PDE Modeler* application. When creating geometry elements in the *PDE Modeler*, each element has its own identifier (e.g., *C* for circle and *P* for polygon). By default, as new elements of each type are created, a sequence number is appended to the corresponding identifier character (e.g., the first circle would be identified by *C1* and the second one by *C2*), etc., as shown in Figure 6.64.

The application model is an *∗.m* file and can be viewed and modified within the MATLAB *EDITOR*. To view this script, the *∗.m* file is opened in the MATLAB *EDITOR*. Figure 6.65 shows a section of this script that is related to the geometry description. The presentation method of the created script by the *PDE Modeler* application is different from what is presented herein and consists of multiple coded lines for each function; however, it can be

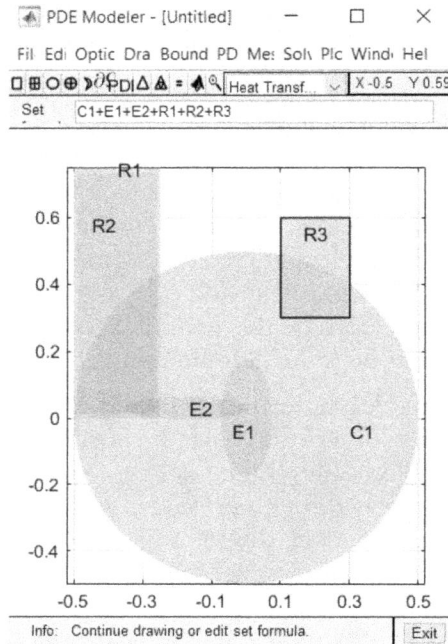

FIGURE 6.64. The 2D geometry created in the MATLAB *PDE Modeler*.

organized to be presented by the script seen in Figure 6.65. This shows that there are six domains (*Circle 1*, *Ellipses 1-2*, and *Rectangles 1-3*), as given in Figure 6.64. These geometry components (domains) are to be assembled to create the model geometry. In this example, they are all added $C1 + E1 + E2 + R1 + R2 + R3$—Equation (152).

$$pdeellip(0,0,0.5,0.5,0,'C1');$$
$$pdeellip(0,0,0.075,0.175,0,'E1');$$
$$pdeellip(-0.175,0.03,0.180,0.034,0,'E2');$$
$$pderect([-0.5 \ -0.25 \ 0.75 \ 0],'R1');$$
$$pderect([-0.49 \ -0.26 \ 0.735 \ 0.015],'R2');$$
$$pderect([0.1 \ 0.3 \ 0.6 \ 0.3],'R3');$$

FIGURE 6.65. Geometry description script to plot the 2D geometry in the MATLAB *PDE Modeler* application.

$$set(findobj(get(pde\_fig,'Children'),'Tag','PDEEval'), \\ 'String','C1+E1+E2+R1+R2+R3') \tag{152}$$

Note that the 2D PDE model physics is introduced to the *PDE Modeler* before the solution can be achieved for the selected physics. Depending on the physics selected, the related PDE specification and coefficients are introduced into the model. For a *Generic System* application, the input parameters are to be fed into $-\nabla.(c\nabla u) + au = f$, which is another form for $-div(c*grad(u)) + a*u = f$, as shown in Figure 6.66. If the *Structural Mechanics* (*Plane Stress* or *Strain*) application is chosen, the input parameters, such as the Young's modulus (E), Poisson ratio (u), density (r), and volumetric force along the x- and y-coordinates (kx, ky), are to be defined (Figure 6.67). For a *Heat Transfer* application, the input parameters are employed in $\rho CT - \nabla.(k\nabla T) = Q + h(Text - T)$, which is another form for $rho * C * T - div(k * grad(T)) = Q + h * (Text - T)$ (Figure 6.68). Note that ρ is the density (kg/m^3), C is specific heat capacity (J/kgK), k is thermal

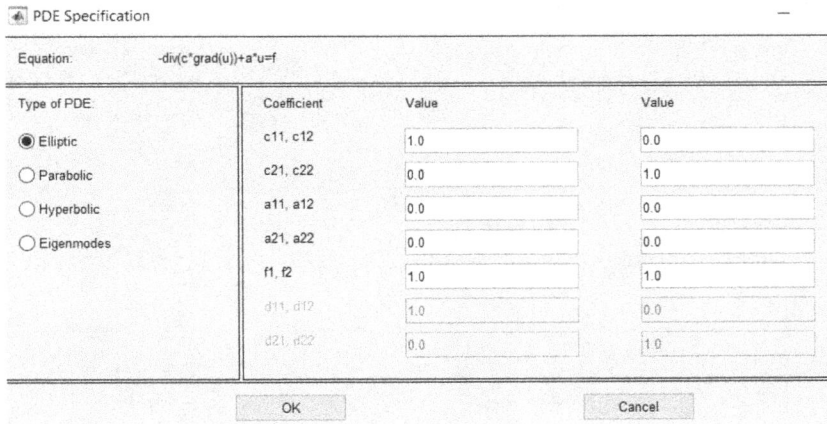

FIGURE 6.66. The *PDE Modeler, PDE Specification* for a *Generic System* model.

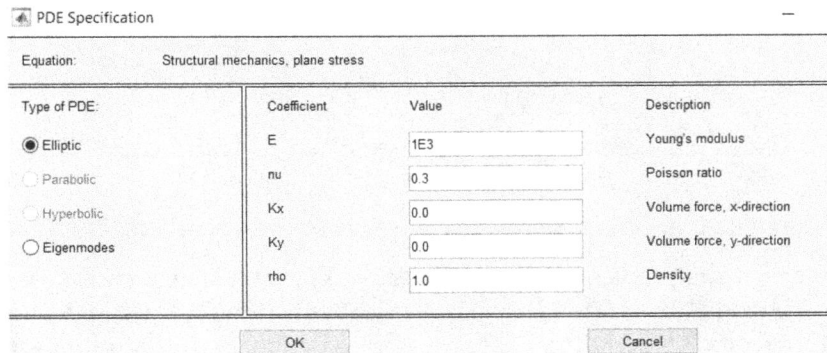

FIGURE 6.67. The *PDE Modeler, PDE Specification* for a *Structural Mechanics, Plane Stress* model.

conductivity (W/mK), Q is internal heat source (W/m³), h is convection heat transfer coefficient (W/m²K), and *Text* is external temperature (°C). As seen in Figure 6.67, *Parabolic Type of PDE* is selected (Figure 6.68). This means the analysis is transient, and therefore all the specified thermal properties above are employed in the equation. If the *Elliptic Type of PDE* were selected, the density and specific heat capacity would not have been needed, and therefore grayed out from the *PDE Specification* menu, creating a steady-state analysis (Figure 6.69).

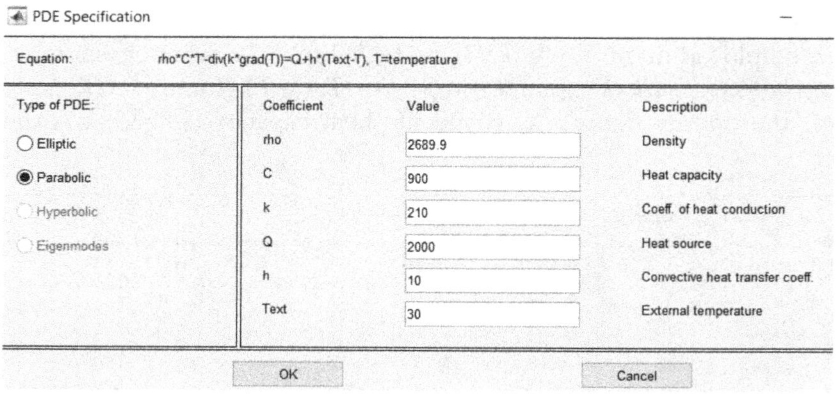

PDE Specification			—
Equation:	rho*C*T-div(k*grad(T))=Q+h*(Text-T), T=temperature		

Type of PDE:	Coefficient	Value	Description
○ Elliptic	rho	2689.9	Density
◉ Parabolic	C	900	Heat capacity
○ Hyperbolic	k	210	Coeff. of heat conduction
○ Eigenmodes	Q	2000	Heat source
	h	10	Convective heat transfer coeff.
	Text	30	External temperature

OK   Cancel

FIGURE 6.68. The *PDE Modeler*, *PDE Specification* for a *Heat Transfer* model, *Parabolic* settings.

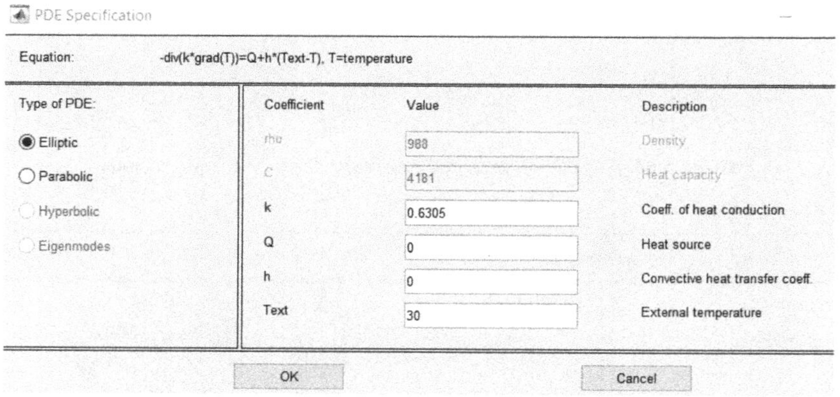

PDE Specification			—
Equation:	-div(k*grad(T))=Q+h*(Text-T), T=temperature		

Type of PDE:	Coefficient	Value	Description
◉ Elliptic	rho	988	Density
○ Parabolic	C	4181	Heat capacity
○ Hyperbolic	k	0.6305	Coeff. of heat conduction
○ Eigenmodes	Q	0	Heat source
	h	0	Convective heat transfer coeff.
	Text	30	External temperature

OK   Cancel

FIGURE 6.69. The *PDE Modeler*, *PDE Specification* for a *Heat Transfer* model, *Elliptic* settings.

Note that the gradient of the function f—$\nabla f(x, y, z)$—is a vector, defined by Equation (153). The divergence of the function f—$\nabla \cdot f(x, y, z)$—is the internal product of the gradient operator by a vector and is therefore a scalar, as shown in Equation (154). When performing an internal (dot) product, the x-, y-, and z-components are multiplied and then added, resulting in a scalar value. The divergence of gradient of a function f—$\nabla.\nabla f(x, y, z)$—is also known as *Laplacian* of that function f—$\nabla f(x, y, z)$—which is also a scalar value and is presented by Equation (155).

$$grad(f) = \nabla f(x,y,z) = \left(\frac{\partial f}{\partial x}, \frac{\partial f}{\partial y}, \frac{\partial f}{\partial z}\right) = \frac{\partial f}{\partial x}\vec{i} + \frac{\partial f}{\partial y}\vec{j} + \frac{\partial f}{\partial z}\vec{k} \qquad (153)$$

$$div(f) = \nabla.f(x,y,z)$$
$$= \left(\frac{\partial}{\partial x}\vec{i} + \frac{\partial}{\partial x}\vec{j} + \frac{\partial}{\partial x}\vec{k}\right).(f_1(x,y,z)\vec{i} + f_2(x,y,z)\vec{j} + f_3(x,y,z)\vec{k}) \qquad (154)$$
$$= \frac{\partial f_1(x,y,z)}{\partial x} + \frac{\partial f_2(x,y,z)}{\partial y} + \frac{\partial f_3(x,y,z)}{\partial z}$$

$$\Delta f(x,y,z) = div(grad(f)) = \nabla.\nabla f(x,y,z)$$
$$= \left(\frac{\partial}{\partial x}\vec{i} + \frac{\partial}{\partial x}\vec{j} + \frac{\partial}{\partial x}\vec{k}\right).\frac{\partial f}{\partial x}\vec{i} + \frac{\partial f}{\partial y}\vec{j} + \frac{\partial f}{\partial z}\vec{k} \qquad (155)$$
$$= \frac{\partial^2 f(x,y,z)}{\partial x^2} + \frac{\partial^2 f(x,y,z)}{\partial y^2} + \frac{\partial^2 f(x,y,z)}{\partial z^2}$$

The physics (*PDE Specification*) should be set for individual subdomains. It is possible to: (a) Select the *PDE Specification* within the *PDE Modeler* application menu; (b) Access the entire domain, consisting of subdomains; (c) Set the model properties (i.e., *Type of PDE* and *Coefficient*), which then will be propagated for the entire domain; (d) Select individual subdomains within the *PDE Mode* available under the *PDE* menu; and (e) Make the required revisions to the PDE data. These data then can be exported from the *PDE Modeler* application to be later used or revised in the MATLAB script. The PDE data variables that can be exported are c, a, f, and d (Figure 6.80), in that order. Inputting any of the aforementioned variables, either in this section or the prior ones, results in viewing the data in the *Command Window* (Figure 6.70 and Figure 6.71).

As mentioned earlier, the application model, which is an *.m file can be viewed and modified within the MATLAB *EDITOR*. To view this script, the *.m file is opened in the MATLAB *EDITOR*. Figure 6.71 shows a section of this script that is related to the PDE specifications. Like the previous case studies, the presentation method of the created script by the *PDE Modeler* application is different from what is presented herein and consists of several coded lines for each function; however, it can be organized to be presented by the script seen in Figure 6.71. This shows that there are two sets of functions (*pdeseteq*, which sets the equation data and

$a = \text{'10!0!10!0!0!0!0!0!0!10!0!0!0'}$

$c = \text{'210!0.6305!210!210!210!0.6305!210!0.6305!0.6305!210!0.6305!0.6305!0.6305'}$

$d = \text{'(2689.9).*(900)!(988).*(4181)!(2689.9).*(900)!(2689.9).*(900)!}$
$(2689.9).*(900)!(988).*(4181)!(2689.9).*(900)!(988).*(4181)!(988).*(4181)!}$
$(2689.9).*(900)!(988).*(4181)!(988).*(4181)!(988).*(4181)'$

$f = \text{'(0)+(10).*(30)!(0)+(0).*(30)!(2000)+(10).*(30)!(4000)+(0).*(30)!(0)+...}$
$(0).*(30)!(0)+(0).*(30)!(0)+(0).*(30)!(0)+(0).*(30)!(0)+(0).*(30)!(10000)+...}$
$(10).*(30)!(0)+(0).*(30)!(-6000)+(0).*(30)!(-6000)+(0).*(30)'$

$pdeseteq(2,c,a,f,d,\text{'0:5000','25','0.0','[0 100]'})$

FIGURE 6.70. The PDE model equation dataset script for the
2D geometry in the MATLAB *PDE Modeler* application.

$aa = \text{'2689.9!988!2689.9!2689.9!2689.9!988!2689.9!988!988!2689.9!988!988!988 '}$

$bb = \text{'900!4181!900!900!900!4181!900!4181!4181!900!4181!4181!4181   '}$

$cc = \text{'210!0.6305!210!210!210!0.6305!210!0.6305!0.6305!210!0.6305!0.6305!0.6305'}$

$dd = \text{'0!0!2000!4000!0!0!0!0!0!10000!0!-6000!-6000   '}$

$ee = \text{'10!0!10!0!0!0!0!0!0!10!0!0!0   '}$

$ff = \text{'30!30!30!30!30!30!30!30!30!30!30!30!30   '}$

$setappdata(pde\_fig,\text{'currparam'},[aa;bb;cc;dd;ee;ff])$

FIGURE 6.71. The PDE model application dataset script for the
2D geometry in the MATLAB *PDE Modeler* application.

setappdata that sets the application data, as shown in Figure 6.70). Solution time (5,000 s) and initial temperature (25 °C) are inputs to this function as well—*pdeseteq(type, c, a, f, d, tlist, u(t0), ut(t0), range)*, with identifying the PDE type—*types* 1 (*Elliptical*), 2 (*Parabolic*), 3 (*Hyperbolic*), and 4 (*Eigenmodes*). *tlist* is the time range, range (e.g., 0 : 5000), *u(t0)* is the initial temperature, *ut(t0)* is the time-derivative of the initial temperature, and *range* is a search string range for the eigenvalue algorithm (on the real axis).

Exported variables (*c, a, f,* and *d*), are the inputs to the former relation (application-defined dataset for the object with handle *h*); while the custom-made variables (*aa, bb, cc, dd, ee* and *ff*) are the inputs to the latter one, *setappdata(h, name, value)* in Figure 6.71. Note that the length of these vectors should be the same. This can be achieved by ensuring the spaces between the single quotation marks (') that are the same for these variables. Blank spaces can be appended to make the lengths the same, as shown in Figure 6.71.

Table 6.3 summarizes the formatting of model parameters withing the *PDE Modeler* script.

TABLE 6.3. Formatting of parameters in the *PDE Modeler* application.

Variable Name	Description	Expression
a and *ee*	Convection heat transfer coefficients	$h_1!h_2!...h_n$
c and *cc*	Thermal conductivities	$k_1!k_2!...k_n$
d	Densities by heat capacities	$(\rho_1.*C_1)!(\rho_2.*C_1)!...(\rho_n.*C_n)$
f	Sum of heat generation and product of convection heat transfer coefficients by external temperatures	$((Q_1)+(h_1).*(Text_1)!(Q_2))((h_2).*(Text_2))!...((Q_n)+(h_n).*(Text_n))$
aa	Densities	$\rho_i!\rho_2!...\rho_n$
bb	Heat capacities	$C_1!C_2!...C_n$
dd	Heat generations	$Q_1!Q_2!...Q_n$
ff	External temperatures	$Text_1!Text_2!...Text_n$

In the boundary mode, the display of *Edge Labels* and *Subdomain Labels* can be activated (Figure 6.72). After the boundary conditions are set, the results consist of the *Neumann* and *Dirichlet* boundary conditions, presented by blue and red colors, respectively. When setting the *Neumann* boundary condition $(n*k*grad(T) + q*T = g)$, the heat flux (g) and heat transfer coefficient (q) are set (Figure 6.73). Note that heat flux is perpendicular to the boundary. When setting the *Dirichlet* boundary condition $(h*T = r)$, the weight (h) and temperature (r) are set (Figure 6.74). In other words, the former condition is related to identifying heat convection and heat flux at the boundaries while the latter one is about setting up constant temperatures at the boundaries. Variables g and b for the decomposed geometry and boundary conditions are exported from the *Boundary* menu in the *PDE Modeler*.

The temperature boundary condition (*Dirichlet, dir*) is applied to the *Edges 2* (45 °C), *6* (35 °C), *18*, and *21* (15 °C), and *22* to *24* and *26* (15 °C). Temperature is assumed constant on the selected edges. Heat flux boundary condition (*Neumann, neu*) is applied to the *Edge 1* at a constant value of

$500 \ W/m^2$. The rest of the edges transfer heat by convection (*Neumann*) to the ambient at 35 °C with the convection coefficient of 10 W/m^2K (Figure 6.72).

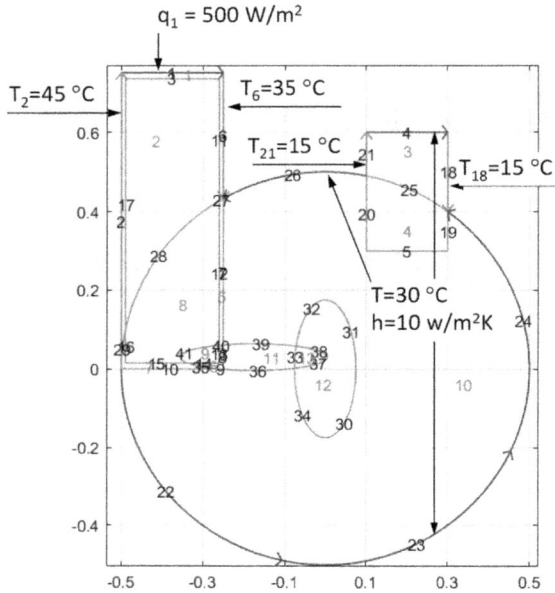

FIGURE 6.72. Boundary conditions applied to the 2D geometry created in the MATLAB *PDE Modeler*.

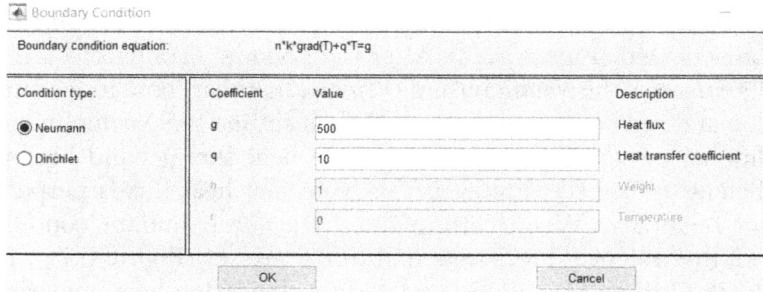

FIGURE 6.73. The *PDE Modeler*, *Boundary Condition* for a *Heat Transfer* model, *Neumann* settings.

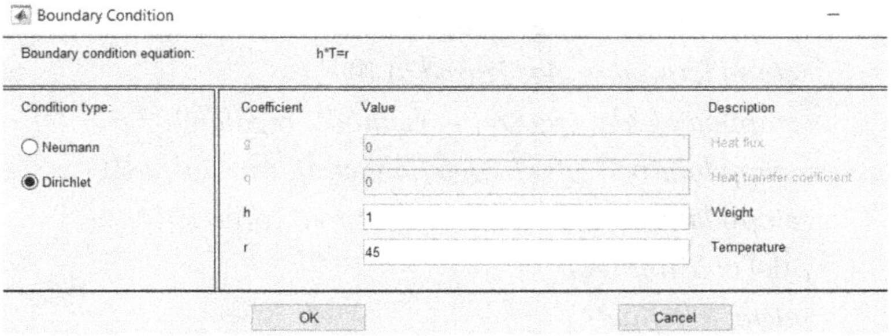

FIGURE 6.74. The *PDE Modeler*, *Boundary Condition* for a *Heat Transfer* model, *Dirichlet* settings.

Opening the script (*.m* file) in the MATLAB *EDITOR*, the boundary conditions can be viewed and modified if required. Figure 6.75 shows the related part of this script. Like the previous case, when describing the geometry script, the presentation method for this section by the *PDE Modeler* application is different from what is presented herein and consists of multiple coded lines for each function; however, it can be organized as displayed in Figure 6.75. This shows that there are two types of the boundary condition types, four *Dirichlet* (*dir*) and six *Neumann* (*neu*) conditions, applied to ten boundaries (edges).

$$p \det ool('change \bmod e', 0)$$

$$pdesetbd(1, 'neu', 1, '10', '500')$$

$$pdesetbd([4, 22:24, 26], 'neu', 1, '10', '0')$$

$$pdesetbd(18, 'dir', 1, '1', '15')$$

$$pdesetbd(21, 'dir', 1, '1', '15')$$

$$pdesetbd(2, 'dir', 1, '1', '45')$$

$$pdesetbd(6, 'dir', 1, '1', '35')$$

FIGURE 6.75. Boundary conditions script to plot the 2D geometry in the MATLAB *PDE Modeler* application.

Mesh properties (triangular), point (*p*), edges (*e*), and triangles (*t*) can be exported from the *Mesh* menu in the *PDE Modeler* within the application editor. The related script is presented in Figure 6.76. The *Maximum Edge Size* (*trisize*) and *Mesh Growth Rate* (*Hgrad*) can be defined for the *PDE Modeler*. If the *Minimum Edge Size* (*Hmin*) is of interest, it can be defined within the script by replacing *Hmax*. The generated mesh is shown in Figure 6.77. The mesh can also be refined (regular and longest methods) and the internal points of the mesh can be jiggled (*optimize minimum and mean* methods) with the *Number of jiggle iterations* identified (<=14).

$setappdata(pde\_fig,'trisize',0.5);$

$setappdata(pde\_fig,'Hgrad',1.5);$

$setappdata(pde\_fig,'refinemethod','regular');$

$setappdata(pde\_fig,'jiggle',char('on','mean','14'));$

$setappdata(pde\_fig,'MesherVersion','preR2013a');$

$pdetool('initmesh')$

$pdetool('refine')$

$pdetool('jiggle')$

FIGURE 6.76. Mesh generation script to plot the 2D geometry in the MATLAB *PDE Modeler* application.

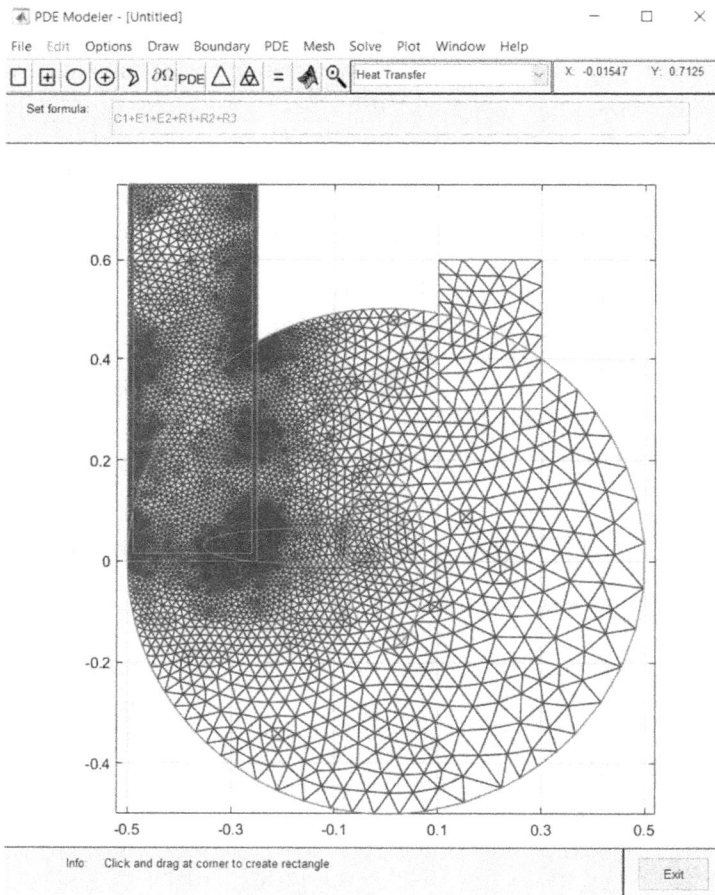

FIGURE 6.77. 2D triangular mesh generated in the MATLAB *PDE Modeler* application.

After all the parameters have been specified using the *pdeseteq* and *setappdata* functions, the solution is executed by the *pdetool('solve')* function. After the solution is obtained, the plot parameters can be set. Properties such as *temperature, temperature gradient*, and *heat flux* can be selected as the output parameters. There is also a *user entry* option available for the property in which the user identifies their own output variable. The arrows (flow variable) can be *temperature gradient, heat flux,* or *user entry* presented in the proportional or normalized forms.

A third dimension (height) can be identified and therefore a 3D plot may be presented. The height can be either *temperature, temperature gradient, heat flux,* or *user entry* property. The output may be animated by identifying *Animation rate* in *fps* and the *Number of repeats. Contour plot levels* can be set, and the mesh may be shown along with the contour plots. Furthermore, color map (e.g., *hot, cool,* and *prism*) may be selected. *Time for plot* can be selected from the drop-down menu, selecting any value within the defined range in 1-s time intervals. After the solution is run, the solution statistics are presented in the *Command Window,* as shown in Figure 6.78. The solution results are presented in Figure 6.79, in which contour plots identify the temperature profiles and arrows represent the heat flux vectors. The *Plot style* is *proportional,* meaning that the magnitude of the heat flux vector determines the size of the arrows.

```
Command Window
>> Pump_Heating_MATLAB_App
83 successful steps
0 failed attempts
168 function evaluations
1 partial derivatives
23 LU decompositions
167 solutions of linear systems
```

FIGURE 6.78. The MATLAB *PDE Modeler* application transient thermal model solution statistics.

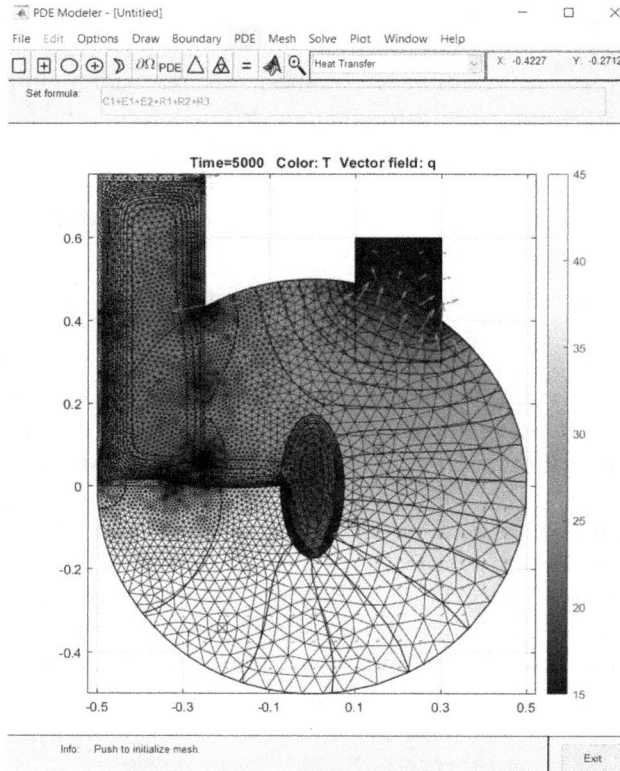

FIGURE 6.79. 2D contour plots generated in the MATLAB *PDE Modeler* application.

6.5.2   PDE Tool Script

With this approach, one can take advantage of the interactive creation of the model geometry using the *PDE Modeler* application and then export the geometry data created to use it within a custom model script. However, one limitation is that 3D geometry creation and export is not supported by the *PDE Modeler*. Within the *PDE Modeler*, the geometry export is achieved by selecting *Export Geometry Description (gd)*, *Set Formula (sf)*, and *Labels* from the *Draw* menu. During the export, the default geometry variable names presented are *gd*, *sf*, and *ns* (Figure 6.80). At this point, these can be changed to the user's preferred names.

The *PDE Tool* script in Figure 6.81 shows how the geometry data saved from the *PDE Modeler* export can be used to create the geometry structure within the *PDE Tool* model object *thermalModelTA*. Before executing this script, the *PDE Modeler* needs to be run and the geometric data exported.

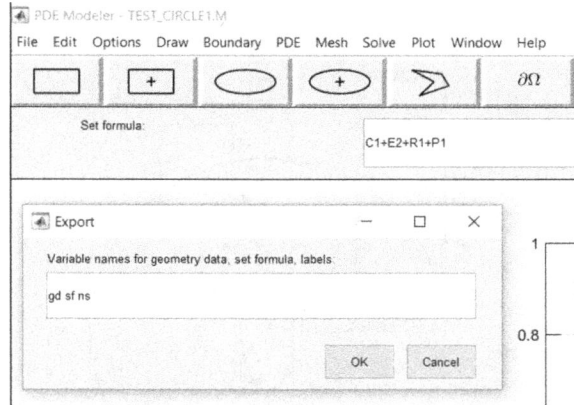

FIGURE 6.80. Variable names for exporting the 2D geometry from the MATLAB *PDE Modeler*.

The example assumes the default geometry data variable names are used (as shown in Figure 6.81). The constructive solid geometry then can be decomposed using the geometry data (*gd, sf, ns*) and the geometry from the edges can be created with the results presented as a figure (Figure 6.81). The *pdegplot* function is employed to plot the geometry to verify its correctness and to identify the edges and faces to which the boundary conditions (and heat sources) are to be applied. Figure 6.82 displays the geometry with edge labels and Figure 6.83, with face labels.

close all; clf ; clc;

thermalModelTA = createpde('thermal','transient');

geometryA = decsg(gd, sf, ns);

geometryFromEdges(thermalModelTA, geometryA);

figure(1);

p deg plot(thermalModelTA,'EdgeLabels','on','FaceLabels','on');

title({'Geometry with edge and face labels displayed';"},'FontWeight','normal');

grid on; grid min or; xlabel('x (m)'); ylabel('y (m)');

x lim([-0.7 0.7]); y lim([-0.6 0.8]);

ax = gca; ax.FontSize = 10;

FIGURE 6.81. Script to plot the exported 2D geometry from the *PDE Modeler*.

Geometry with edge labels

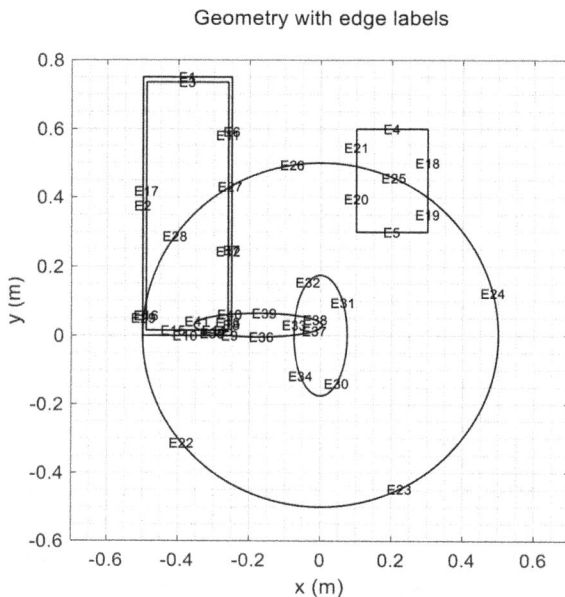

FIGURE 6.82. The 2D geometry plotted using the *PDE Toolbox* script with the edge labels on.

Geometry with face labels

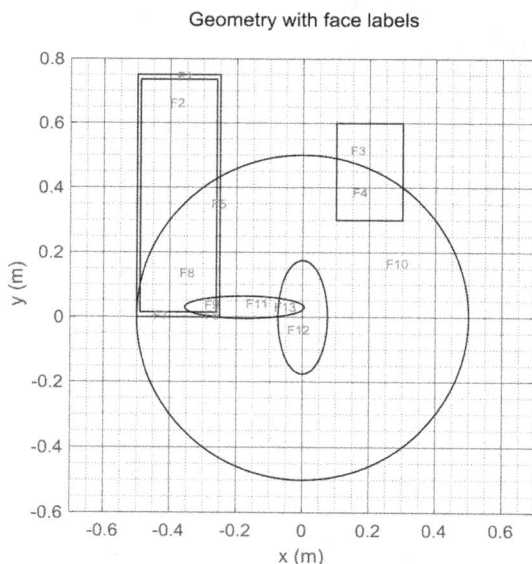

FIGURE 6.83. The 2D geometry plotted using the *PDE Toolbox* script with the face labels on.

The axes limits (x- and y-coordinates) are also to be set in this step using the $xlim(x_{min}, x_{max})$ and $ylim(y_{min}, y_{max})$ functions. Equal spacing between the axes lower and upper limits may be set using the axis equal command.

Note that the edge and face labels are the same for the geometry created in the *PDE Modeler* and *Toolbox*. It is possible to set the physics (*Application Options*) or PDE type independently from the physics selected when setting the model using the *PDE Modeler* (e.g., *Structural Mechanics* in the former case and *Elliptic* in the latter case). In this case study, however, two identical models are created for comparison purposes.

The next step is to specify the thermophysical properties (i.e., thermal conductivity for both steady-state and transient problems and thermal conductivity, density, and heat capacity for transient problems), as shown in Figure 6.84. All faces should be assigned the appropriate properties.

%Water

$k15 = 0.6305$; *% thermal conductivity (W/mK)*

$rho15 = 988$; *% density (kg/m3)*

$cp15 = 4181$; *% heat capacity (J/kgK)*

$tp15 = thermal\Pr operties(thermalModelTA,'ThermalConductivity',k15,...$
    $'MassDensity',rho15,'SpecificHeat',cp15,'Face',[2,8,9,11,12,13]);$

%Aluminum

$k25 = 210$; *% thermal conductivity (W / mK)*

$rho25 = 2698.9$; *% density (kg/m3)*

$cp25 = 900$; *% heat capacity (J/kgK)*

$tp25 = thermalProperties(thermalModelTA,'ThermalConductivity',k25,...$
    $'MassDensity',rho25,'SpecificHeat',cp25,'Face',[1,3:7,10]);$

FIGURE 6.84. Thermophysical properties for the 2D geometry plotted in Figure 6.82.

The next step is to assign boundary conditions (e.g., convection, temperature, and heat flux), as shown in Figure 6.85. If the boundary conditions are not assigned to an Edge, the edge is assumed to be insulated.

$BC51 = thermalBC(thermalModelTA, 'Edge', 6, 'Temperature', 35);$

$BC52 = thermalBC(thermalModelTA, 'Edge', 2, 'Temperature', 45);$

$BC53 = thermalBC(thermalModelTA, 'Edge', [21,18], 'Temperature', 15);$

$BC54 = thermalBC(thermalModelTA, 'Edge', [1,4,22:24,26], ...$

$\qquad\qquad 'ConvectionCoefficient', 10, ...$

$\qquad\qquad 'AmbientTemperature', 30);$

$BC55 = thermalBC(thermalModelTA, 'Edge', 1, 'HeatFlux', 500);$

FIGURE 6.85. Boundary conditions assigned to the 2D geometry plotted in Figure 6.82.

The heat source (internal heat generation) can be assigned to any region (*Face*). The regions can be either listed as an array, if they all have the same setting, or assigned values individually (Figure 6.86).

$ih51 = internalHeatSource(thermalModelTA, 2000, 'Face', 3);$

$ih52 = internalHeatSource(thermalModelTA, 4000, 'Face', 4);$

$ih53 = internalHeatSource(thermalModelTA, 10000, 'Face', 10);$

$ih54 = internalHeatSource(thermalModelTA, -6000, 'Face', [12,13]);$

FIGURE 6.86. Heat sources (internal heat generation) assigned to the 2D geometry plotted in Figure 6.82.

Assigning the initial conditions is the next phase, as shown in Equation (156). This can be set individually for each region (*Face*). Note that it is possible to set the initial temperature for each region; in this case, the *RegionID* and *RegionType* should be identified, as shown in Equation (157). The *RegionType* can be *Vertex*, *Edge*, and *Face* for the 2D and 3D models and *Cell* for the 3D models

$$IC56 = thermalIC(thermalModelTA, 25); \qquad (156)$$

$$IC\_Regions = thermalIC(thermalModelTA, T0, 'RegionType', RegionID); (157)$$

The next phase is the meshing of the geometry using the specified parameters, as shown in Figure 6.87. In this case study, the grid size (*grid_size5*) and growth rate (*Hgrad*) are defined. It is also possible to specify the relative ($1E-5$) and absolute ($1E-9$) tolerances. The solution total time (*tfinal5*) is set in this step, as well as the frequency (*tlist5*) of the data output. Note that *quadratic* elements are selected for this analysis, which are more accurate. To save memory space, *linear* elements may be employed. Mesh properties may be confirmed by entering *mesh5* on the command line, as presented in Figure 6.88. The plot of node locations is presented in Figure 6.89.

$grid\_size5 = 0.05;$

$mesh5 = generateMesh(thermalModelTA,'H\max',grid\_size5,'Hgrad',1.2,...$

$'GeometricOrder','quadratic');$

$tfinal5 = 5000;$

$tlist5 = 0:100:tfinal5;$

FIGURE 6.87. Mesh parameters and thermal model solver options assigned to the 2D geometry plotted in Figure 6.82.

```
Command Window
   mesh5 =

     FEMesh with properties:

                 Nodes: [2×2374 double]
              Elements: [6×1143 double]
        MaxElementSize: 0.0500
        MinElementSize: 0.0250
         MeshGradation: 1.2000
        GeometricOrder: 'quadratic'
```

FIGURE 6.88. Mesh properties.

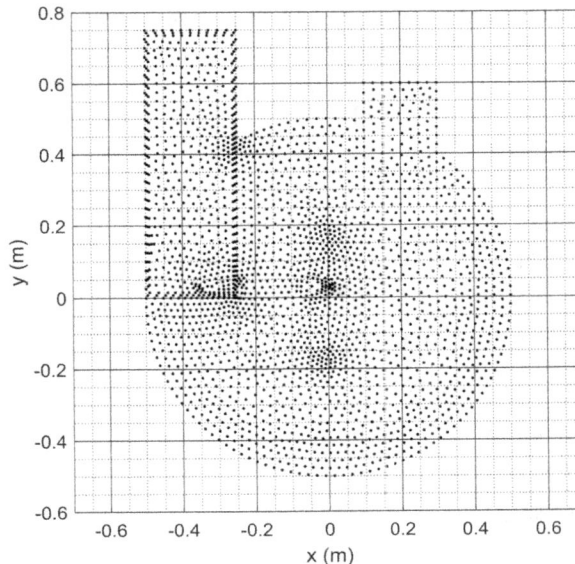

FIGURE 6.89. Node locations.

The model is to be solved next, given the solution parameters. Any changes made to the boundary or initial conditions as well as the solution parameters result in new sets of solutions. Note that if the solution is to be rerun, the solution of the problem remaining in the MATLAB *Workspace* will be used as the initial condition unless the initial condition and the iteration solution first *guess* is to be set first. This is an *educated* guess when using coding languages such as Fortran or C++ and when numerical methods such as *Gauss-Sidle* elimination methods are employed. Note that this is not to be taken as a built-in step when setting PDE models in the MATLAB environment. Therefore, clearing the MATLAB *Workspace* beforehand ensures unknown initial iteration values and residual errors are eliminated.

It is always possible to create "pretty" images; however, how accurate or precise they are of interest to a modeling scholar. Note, however, that as mentioned earlier, in case the inputs of another program (e.g., geometry data imported from the MATLAB *PDE Modeler* application are needed as the inputs to the PDE model, the MATLAB *PDE Modeler* application should be run first for the data to be created. The data then should be exported from the MATLAB *PDE Modeler* application to the MATLAB *Workspace*, and then inputted to the MATLAB script.

The temperature ($T51$), temperature gradient ($Tx51$), heat flux ($qx51$) magnitudes for the model nodes ($xy51$), or any other combination of these variables can be valuated as the next step, as shown in Figure 6.90. Note that T_st51 is the solution times and is a vector; it is the same as the tlist5 (Figure 6.87). The results' properties can be displayed by entering on the command line the name of the results object (*results5*), which contains the solution. The results object is created on the first line in Figure 6.90. The transient temperature results object's properties are presented in Figure 6.91. The information includes sizes of the arrays within the results object (e.g., the temperature array and solution times array).

$results5 = solve(thermalModelTA, tlist5);$

$point = thermalModelTA.Mesh.Nodes;$

$nodesIndex = find(point);$

$xy51 = point; x51 = xy51(1,:); y51 = xy51(2,:);$

$T51 = results5.Temperature;$

$Tx51 = results5.XGradients;$

$Ty51 = results5.YGradients;$

$T\_st51 = results5.SolutionTimes;$

$[qx51, qy51] = evaluateHeatFlux(results5, x51, y51, 1 : length(tlist5));$

FIGURE 6.90. Initial conditions assigned to the 2D geometry plotted in Figure 6.82.

```
Command Window
    results5 =

    TransientThermalResults with properties:

            Temperature: [2374×51 double]
          SolutionTimes: [1×51 double]
             XGradients: [2374×51 double]
             YGradients: [2374×51 double]
             ZGradients: []
                   Mesh: [1×1 FEMesh]
```

FIGURE 6.91. Solution properties for the 2D geometry plotted in Figure 6.82.

The solution's statistics can be displayed by activating the *ReportStatistics* feature within the *SolverOptions* (Figure 6.92 and Figure 6.93).

thermalModelTA.SolverOptions.ReportStatistics = 'on';

modelD.SolverOptions.RelativeTolerance = 1E-5;

modelD.SolverOptions.AbsoluteTolerance = 1E-9;

FIGURE 6.92. Solution statistical data.

```
Command Window
    thermalModelT =

    ThermalModel with properties:

                    AnalysisType: 'transient'
                        Geometry: [1×1 AnalyticGeometry]
              MaterialProperties: [1×1 MaterialAssignmentRecords]
                     HeatSources: [1×1 HeatSourceAssignmentRecords]
          StefanBoltzmannConstant: []
              BoundaryConditions: [1×1 ThermalBCRecords]
               InitialConditions: [1×1 ThermalICRecords]
                            Mesh: [1×1 FEMesh]
                   SolverOptions: [1×1 pde.PDESolverOptions]

    109 successful steps
    0 failed attempts
    220 function evaluations
    1 partial derivatives
    25 LU decompositions
    219 solutions of linear systems
```

FIGURE 6.93. The MATLAB script transient thermal model solution properties and statistics.

Note that it is also possible to create the result animation by exporting *j* number of image frames and playing them *n* times. To achieve this, the *getframe* function is employed (Figure 6.94). This function may be used either right after any plot functions or specifically as a function, calling frames for a specific figure *getframe(figure(n))*, where *n* is the figure ID (number).

for j = 1:5:*length(tlist5)*

    pdeplot(thermalModelTA,'XYData',T51(:, j),'colormap','hot',...

    'Contour','on','Levels',24);

    axis equal; grid on; grid minor; xlabel('x (m)'); ylabel('y (m)');

    *x*lim([-0.7 0.7]); *y*lim([-0.6 0.8]); *ax* = *gca; ax.FontSize* = 20;

    hold all;

    quiver(x51(:),y51(:),qx51(:, j),qy51(:, j),'-','LineWidth',0.5,...

    'Color','black',AutoScale','on','ShowArrowHead','on');%,'AutoScaleFactor',3

    axis equal; grid on; grid minor; xlabel('x (m)'); ylabel('y (m)');

    *x*lim([-0.7 0.7]); *y*lim([-0.6 0.8]); *ax* = *gca; ax.FontSize* = 20;

    title({['Temperature contours'];['and heat flux arrows (t = ',...

    num2str(round(tlist5(j),2)),' s)'];''},'FontWeight','normal')

    M5(j) = *getframe;*

end

FIGURE 6.94. Animated data setting for the 2D geometry plotted in Figure 6.82.

The collected data frames then can be called back and presented in the form of a movie using the *movie* function (Figure 6.95). The script provided herein plays the time-array presented by the vector 1:5:*length(tlist5)* ten times, which displays the thermal data contours (image frames) for every fifth frame; 14 frames per second, played once, are played seven times in this animated data. The *for* loop is used for retrieving the data frames and animating them. The *movie(gca, M5, [n, k], fps)* function, where *gea* is the graphic object handle of the current axis, *M5* is the movie frames array, [*n k*] in this scenario represents the number of times each frame plays (*n*) and selected time-frames (*k*), and *fps* is the number of frames per second that the animation plays. One of the challenges of these diagrams is how to include titles so that they represent the selection parameters used as the time step progresses. The method of the presentation of title for Figure 6.94 and Figure 6.95, with loop variables (*j* and *k*) is included in the script.

$for\ k\ =\ 1:5: length(tlist5)$

  $title(\{['Temperature\ contours'];['and\ heat\ flux\ arrows\ (t\ =\ ',...$
  $num2str(round(tlist5(k),2)),'\ s)'];''\},'FontWeight','normal')$
  $axis\ equal;\ grid\ on;\ grid\ \min or;\ xlabel('x\ (m)');\ ylabel('y\ (m)');$
  $x\lim([-0.7\ 0.7]);\ y\lim([-0.6\ 0.8]);\ ax\ =\ gca;\ ax.FontSize\ =\ 20;$
  $movie(gca,M5,[7\ k],14);$

end

FIGURE 6.95. Animated data setting for the 2D geometry plotted in Figure 6.82.

In this case study, the solution results are presented in the form of temperature contour plots, including the heat flux arrows (Figure 6.96), as well as the temperature gradient contour plots with respect to the x- and y-coordinates, including the temperature gradient arrows (Figure 6.97).

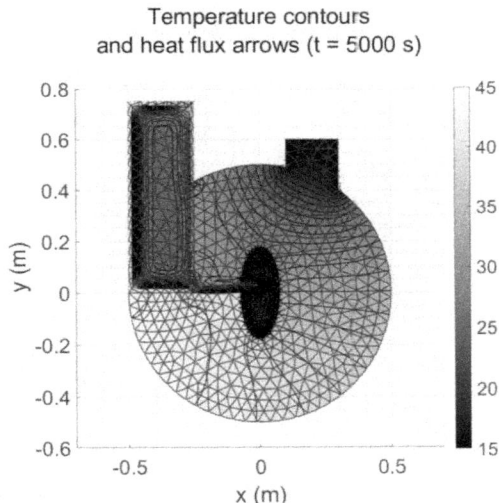

FIGURE 6.96. Temperature contour plots and heat flux vectors for the script presented in Figure 6.94.

The color map can be set using the $cmap\ =\ colormap(target,\ map)$. map identifies the color map (e.g., $gray$, $turbo$, and $spring$) and target is the figure, axes or graphics object (e.g., gca or $figure(n)$), where n is the figure ID (number). In Figure 6.96, the $gray$ color map is selected, while for Figure 6.97, the $turbo$ color map is selected.

Temperature gradient contours with respect to
x (dT/dx) and their arrows (t = 5000 s)

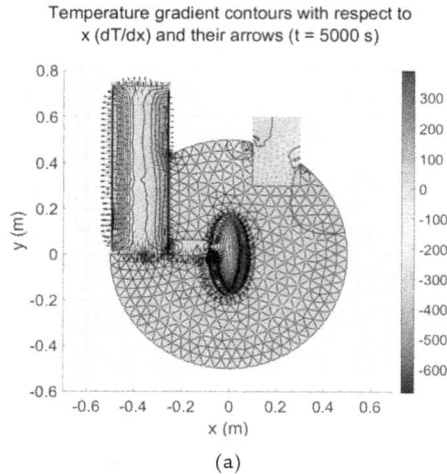

(a)

Temperature gradient contours with respect to
y (dT/dy) and their arrows (t = 5000 s)

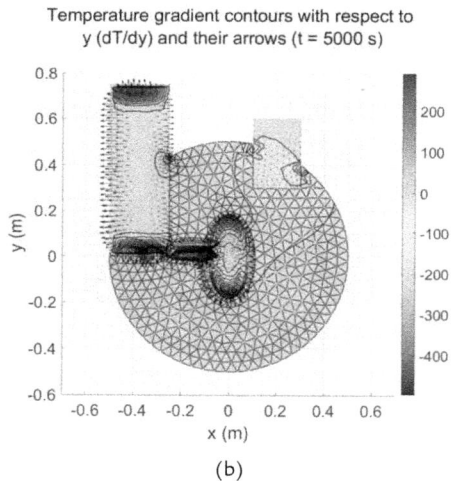

(b)

FIGURE 6.97. Temperature gradient contour plots and vectors with respect to:
(a) x (dT/dx), (b) y (dT/dy).

It is possible to enquire about a node at the selected coordinate. However, data may not be available at the exact location; therefore, the closest location is queried. If the desired coordinate is located at the point $A(x, y)$ the closest data is then defined as $A\_data = @(p, x, y)min(x51 - x)^2 + (y51 - y)^2$. This is the minimum distance between the desired point and available mesh points, where the $x51 = point(1,:)$ and $y51 = point(:,1)$ are x- and y-coordinates of the solution nodes calculated from the $point = thermalModelTA.Mesh.Nodes$.

For example, assume the temperature data at the center of the left edge is of interest—$A(-0.5, 0)$. This coordinate is to be replaced in the function, which can be called to get the specific node near the center of the left edge $[\sim, nid1] = data(mesh5.Nodes, -0.5, 0)$, where nid is the node ID. The node ID ($nid1$) for this point is 15. In another example, the temperature data at a point located at $B(0.2, -0.4)$ is of interest. The node ID ($nid2$) for this point is 1,754. The third example, the node ID ($nid3$) for the point located at $C(-0.1, 0.2)$ is 2,098. The related script is presented in Figure 6.98. The transient temperatures for these selected nodes are plotted in Figure 6.99.

$n = length(tlist5);$

$data = @(p,x,y)min((x51-x).^2 + (y51-y).^2);$

$[\,,nid1] = data(mesh5.Nodes,-0.5,0);$

$[\,,nid2] = data(mesh5.Nodes,0.2,-0.4);$

$[\,,nid3] = data(mesh5.Nodes,-.1,0.2);$

$A = nid1; \quad B = nid2; \quad C = nid3;$

$plot(tlist5(1:n),T51(A,[1:n]),'-<','MarkerSize',5,'MarkerIndices',...$

$\quad 1:5:length(tlist5(1:n)),'MarkerFaceColor',[1\ 1\ 1]);$

$plot(tlist5(1:n),T51(B,[1:n]),'-d','MarkerSize',5,'MarkerIndices',...$

$\quad 1:5:length(tlist5(1:n)),'MarkerFaceColor',[0.5\ 0.5\ 0.5]);$

$plot(tlist5(1:n),T51(C,[1:n]),'->','MarkerSize',5,'MarkerIndices',...$

$\quad 1:5:length(tlist5(1:n)),'MarkerFaceColor',[0\ 0\ 0]);$

FIGURE 6.98. Script to generate the 2D geometry and plot in Figure 6.99.

Geometry with edge labels

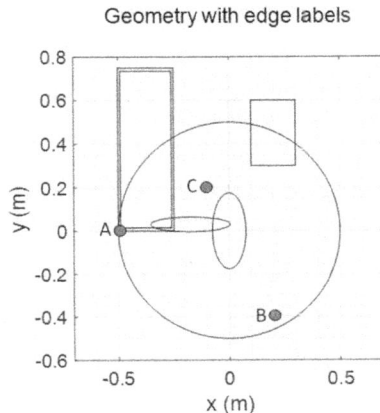

(a)

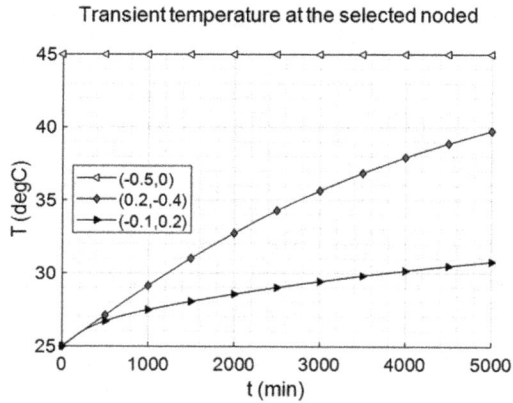

(b)
FIGURE 6.99. Getting data at the query points:
(a) Point locations, (b) Transient temperature at the selected points.

End Notes

[55] http://www.matweb.com/search/datasheet_print.aspx?matguid=9a0c88f81df945
      218033319fa4dd1cb6
[56] COMSOL Material database
[57] http://www.matweb.com/search/datasheet_print.aspx?matguid=9a0c88f81df945
      218033319fa4dd1cb6
[58] COMSOL Material database

CHAPTER 7

THE COMSOL MULTIPHYSICS MODELS

T he COMSOL Multiphysics modeling has been discussed in detail in the author's previous works [2,3,4]. However, for the purpose of this publication, a summary of some of the main concepts is presented herein. The reader is encouraged to learn more about heat transfer and thermal modeling by studying the previous works on the extended surfaces using analytical methods and numerical analysis, with the focus on COMSOL Multiphysics as the FEM analysis tool in combination with CAD tools to generategeometry [2], geometry import and creation using COMSOL Multiphysics [4], and a complete review of heat transfer science, thermal analysis modeling methods, and multidisciplinary case studies involving heat and flow [3].

The process of heat transfer analysis using FEM is carried out in three stages: (a) model setup or pre-processing, (b) solution, and (c) post-processing. FEA tools share these same steps, and their organization methods (model tree) are very similar among the variety of the tools with the most used ones, such as ANSYS, ABAQUS FEA, and COMSOL Multiphysics. The modeling setup steps are generally as follows:

(1) Evaluate the available resources (machine, tool, and human).

(2) Identify the number of spatial dimensions (zero to three) defining the physical phenomena.

(2) Consider the possibility of representing the model using fewer dimensions (e.g., 1D or 2D), fewer geometrical features, or by a simplified version that takes advantage of any symmetry.

(4) Generate the model geometry by: (a) Importing the model from a dedicated CAD tool, (b) Creating it using the FEM built-in tools, (c) Creating from a mesh, or (d) Importing from another FEA tool.

(5) Identify the dominant physics, the domains to which they apply, and whether single or multiple physics apply (e.g., heat transfer and structural mechanics).

(6) Identify the boundary conditions at which the boundaries interact with their environment(s). For heat transfer modeling, these conditions can be expressed by the temperature, heat flux, heat rate, and heat generation, which are derivatives of the dependent variable (temperature). They can be either constant or functions of space and time (e.g., zero gradient and constant value).

(7) Decide whether the physical system will be modeled as time-dependent or stationary by considering the given physics and available resources. This step is one of the most challenging ones, for the time-dependency should be correctly selected to represent the system's behavior given the operational conditions. Time and space increments work closely together and, on many occasions, the combination of them is the determining factor in generating accurate solutions; therefore, care should be taken to ensure the proper selection.

(8) Identify the initial conditions if a transient analysis (time-dependent) is selected. These are either applied to certain regions within the modeling domain or portions of it, meaning that there would be either single or multiple subdomains within the model. If a steady-state analysis is preferred, the initial guess at which the solution starts (especially for the iterative-elimination methods) should be set.

(9) Mesh the model geometry by means of 1D, 2D, or 3D entities, depending on the models' dimension. The order of the mesh (*quadratic* versus *linear*) determines the level of accuracy achieved, which is mainly determined by the available resources (e.g., time and machine).

(10) Select the numerical technique(s) to produce solutions that converge, given the acceptable relative and absolute tolerances. On occasions, the solution may need to be divided into multiple steps, each step adopting the numerical method that better suits the conditions. The inputs (or initial guesses) to the solutions in such cases are usually the outputs of the previous solution steps.

(11) Specify the solution parameters such as the duration of the interval simulated and the time intervals at which the solution data are saved.

(12) Carry out solution post-processing as the last step. This may involve solution data in the wide variety of forms (e.g., diagrams and animations) required in the internal or external reports. This may include steps such as: (a) Extracting the solution using output commands; (b) Evaluating the solution by taking integrals or averages over regions; (c) Customizing the report templates to follow certain formats or add the input commands to make interactive and user-friendly Web-based applications; and (d) Processing and visualizing the solution output by means of tables, diagrams (1D, 2D, and 3D), contour and surface plots, and spatial and temporal probes.

This chapter discusses briefly how to work with models in COMSOL Multiphysics Version 5.6; the software can be updated by checking under *File > Help*. The first section below presents considerations pertinent to setting up a heat transfer model. The next section focuses on the geometry creation process, importing the geometry as an independent part or assembly from a dedicated CAD tool such as Solid Edge or SOLIDWORKS, and creating it using the built-in geometry creation tools. This is followed by a summary of all the steps involved in carrying out an analysis. Finally, a section is dedicated to introduction of the COMSOL Multiphysics *LiveLink for MATLAB Module*; it shows how one can use this module to take advantage of the MATLAB's computational and data processing capabilities while also benefiting from the COMSOL Multiphysics' capabilities as a dedicated FEM tool.

7.1 Heat Transfer Modeling Considerations

To model a heat transfer problem using any tool, including COMSOL Multiphysics, the modes of heat transfer that should be included in the model need to be selected. In most cases, heat transfer phenomena include all three main heat transfer modes of conduction, convection, and radiation. However, it is often the case that one or more of these can be ignored to simplify the model. For example, the radiation mode may have a negligible effect at relatively low temperatures.

Methods of setting up models are different for the solid and fluid domains. *Conjugate Heat Transfer* physics is a thermal-flow modeling system that combines heat transfer in both solids and fluids, while considering their interaction. Gravity effects may play a more important role where the flow in the vicinity of a vertical surface is examined, for instance, in the free convection case, where the ratio of the buoyancy to viscosity forces is dominant. Radiation heat transfer also can be combined with the previous scenario and make the model more complex. Radiation sources can include solar radiation or another intense heat source such as fire. The former one introduces the radiation heat flux for it hits the domain surface (W/m^2), while the latter can either hit the domain surface (W/m^2) or be generated within the domain (W/m^3). Both heat intensities can have spatial and temporal dependency.

Among the important factors to be considered when setting up the FEM models are: (a) methods used to accurately capture the temporal and spatial variations of the thermophysical properties; (b) physics to be used; (c) accurate estimation of the convection heat transfer coefficient; and (d) how different physics, if present, interact with one another. The choice of the transient versus the stationary solutions should also be carefully considered. For some models, a stationary study may result in a solution, while a transient one will not produce a solution due to the convergence issues; the opposite may be valid in some cases.

7.2 Creating a Model in COMSOL Multiphysics

To create a new FEM model, *File > New* is selected; a window opens as the result, which offers the choice to create a *Blank Model* or *Model Wizard* (Figure 7.1). If the analyst selects the *Blank Model*, a new empty model is created. However, in most cases, it is much simpler to set up the model from the start based on the applicable physics. For this purpose, *Model Wizard* should be activated (Figure 7.1a) which will take you through a few basic steps for model setup.

First, a new window opens where you can select the appropriate space dimension (e.g., *1D* or *2D Axisymmetric*) (Figure 7.1b). Now you need to decide which physics to include in the model (e.g., heat transfer, fluid flow, or a combination of multiple physics). Figure 7.2 shows the physics selection when a specialized add-on module is available (e.g., *Heat Transfer*). In this case, there are twelve different physics available. *Heat Transfer in Solids* is highlighted in the image. Without the add-on module, only a few of the most basic physics would normally be available. After the physics is added, its dependent variable (e.g., temperature, T) is set. The dependent variable name can be changed and may include subscripts (e.g., $T1$). After completing this step, the *Study* type is selected (Figure 7.3).

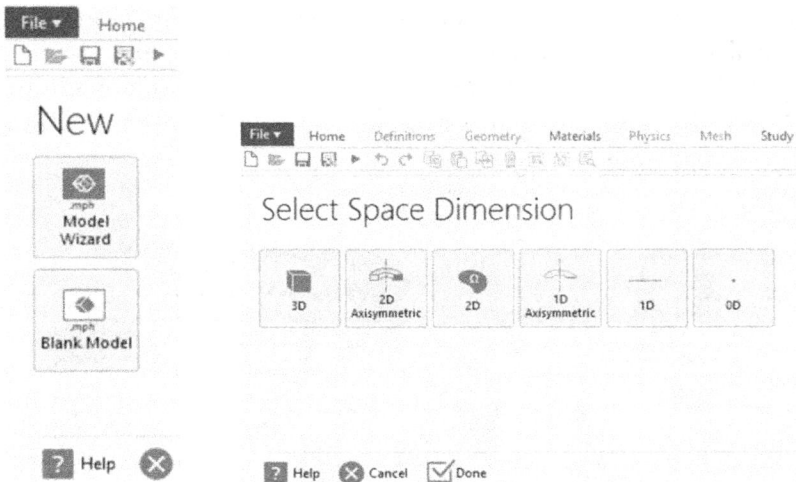

(a)   (b)

FIGURE 7.1. (a) Setting up a new model, (b) Selecting the space dimension.

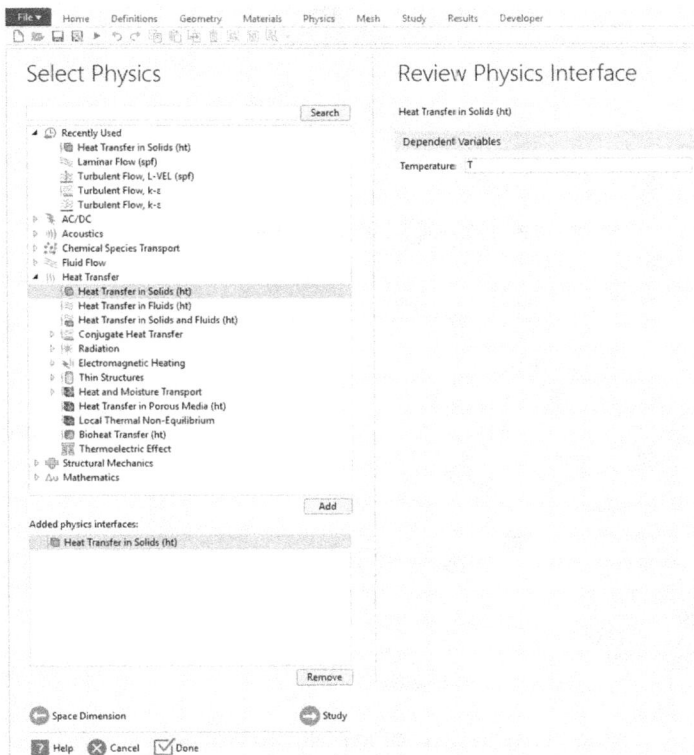

FIGURE 7.2. Selecting physics—*Heat Transfer, Heat Transfer in Solids* (*ht*).

Clicking on the *Done* button after the previous steps are completed takes the analyst to the modeling window, a home for the brand-new model (Figure 7.4). Here, the analyst has multiple regions or windows (four in the provided example). Most of the model operations are done in the *Model*

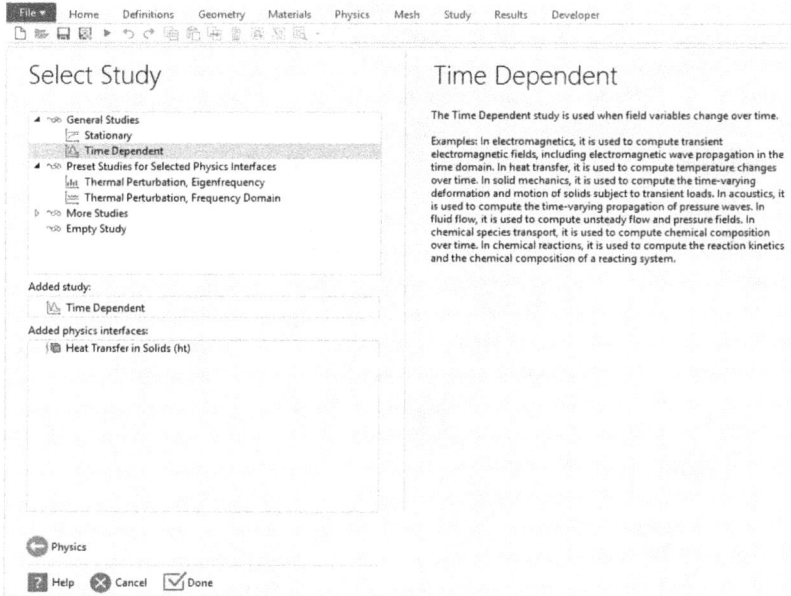

FIGURE 7.3. Selecting study—*General Studies, Time Dependent*.

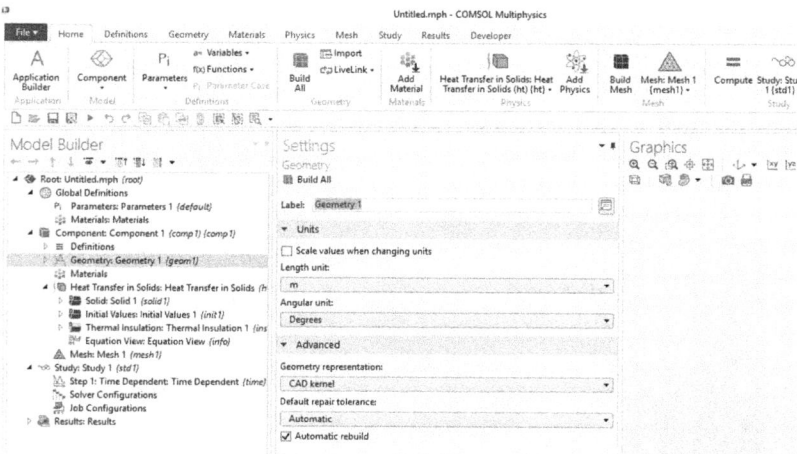

FIGURE 7.4. The COMSOL Multiphysics model tree window.

Builder window. It is a tree structure starting with the *Root* named after the model, with the main branches of *Global Definitions*, *Component(s)*, *Study*, and *Results*. *Model Builder*'s neighbor to the right shows *Settings* pertaining to the selection made in the *Model Builder* window.

7.3 Creating Geometry

This section presents a brief overview of the geometry creation in COMSOL Multiphysics. More detail is available in the author's publication *COMSOL Multiphysics Geometry Creation and Import* [3]. In the COMSOL Multiphysics model file, the geometry is created within the *Geometry* node found under each *Component*. The parameters used for the geometry may be defined either on the local or global level. The two most common ways to create a geometry are: (a) creating and manipulating the elementary geometric entities and (b) importing the geometry from a third-party CAD tool.

To import from the external CAD tools (e.g., SOLIDWORKS or Solid Edge), the analyst needs to have either a *CAD Import Module* or *LiveLink Module*, the latter being associated with a specific CAD tool. Any of the software-specific *LiveLink Modules* include the *CAD Import Module* functionality. Using the *LiveLink Module* allows one to update the model geometry in COMSOL Multiphysics as soon as the changes are made in the CAD software.

In the example shown in Figure 7.5, *Geometry* is highlighted. In this case, the *Settings* window presents the related geometrical characteristics such as the *Length unit* (e.g., m, nm, and GM), *Angular unit* (Figure 7.6), *Geometry representation* kernel (Figure 7.7), and *Default repair tolerance* method (Figure 7.8). A kernel is the fundamental geometrical language used to describe the model geometry. In this example, two geometrical kernels are available—CAD and COMSOL Multiphysics; the former is only available with the optional *CAD Import Module* while the latter is part of the base package.

Default repair tolerance is applied when the geometry is imported or when *Boolean* operations are performed. It defines a threshold below which the geometry entities may be considered coincident and appropriate repairs are made to avoid, for example, in the cases of vertices which are very close to one another. Selection of the *Relative* setting expresses the tolerance as a ratio between the error dimension and the maximum model coordinate. *Absolute tolerance* is expressed in the length units of the model.

Automatic tolerance (the default choice) sets it at a relative value of *1e*-6 and takes adjustment steps if needed. The following sections provide an overview of ways to generate geometry.

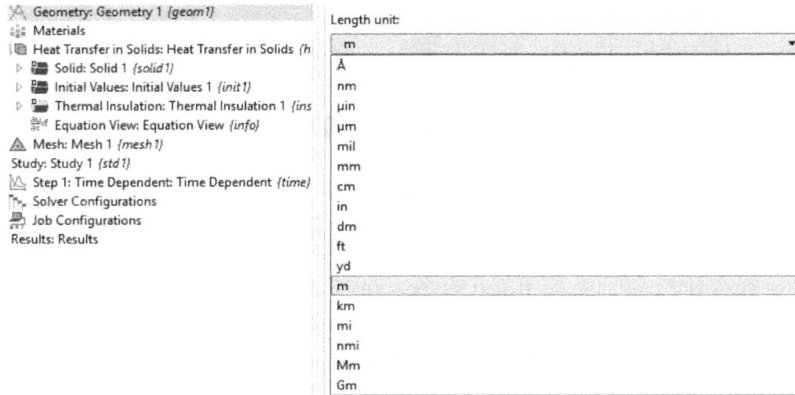

FIGURE 7.5. Geometry—*Length unit* options.

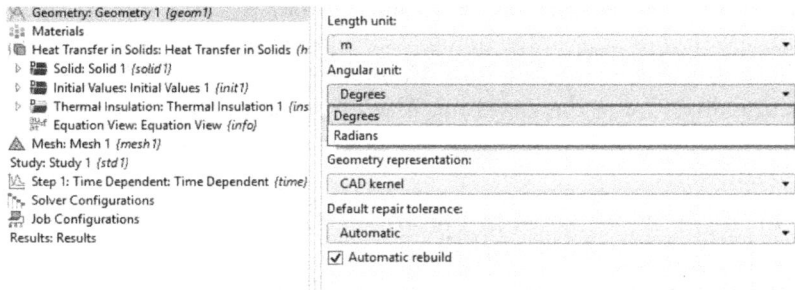

FIGURE 7.6. Geometry—*Angular unit* options.

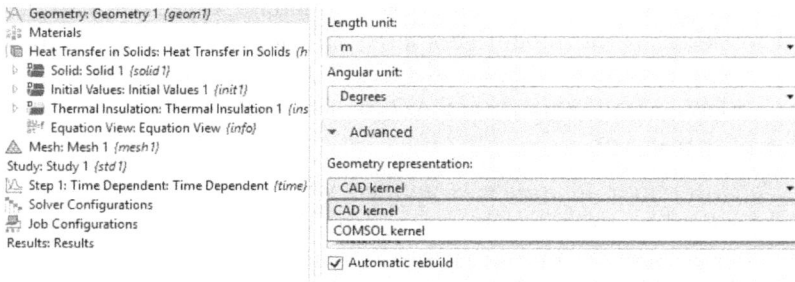

FIGURE 7.7. Geometry—*Geometry representation* kernels options.

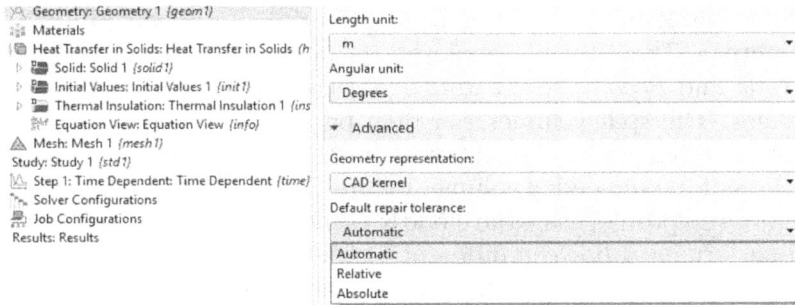

FIGURE 7.8. Geometry—Default repair tolerance options.

7.3.1 Using Elementary Geometric Entities

In this method, the geometry is created by defining the basic geometrical entities, such as: (a) the intervals and points (1D); (b) lines, curves, rectangles, and circles (2D); and (c) blocks, cylinders, and spheres (3D). Various transformations can then be applied to these entities; these include the *Booleans* and *Partitions* operations (e.g., union and difference), *Transforms* operations (e.g., copy, mirror, rotate, and scale), and uncategorized ones such as the extrude, revolve, sweep, fillets, and chamfers. Geometries are created in steps; each step can be disabled (or enabled) when needed without affecting the following unrelated sequential steps. This makes it possible to temporarily exclude geometry-creation steps without affecting the future steps. Furthermore, the steps can be duplicated, and copies revised to facilitate adding new steps. For example, to create the 3D ring shown in Figure 7.9, the following sequence may be used: (a) cylinder of radius 0.03 m and height of 0.015 m (*cyl*1); (b) cylinder of radius 0.024 m and height of 0.015 m (*cyl*2); and (c) *Difference*, *cyl*2 – *cyl*1(*dif*1).

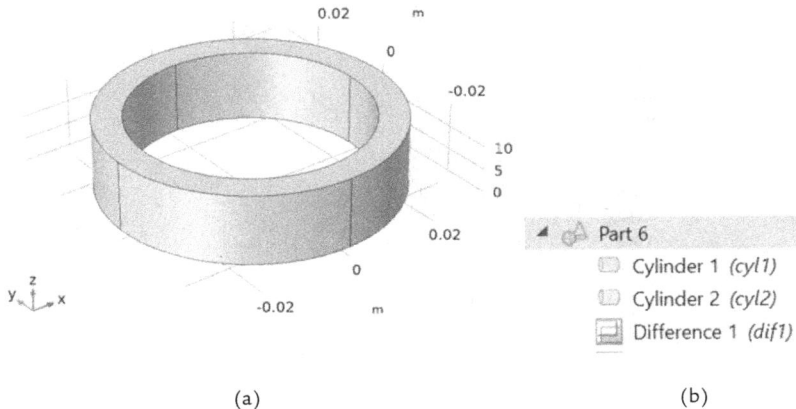

(a)   (b)

FIGURE 7.9. (a) An example of a 3D ring geometry, (b) The geometry sequence shown.

Use of the *Work Planes* can be a powerful tool when processing 3D geometries. They can be employed to: (a) process 2D sketches such as *Extrude* and *Revolve* to create 3D objects and (b) partition or split 3D domains. The split volumes can then be used for subsequent modeling steps, allowing for setting different material and mesh properties, and deletion of the unneeded volume. Figure 7.10 shows the method to create the revolved profile, using the circle (*c1*) and rectangle (*r1*) 2D shapes (left) and performing a *Boolean* difference (*dif1*), where the circle is subtracted from the rectangle (right). The *Revolve* operation is then performed, which uses this profile and defines a revolution axis to create the 3D shape shown.

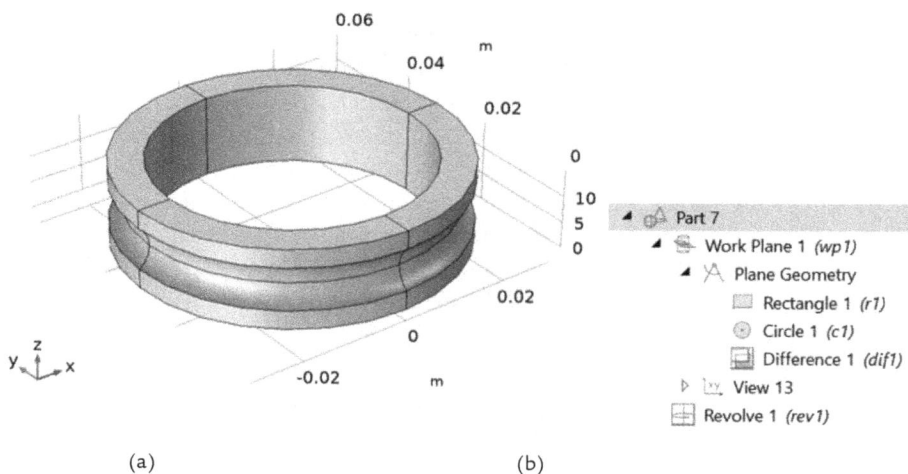

(a)   (b)

FIGURE 7.10.   (a) A 3D ring with a groove created by revolving the 2D shapes;
(b) The geometry sequence shown.

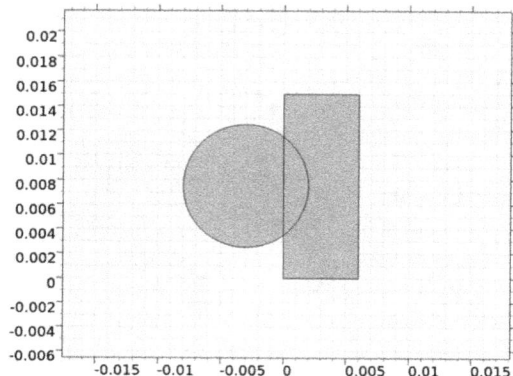

(a)

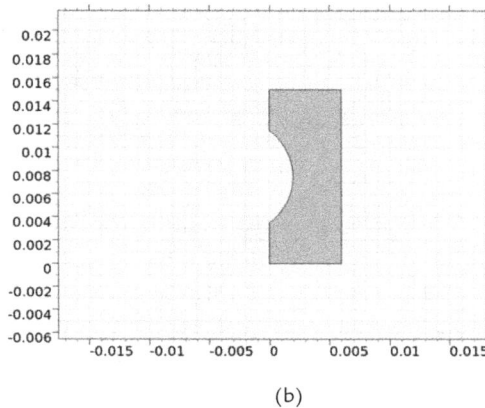

(b)

FIGURE 7.11. Geometry components used in Figure 7.10: (a) *Work Plane* used to define input to the 2D shapes, (b) Finished profile for a 3D ring with a groove.

7.3.2 Importing Geometry

The geometry import is performed by right-clicking on *Geometry* and selecting *Import*. Sources available for bringing the geometric entities into the file are the: (a) geometry sequence, (b) mesh, (c) *.stl; (3D geometries only), (d) COMSOL Multiphysics file, and (e) 3D CAD files (3D geometries only).

The *Geometry sequence* option allows one to bring a set of geometry creation steps from another component of compatible dimensions (1D, 2D, or 3D) within the same model file into the current component. This option allows one to reuse the geometry created in one component within another component.

Mesh import takes a mesh from another component within the same model file and brings it as a geometry (shape) into the current component. The two components must have the same number of spatial dimensions (e.g., there is a 2D component, and so only a mesh from another 2D component can be used).

Stereolithography (*.stl) file import only applies to 3D components. It allows the analyst to introduce a geometry defined using an *.stl file, which is a common way to store information used for 3D printing of objects. The *.stl geometries describe the surfaces of 3D objects by flat triangular surface patches, with the coordinates of all triangle vertices stored in the *.stl file. Thus, curved surfaces will be approximated by these flat patches. For curved surfaces with small radii, small triangles are needed to have an accurate surface representation, possibly leading to very large file sizes for highly complex objects.

The *COMSOL Multiphysics file* option allows the analyst to import into the definition of the current component a geometry extracted from another COMSOL Multiphysics file. Such a source geometry file can be created in the first place by right-clicking on the *Geometry* and selecting *Export*. A file in a *.mphbin* format is then created, which can be selected for import via this option.

The *3D CAD File* option is only available for 3D components and will only be visible if the analyst has one of several optional add-ons, such as the *CAD Import Module* or one of the *LiveLink Modules*. *CAD* files of several different formats can be imported. These include non-proprietary geometry exchange formats, such as *.step*, *.sat*, and *.iges*, and proprietary ones, such as the SOLIDWORKS part (*.sldprt*) and assembly (*.sldasm*).

7.4   Adding Materials

After the model geometry is created or imported, the next step is to add one or more materials to the entire domain (consisting of subdomains). A material needs to be assigned to each domain of the model. Materials can be added from: (a) the list of the built-in basic ones that are included with the core COMSOL Multiphysics package, (b) the optional *Material Library* add-on *Module*, or (c) by introducing a blank material and providing the input data by filling the related fields (e.g., mechanical and thermophysical properties).

Items such as variables, parameters, and materials may be added either at the: (a) *local* (directly under the subcomponent, such as tree leaf, local) level or (b) *global* (under the upper-level component, such as tree trunk, global) level. In the latter case, a link is to be created, connecting the local property or entity (*child*) to the global property or entity (*parent*). Any of the preset parameters (or variables), including the material properties and solution control options, may be revised at any time.

7.5   Adding or Revising Physics

Even after the physics and study selections have been made during the model setup, it is possible to add a new physics under the current *Component* or a new study under the *Root* (top-level tree). It is also possible to add a new *Component* (of any dimension) under the *Root*. The user may add the *Study Steps* as well as a variety of sweeps (e.g., *Parametric*, *Function*, and *Material Sweep*). *Study Extensions* may also be activated under each of the

Study Steps. The latter two features make it possible to perform sensitivity analysis for the selected parameter (or variable).

The sub-physics and conditions (e.g., boundary conditions such as inflow, symmetry, heat flux, and loads) are added under the main physics. The user should ensure that these sets of input data are provided so that the problem solution can be attempted. In case a boundary condition is missing, the program may employ the default conditions for the missing regions (e.g., lines, areas, volumes) if the user has not excluded them from the physics. For example, when defining the material for the first time, its properties are propagated to the entire geometry to be analyzed. However, this is not the case if the user decides to exclude parts of the geometry from the physics or material definition. Leaving the properties of the materials the same as the default values is advantageous in the sense that errors due to the missing information (e.g., options, properties, and their values) are avoided. The disadvantage of this method is when the user inadvertently neglects setting up the options that are not shared among all the features such as radiation properties.

7.6 Solution

If a model consists of multiple components and physics, they are solved in the order in which they were defined. The user can choose which component-physics to solve by placing a checkmark in the corresponding box (under *Study Step > Physics and Variable Selections*). Additionally, the component mesh for the selected physics can be confirmed so that the user can ensure that the correct physics and component are selected. This means the user can exclude an analysis if they are not interested in solving it by not placing the checkmark in its box. This may be done either to perform the analysis in multiple steps, where the output of one analysis is the input to the next analysis (in whatever order desired) or when there is no interest in performing the analysis steps simultaneously; for example, when the effect of including or excluding certain features (e.g., heat flux versus the convection boundary condition) is to be studied individually using the *Parameter Sweep*.

Another application of a sweep feature is when it is required to run multiple solutions for different combinations of model materials. For example, if there are several different fluids (n) to be modeled with one region and several solids (m) with another region, one can explore the effect of having them in the model in all possible combinations. Then, choosing the *All combinations* setting in the *Parametric* (or *Material*) *Sweep* setup, results in $m \times n$ scenarios in total that are solved for.

7.7 The COMSOL *LiveLink* for MATLAB

A connection between COMSOL Multiphysics and the MATLAB scripting environment is possible by means of *LiveLink* for MATLAB, a COMSOL add-on module. This module makes it possible to create a COMSOL model using the MATLAB script that includes the use of the COMSOL API. The API (Application Programming Interface) provides access to the internal COMSOL commands from the MATLAB environment. These are the same commands that are executed in the background when working with the COMSOL Multiphysics graphical user interface.

All actions taken when creating the COMSOL Multiphysics model are recorded in the model history as a Java code. When the model is saved as the *Model File for Java* (*.java), this code is exported and can be reviewed later. In other words, this code gives access to the building blocks of the FEM model. This is similar to the way macros can be created when working in Excel, with all the steps taken, such as selections, executing mathematical operations, and adding text, recorded to be later retrieved and modified, if needed, using the built-in *Visual Basic* editor.

This means that any operation performed in COMSOL Multiphysics creates a command (simplified Java script-based) which is added to the MATLAB script. Therefore, using this feature, one can define material properties or boundary conditions as the MATLAB script, which is then evaluated when the script is run. Incorporating the COMSOL Multiphysics commands within a script makes it possible to implement, for example, nested loops (e.g., *for* or *while*) and conditional model settings using the *if –switch* statements. This link is bidirectional, meaning that if changes are made to the MATLAB script or within the MATLAB environment, they become simultaneously available in COMSOL Multiphysics; the opposite is true as well.

Figure 7.12 shows a Java script created in the COMSOL Multiphysics environment. A 3D model of a 150-mm long pipe with exterior diameter of 65 mm and interior diameter of 55 mm is generated by creating two concentric cylinders with center of one end positioned at $(0, 0, 0)$. Reviewing the code, the analyst can see that the file name is *HollowCylin_3D_Conj* located on the G drive. One component (*comp*1) is created and added to the

```
/*
 * HollowCylin_3D_Conj.java
 */

import com.comsol.model.*;
import com.comsol.model.util.*;

/** Model exported on Apr 23 2021, 10:05 by COMSOL 5.6.0.341. */
public class HollowCylin_3D_Conj {

  public static Model run() {
    Model model = ModelUtil.create("Model");

    model.modelPath("G:\\");

    model.component().create("comp1", true);

    model.component("comp1").geom().create("geom1", 3);

    model.component("comp1").mesh().create("mesh1");

    model.component("comp1").geom("geom1").create("cyl1",
"Cylinder");
model.component("comp1").geom("geom1").feature("cyl1").set("r",
0.065);

model.component("comp1").geom("geom1").feature("cyl1").set("h",
0.15);
    model.component("comp1").geom("geom1").run("cyl1");

model.component("comp1").geom("geom1").feature().duplicate("cyl2"
, "cyl1");

model.component("comp1").geom("geom1").feature("cyl2").set("r",
0.055);
    model.component("comp1").geom("geom1").run("cyl2");

    return model;
  }

  public static void main(String[] args) {
    run();
  }
```

FIGURE 7.12. Java script created in COMSOL Multiphysics to set up the model geometry.

model. Within this component there is a single geometry (*geom*1), consisting of two cylindrical features (*cyl*1 and *cyl*2). *Cylinder 1—geom("geom*1"). *feature("cyl*1", "*Cylinder")*—has its dimensions of radius and height defined by *feature("cyl*1").*set("r"*, 0.065), *feature("cyl*1").*set("h"*, 0.15).

Cylinder 2 is created by copying the *Cylinder 1* and then changing its radius: *feature().duplicate("cyl*2", "*cyl*1"). COMSOL Multiphysics has a command called *Compact History*, which can be found under the *File* menu. Applying this command before saving the file into a *∗.java* deletes unnecessary steps that were carried out at the time of model creation.

Figure 7.13 is the Java script presented earlier but after the *Compact History* command has been applied. Comparing the two versions, one can see that most of the code is the same, except for the duplication step to create the second cylinder—*geom("geome1").feature("cyl2", "Cylinder")*. The redundant model creation steps are deleted in the compacted script; however, the disadvantage is that some of the steps that were employed to create the model by the user may be lost and thus one may not be able to learn about the techniques employed.

```java
/*
 * HollowCylin_3D_Conj_2.java
 */

import com.comsol.model.*;
import com.comsol.model.util.*;

/** Model exported on Apr 23 2021, 10:24 by COMSOL 5.6.0.341. */
public class HollowCylin_3D_Conj_2 {

  public static Model run() {
    Model model = ModelUtil.create("Model");

    model.modelPath("G:\\");

    model.label("HollowCylin_3D_Conj.mph");

    model.component().create("comp1", true);

    model.component("comp1").geom().create("geom1", 3);

    model.component("comp1").mesh().create("mesh1");

    model.component("comp1").geom("geom1").create("cyl1",
"Cylinder");

model.component("comp1").geom("geom1").feature("cyl1").set("r",
0.065);

model.component("comp1").geom("geom1").feature("cyl1").set("h",
0.15);
    model.component("comp1").geom("geom1").create("cyl2",
"Cylinder");

model.component("comp1").geom("geom1").feature("cyl2").set("r",
0.055);

model.component("comp1").geom("geom1").feature("cyl2").set("h",
0.15);
    model.component("comp1").geom("geom1").runPre("fin");
    model.component("comp1").geom("geom1").runPre("fin");

    return model;
  }

  public static void main(String[] args) {
    run();
  }

}
```

FIGURE 7.13. Java script created in COMSOL Multiphysics to set up the model geometry, after applying the *Compact History*.

Adding material to the model is the next step. Water is assigned to the interior cylinder (*Domain 2*) in this case and its properties are chosen from the built-in library. Figure 7.14 shows the Java script for the material block. Note that there are eight properties defined for water in the built-in library (e.g., heat capacity, density, and bulk viscosity). These properties are defined within the library using different functions (e.g., piecewise, interpolation, and analytic). Among these properties, only the thermal conductivity (*k*) is selected for this exercise. Thermal conductivity—*k* in W/(mK)—for water is a *Piecewise* function with the temperature—*T* in K—used as the argument (*arg*), meaning that the property is temperature-dependent. This method makes it possible to use the MATLAB functions to set the model properties, and then feed them into COMSOL Multiphysics. This approach provides the freedom to define any property relationship using the very versatile the MATLAB tools.

```
model.component("comp1").material().create("mat1", "Common");
model.component("comp1").material("mat1").selection().set(2);

model.component("comp1").material("mat1").propertyGroup("def").fu
nc().create("k", "Piecewise");

model.component("comp1").material("mat1").label("Water");
model.component("comp1").material("mat1").set("family", "water");

model.component("comp1").material("mat1").propertyGroup("def").fu
nc("k").set("arg", "T");

model.component("comp1").material("mat1").propertyGroup("def").fu
nc("k")
        .set("pieces", new String[][]{{"273.15", "553.75",
"-0.869083936+0.00894880345*T^1-1.58366345E-5*T^2+7.97543259E-9
*T^3"}});

model.component("comp1").material("mat1").propertyGroup("def").fu
nc("k").set("argunit", "K");

model.component("comp1").material("mat1").propertyGroup("def").fu
nc("k").set("fununit", "W/(m*K)");

model.component("comp1").material("mat1").propertyGroup("def")
        .set("thermalconductivity", new String[]{"k(T)", "0",
"0", "0", "k(T)", "0", "0", "0", "k(T)"});
```

FIGURE 7.14. Java script created in COMSOL Multiphysics to define materials.

Physics *physics().create("ht", "HeatTransferInSolidsAndFluids", "geom1")* is added as the next step, followed by the boundary conditions (boundary heat source, convective boundary, temperature, inflow, and outflow) applied to the selected boundaries and domains (Figure 7.15). The *convective heat flux* boundary condition, *feature("hf1").set("HeatFluxType",*

"*ConvectiveHeatFlux*"), is applied to the surfaces (1, 2, 9, 12) of *Domain 2*, with heat transfer coefficient and external temperature, defined by ("*h*",

```
model.component("comp1").physics().create("ht",
"HeatTransferInSolidsAndFluids", "geom1");

model.component("comp1").physics("ht").feature("fluid1").selectio
n().set(2);
    model.component("comp1").physics("ht").create("hf1",
"HeatFluxBoundary", 2);

model.component("comp1").physics("ht").feature("hf1").selection()
.set(1, 2, 9, 12);
    model.component("comp1").physics("ht").create("bhs1",
"BoundaryHeatSource", 2);

model.component("comp1").physics("ht").feature("bhs1").selection(
).set(3, 4);
    model.component("comp1").physics("ht").create("temp1",
"TemperatureBoundary", 2);

model.component("comp1").physics("ht").feature("temp1").selection
().set(8);
    model.component("comp1").physics("ht").create("ifl1",
"Inflow", 2);

model.component("comp1").physics("ht").feature("ifl1").selection(
).set(8);
    model.component("comp1").physics("ht").create("ofl1",
"ConvectiveOutflow", 2);

model.component("comp1").physics("ht").feature("ofl1").selection(
).set(7);

  "1.3799566804-0.021224019151*T^1+1.3604562827E-4*T^
2-4.6454090319E-7*T^3+8.9042735735E-10*T^4-9.0790692686E-13*T^5+
3.8457331488E-16*T^6"}, {"413.15", "553.75",
"0.00401235783-2.10746715E-5*T^1+3.85772275E-8*T^2-2.39730284E-11
*T^3"}]);

model.component("comp1").physics("ht").feature("hf1").set("HeatFl
uxType", "ConvectiveHeatFlux");

model.component("comp1").physics("ht").feature("hf1").set("h",
"10[W/(m^2*K)]");

model.component("comp1").physics("ht").feature("hf1").set("Text",
"300[K]");

model.component("comp1").physics("ht").feature("bhs1").set("Qb",
"100[W/m^2]");

model.component("comp1").physics("ht").feature("temp1").set("T0",
"500[K]");

model.component("comp1").physics("ht").feature("ifl1").set("Tustr
", "300[K]");
```

FIGURE 7.15. Java script created in COMSOL Multiphysics to define boundary conditions.

"10[W/(m^2*K)]") and ("*Text*", "300[K]"), respectively. The *boundary heat source*, ("*bhs*1", "*BoundaryHeatSource*", 2), is applied to the *Surfaces* (3, 4) in *Domain* 2, ("*Qb*", "100[W/m^2]"). The *temperature* boundary condition of 500 K is applied to *Surface* 8 in *Domain 1*, ("*temp*1", "*Temperature-Boundary*", 2), by ("*T0*", "500[K]"). The temperature boundary condition is overwritten by the *Inflow* condition, which is applied to the *Surface* 8 in *Domain* 1, ("*ifl*1", "*Inflow*", 2). The upstream temperature is 300 K, ("*Tustr*", "300[K]"). The *outflow* condition is applied to *Surface* 7 in *Domain* 1, ("*ofl*1", "*ConvectiveOutflow*", 2), which is where the flow leaves under the atmospheric conditions.

Mesh (*mesh*1) and study blocks (*std*1) are presented in Figure 7.16. The model is set as transient, *model.study*("*std*1").*create*("*time*", "*Transient*"). The results are stored in solution 1 (*sol*1). The fully coupled direct iterative technique is used to solve the model, ("*fc*1", "*FullyCoupled*"), ("*d*1", "*Direct*"), and ("*i*1", "*Iterative*"), with the iteration as multigrid, ("*mg*1", "*Multigrid*"). The solution parameters of the initial time, time step, and solution time are defined under the related node ("*tlist*", "*range*(0, 1.5, 30)*").

```
model.component("comp1").mesh().create("mesh1");
model.study().create("std1");
model.study("std1").create("time", "Transient");
model.sol().create("sol1");
model.sol("sol1").study("std1");
model.sol("sol1").attach("std1");
model.sol("sol1").create("st1", "StudyStep");
model.sol("sol1").create("v1", "Variables");
model.sol("sol1").create("t1", "Time");
model.sol("sol1").feature("t1").create("fc1", "FullyCoupled");
model.sol("sol1").feature("t1").create("d1", "Direct");
model.sol("sol1").feature("t1").create("i1", "Iterative");
model.sol("sol1").feature("t1").feature("i1").create("mg1",
"Multigrid");
model.study("std1").feature("time").set("tlist",
"range(0,1.5,30)");
```

FIGURE 7.16. Java script created in COMSOL Multiphysics to define mesh and solution.

Using the same methodology, the model results and the data related to selected nodes can be extracted in the Java format and analyzed in MATLAB. The MATLAB functions such as *evaluateHeatFlux*, *evaluateHeatRate*, *evaluateTemperatureGradient*, and *interpolateTemperature* then can be used to process the extracted data to obtain the results at arbitrary coordinates. Mesh data can be extracted and manipulated. To sum up, the extracted data are available as the MATLAB variables ready to be used with any MATLAB functions.

Postprocessing the COMSOL Multiphysics data is also possible, using the same technique. Figure 7.17 presents the Java script developed related to a point graph. The *Cut Point* coordinates are defined by $cpt1(0, 0, 0.075)$, represented by (*"pointx"*, 0), (*"pointy"*, 0), and (*"pointz"*, 0.075) functions. The *1D point* diagram is associated with the *cpt1* and (*"cpt1"*, *"CutPoint3D"*) functions, and shows the transient temperature versus the time at that point. The axis titles are also defined in this script—(*"xlabel"*, *"Time(s)"*).

```
model.result().dataset().create("cpt1", "CutPoint3D");
model.result().dataset("cpt1").set("pointx", 0);
model.result().dataset("cpt1").set("pointy", 0);
model.result().dataset("cpt1").set("pointz", 0.075);
model.result("pg3").feature("ptgr1").set("data", "cpt1");
model.result().create("pg3", "PlotGroup1D");
model.result("pg3").create("ptgr1", "PointGraph");
model.result("pg3").set("xlabel", "Time (s)");
model.result("pg3").set("ylabel", "Temperature (degC)");
model.result("pg3").set("xlabelactive", false);
model.result("pg3").set("ylabelactive", false);
model.result("pg3").feature("ptgr1").set("data", "cpt1");
model.result("pg3").feature("ptgr1").set("looplevelinput", new
String[]{"all"});
model.result("pg3").feature("ptgr1").set("unit", "degC");
```

FIGURE 7.17. Java script created in COMSOL Multiphysics to display transient temperature at a point.

Figure 7.18 presents the Java script that creates a line graph. The *Cut Line 3D* is defined by $cln1(0, 0, 0; 0, 0, 0.075)$, represented by (*"genpoints"*, *new double*[][]{{0, 0, 0}, {0, 0, 0.15}}). The 1D line diagram is associated with the *Cut Line 3D cln1*, (*"cln1"*, *"CutLine3D"*). The plot shows the variation of temperature along this line at the given time in the case of transient analysis.

```
model.result().dataset().create("cln1", "CutLine3D");
model.result().dataset("cln1").set("genpoints", new double[][]
{{0, 0, 0}, {0, 0, 0.15}});
model.result("pg4").feature("lngr1").set("data", "cln1");
model.result("pg4").create("lngr1", "LineGraph");
model.result("pg4").feature("lngr1").set("xdata", "expr");
model.result("pg4").feature("lngr1").set("data", "cln1");
model.result("pg4").feature("lngr1").set("looplevelinput", new
String[]{"manual"});
model.result("pg4").feature("lngr1").set("looplevel", new
String[]{"8, 15, 21"});
model.result("pg4").feature("lngr1").set("unit", "degC");
model.result("pg4").feature("lngr1").set("xdataexpr", "z");
model.result("pg4").feature("lngr1").set("xdataunit", "m");
model.result("pg4").feature("lngr1").set("xdatadescr", "z-
coordinate");
model.result("pg4").feature("lngr1").set("linecolor", "black");
model.result("pg4").feature("lngr1").set("linemarker", "cycle");
model.result("pg4").feature("lngr1").set("resolution", "normal");
```

FIGURE 7.18. Java script created in COMSOL Multiphysics to display temperature variation along a line.

CHAPTER 8

THE *COMSOL HEAT TRANSFER PROBLEM CASE STUDIES*

T his chapter presents several case studies related to heat transfer in pipes. These studies employ thermal-fluid models created using the COMSOL Multiphysics FEM commercial package with two add-ons: the *Heat Transfer Module* and the *CAD Import Module*. These studies should be helpful in learning about the software; they range from less complex cases (a cylindrical pipe) to more complex ones (a cylindrical pipe with internal-external extended surfaces). All models are represented in 3D; however, in one case, a validation is conducted to compare the results of the base condition (a 3D cylindrical pipe) with a 2D axisymmetric model of the same geometry.

The physical system modeled is the same as that of the metal (copper) pipe through which hot water is flowing. The pipe's exterior is exposed to atmospheric conditions with natural convection. The model considers both the flow of water and the heat transfer between water and the pipe solid. This interaction between a non-isothermal flow and a solid requires addition of a *Conjugate Heat Transfer* physics into the model. A stationary model is solved representing the state of the system after it reaches equilibrium. These cases are comparable to the case investigated within MATLAB with nominally one-inch diameter copper pipe (Section 6.1). However, the MATLAB model involved a greater degree of approximation.

The COMSOL Multiphysics model files are made available for most of the case studies presented in the following sections. The reader is encouraged to think about heat transfer concepts in these case studies, review the model files, and make new scenarios by changing the variables, which they can then solve and post-process using the guidelines provided in this book. True learning happens only by patience and practice.

The next section introduces the overall process of modeling for these case studies; following that, the five case studies that investigate the effect of the extended surfaces are presented. The last section presents a comparison of the results.

8.1 Modeling Heat Transfer in a Pipe—Overview of the Case Studies

The purpose of these studies is to compare the heat loss to the environment for the selected pipes. It is expected that pipes with greater area of extended surfaces will show a lower average temperature on the pipe wall. Effort was made to use pipe geometries with similar dimensions to make the comparisons possible. The base case scenario is a simple pipe (a hollow cylinder). The first enhancement is to make an internally finned pipe; next, an externally finned pipe, and then an internally-externally-finned pipe. The final geometry is an externally-twisted-finned, rotini-shaped pipe.

For each geometry, the exterior surface area and other geometrical parameters are calculated to investigate the effect of adding the extended surfaces in the form of internal, external, and internal-external extended fins.

8.1.1 Model Geometry

All pipe geometries were created in a dedicated CAD software (SOLIDWORKS) and exported as the *.stp* files. These files were then imported into COMSOL Multiphysics, with import applied to both solids and surfaces. The absolute import tolerance was *1e*-5, unless otherwise stated. The imported objects were checked and repaired for errors based on the above tolerance. A *CAD Import Module* or a specific CAD tool *LiveLink Module* are required to enable CAD model import.

Care is to be taken when importing the geometry data, ensuring the units are correctly set. The analyst will also have an option to scale the geometry dimensions when changing units or to keep the size unchanged. The *Angular unit* of degrees and *Length unit* of *m* were employed when importing the geometries. The analyst can always confirm any of the model's dimensions using the *Measure* feature under the *Geometry* node.

If required, it is possible to modify the imported geometry, taking advantage of the built-in features available in the COMSOL Multiphysics base module. For example, the geometry may be cut by a plane to remove one of the resulting volumes. To carry out more complex geometry operations, such as creation of fillets or advanced defeaturing, the *Design Module* is required.

The imported part is a hollow pipe, with no material defined within its interior. However, to create the flow model within the pipe's interior, a domain needs to exist there. This interior domain is created by carrying out a capping operation on the two ends of the pipe. To do this, the edges around the pipe opening are selected within the *Geometry, Cap Faces* node.

The pipe geometry is oriented so that the pipe axis is aligned with the z-coordinate, with flow along the positive z-coordinate. Gravity is assumed to act along the negative y-coordinate. The x- and z-coordinates form the horizontal plane.

In each study, the domain volume and convective surface area are measured using the COMSOL built in geometry measuring tool. The pipe volume can be obtained by selecting the *Measure* feature by right-clicking on the *Geometry* node. Within the tool, the *Geometric entity level* is set to *Domain,* and the pipe (solid) domain is selected. The pipe volume refers to the volume of the pipe itself and not its interior space.

The convective surfaces include all exterior surface excluding the end surfaces, since these surfaces are assumed insulated. The convective surfaces can be measured by selecting the *Measure* as above. Within the tool, the *Geometric entity level* is set to *Boundary,* and the pipe exterior surfaces are selected.

8.1.2 Material Properties

The model includes two materials: the copper of the pipe and the water that flows through it. These materials are available in the built-in material database of COMSOL, and thus were added to the model from there. Their properties include, for example, temperature-dependent density for water. Materials can be added to the model at the global level or at the component level. In the former case, the material would then be referenced by each of the components. Thus, if any customization is done to it, the effect would propagate to all components. In this case, materials were added on the component level. Here, the *component* means the 3D model of the pipe geometry. Water is linked to the material properties of the fluid domain inside the pipe and copper is linked to the pipe (shell) domain.

8.1.3 Model Physics

Conjugate Heat Transfer physics in the COMSOL Multiphysics *Heat Transfer Module* is selected to define and solve these problems, which is a hybrid model consisting of *Heat Transfer in Solids and Fluids (ht)* and *Laminar/Turbulent Flow (spf)* physics. The former physics are employed to model heat transfer within the solid domain (i.e., pipe with fins and the conductive shell containing the fluid), while the latter physics are selected to model the flow inside the channel. The two physics are interfacing under the *Multiphysics* node and create the *Nonisothermal Flow* model.

8.1.4 Boundary and Initial Conditions

All pipe models are set up to represent the following conditions. The exterior surfaces are exposed to a constant ambient temperature (25 °C); the pipe initial temperature is 80 °C. Water flows within the pipe at a constant average speed of 50 mm/s and at initial temperature of 80 °C at the inlet. All the external (exposed) surfaces transfer heat by convection mechanism with the convection heat transfer coefficient of 10 W/m²K. The external end surfaces at the inlet and outlet areas are insulated (heating flux of 0 W/m²). There is no internal heat source within the (pipe) solid domain (heat rate of 0 W/m³). All internal interfaces (the interface between the solids and fluids) are treated as walls in the flow models.

The Reynolds number was considered when setting up the problem by selecting a flow velocity sufficiently low that a laminar flow will result. These studies focus on the heat exchange between the fins and the exterior, and thus complex turbulent fluid flow in the pipe interior is not considered here. The inlet velocity of 50 mm/s results in a Reynolds number of 1,420 (< 2,300), which is associated with a laminar regime. The input parameters variable names and values are listed in Table 8.1.

TABLE 8.1. Thermal model input parameters.

Parameters			
Name	**Expression**	**Value**	**Description**
density	1000[kg/m^3]	1000 kg/m³	fluid density
diam	0.995[in]	0.025273 m	pipe diameter
dyn_vis	(8.90/10000)[Pa*s]	8.9E-4 Pa.s	fluid dynmaic viscosity
hc	10[W/(m^2*k]	10[W/m²K]	heat convection coefficient
qf	0[W/m^2]	0 W/m²	heat flux
qfunc	0[W/m^3]	0 W/m³	heat generation

Parameters			
Name	**Expression**	**Value**	**Description**
Reynolds	density*velocity*diam/dyn_vis	1,419.8	fluid Reynolds number
T_ambient	25[degC]	298.15 K	ambient temperature
T0_aluminum	25[degC]	298.15 K	aluminium initial tempeature
T0_water	80[degC]	353.15 K	water initial temperature
time	120[s]	120 s	solution time
toler	0.05	0.05	solution toleance
velocity	50[mm/s]	0.05 m/s	flow velocity

The conditions described above are implemented within each physics node described below.

Heat Transfer in the Solids and Fluids (ht) Node

(1) *Node root*—Applied to the solid and fluid domains, and the reference temperature is set to *T_ambient*.

(2) *Solid*—Applied to the solid (pipe and fins) domain. Relevant thermal material properties are taken from the materials added to the component as described above.

(3) *Fluid*—Applied to the fluid domain (water) inside the pipe. Relevant thermal properties are taken from the materials added to the components. See Note 1.

(4) *Initial values*—The initial temperature is set to *T0_water* for both solid and fluid domains.

(5) *Thermal insulation*—Applied to both end surfaces of the pipe solid domain.

(6) *Heat flux*—Convective heat flux of *hc* is applied to all exterior surfaces, with the ambient temperature set to *T_ambient*.

(7) *Inflow*—The end surface of the fluid domain when the lower *z*-coordinate value is selected and *T0_water* upstream temperature is applied to it.

(8) *Outflow*—The end surface of the fluid domain when the $z = 0$ value is selected.

Laminar/Turbulent Flow (spf) Node

(1) *Node root*—Applied to the fluid domain (water). Compressibility is set to Weakly compressible flow (see Note 2). *Include gravity* is selected. *Discretization of fluids* is selected as *P*1 + *P*2 (see Note 3).

(2) *Fluid properties*—Applied to the fluid domain and the applicable material properties are taken from the materials added to the component as described above.

(3) *Initial values*—Applied to the fluid domain uses a velocity field with a *z*-component equal to the *velocity* variable and pressure is set to that of the ambient (zero value).

(4) *Wall*—Applied to all interior pipe walls with a *No slip* wall condition.

(5) *Gravity*—This volumetric force is introduced as the Acceleration of gravity that can have *x*-, *y*-, and *z*-components. The gravity is assumed to act perpendicular to the pipe's longitudinal axis (*z*-coordinate), along the negative *y*-coordinate and thus a –*g_const* variable value is entered under the *y*-component.

(6) *Inlet*—Applied to the fluid domain end where the *z*-coordinate value is lower. The inlet boundary condition can be defined as one of *Velocity, Pressure, Mass flow, and Fully developed flow*. For the problems herein, the *Velocity* option is selected, and the *Normal inflow velocity* is set to the *velocity* variable value.

(7) *Outlet*—Applied to the fluid domain end where the *z*-coordinate value is zero. The outlet boundary condition can be defined as one of *Pressure, Velocity, and Fully developed flow*. For the problems herein, the *Fully developed* flow option is selected, and the *Average pressure* is set to that of the ambient (zero value). Also, the *Compensate for hydrostatic pressure approximation* option is selected since gravity is acting transverse to the flow direction.

Notes (1) *A Fluid node is also included in the solid physics model in the Conjugate Heat Transfer cases. This means that both domains (solids and fluids) are included in the main physics (Heat Transfer in Solids and Fluids). Thus, the fluid domains are selected within the Fluid node above and in the Laminar Flow physics. In heat transfer models, temperature can be discretized using linear, Lagrange or Serendipity (quadratic, cubic, quantic, and quintic) elements for quadratic or higher order discretization. The Isothermal Domain Interface in Heat Transfer in Solids and Fluids node are included; furthermore, the interface type can be selected as Continuity, Ventilation, Convective heat flux, Thermal insulation, and Thermal contact. However, in this case, the applicable boundaries become visible, and the rest are grayed out (not selectable) and since all of them are greyed out, the isothermal domain is not applicable.*

(2) *When setting up the Laminar Flow model, the Compressibility condition can be selected as either Incompressible, Weakly compressible, or Compressible with Ma < 0.3 flows. Choosing an appropriate option may be able to improve the solution convergence in case difficulties are encountered. In this study, the Weakly compressible flow is assumed.*

(3) *Fluid domains can be discretized using the first to the third order elements for the velocity and pressure fields. This can be controlled by selection of any combination of the Pi + Pj, where $1 \leq i, j \leq 3$; i and j are the order of the elements for the velocity components and pressure fields, respectively. For example, P1 + P3 presents the linear elements for the velocity components and the third order elements for the pressure fields.*

8.1.5 Meshing

Physics-Controlled mesh is selected for the problems presented herein with the element size setting that varies from *Coarse* to *Finer*, depending on the model. The choice of the mesh element size determines the number of nodes and elements in the model. Creating the mesh by selecting the *Finer* element size may result in about one-million elements (for the models presented herein), while selecting *Coarse* element size may result in two-hundred thousand elements.

Choosing the *Physics-Controlled* mesh setting means that all the mesh parameters will be set automatically based on the physics used in the model, region type, model geometry (narrow regions, walls), and the element size selection from one of nine levels, ranging from *Extremely Fine* to *Extremely Coarse*.

To examine the mesh settings chosen by COMSOL, select the *Sequence type* setting and choose the *User-controlled mesh* instead. This will reveal a set of nodes under the *Mesh* node that define all the mesh parameters in detail. Any one of them can then be adjusted by the user. For each case study below, the maximum and minimum overall mesh size limits are reported; special mesh limits imposed on the fluid domain are reported, as well.

The purpose of these case studies is to familiarize the reader with the FEM techniques and establish a reference to compare heat transfer rates between different geometries. Thus, there is no need to create models with an excessively high number of nodes that may lead to very long solution times. To choose an appropriate setting, element size sensitivity analysis may be carried out with gradually decreasing size, starting from a coarse setting. As the element size is progressively decreased, selected solution output value can be compared between iterations; once no change is detected, the appropriate element size has been reached.

8.1.6 Solution Settings

Heat Transfer in Solids and Fluids (ht) and *Laminar Flow (spf)* physics are connected through the *Multiphysics* node, which assembles them. The *Nonisothermal* node created under the *Multiphysics* node couples the solid-fluid interfaces by identifying the *Flow (spf)* and *Heat transfer (ht)* models. In these case studies, the *Stationary Solver* is employed. If *Show Default Solver* under the *Stationary* solution node is activated, the analyst can review the solution settings and implement any desired changes.

For these analyses, *Direct Nonisothermal flow (merged)* has been selected from the *Linear Solver* menu, *AMG Nonisothermal flow*, or iterative option, is also available. The disadvantage of the direct method is that all the matrices are solved simultaneously, and therefore it requires large computational memory resources compared to the iterative method. However, the latter approach is more likely to result in convergency issues; the *Direct Solver* approach is generally better in that regard. For the presented analyses, the former approach is helpful in obtaining solution convergence.

If the *Transient Solver* is chosen, time-dependent solver settings such as *Output times* and *Relative tolerance* should be specified. The *Output times* can be defined by *range(initial time, time step, solution time)*. The initial values of variables solved for may be chosen either as the *Physics controlled, User controlled* settings with the *Initial expression, Solution* of *Zero solution*, or the previously-run *Study*. As part of the settings for the *Study* node, the *Information* tab identifies the last computation time. The default *Relative tolerance* in the convergence setting is 0.005; however, this can be revised after reviewing the convergence plot. This plot shows the *reciprocal of step size* versus the *time step* for the time-dependent solver. The larger the step size is, the more the solution can advance, resulting in a smaller reciprocal step size. This means the smaller values are associated with better solution convergence. If the reciprocal step size remains large, it means that the solution is not converging. This can also be due to the accumulation of errors (residual errors) at each iteration step to the point of solution divergence. When a flow model is combined with the heat transfer model, convergence tolerance values may be affected.

8.2 Case Study 1—Pipe

8.2.1 3D Model Setup and Results

Case study 1 is the base study, where a pipe does not include any extended surfaces. The model parameters are presented in Table 8.1. The geometry used in this study is presented in Figure 8.1. It represents a copper pipe of 0.2-m length with nominal one-inch diameter. The inlet, where water enters the pipe, is located at $z = -0.2$-m and the outlet is at the opposite end at $z = 0$. The gravity vector points along the negative y-coordinate, which is perpendicular to the flow direction (Figure 8.2).

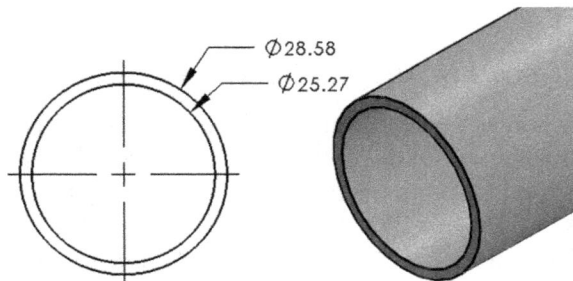

FIGURE 8.1. Geometry of the not-finned pipe (dimensions in mm).

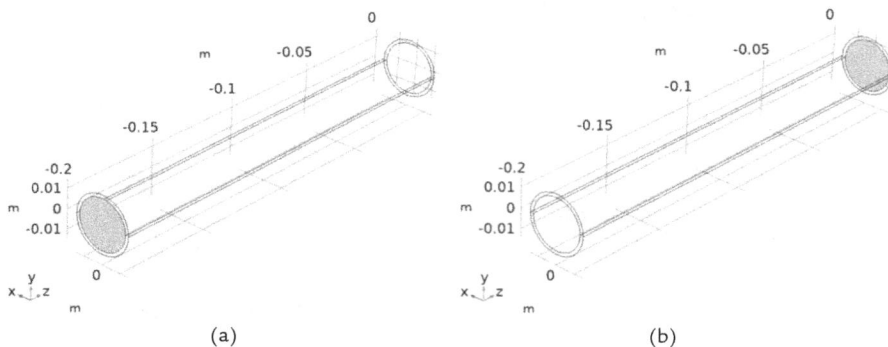

FIGURE 8.2. (a) Inlet, (b) Outlet.

Figure 8.3 presents the meshed geometry. The generated mesh had the overall maximum and minimum element sizes of 0.016 m and 0.002 m, respectively. The fluid domain and boundary maximum element sizes were 0.00195 m and 5.81E-4 m, respectively, with the maximum element growth rate of 1.45.

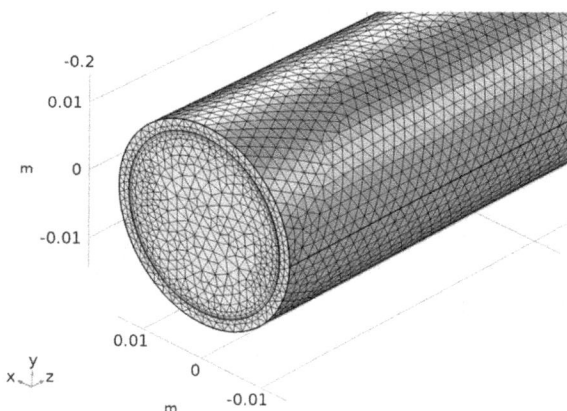

FIGURE 8.3. Mesh distribution for the not-finned pipe.

Figure 8.4 presents the mesh statistics for this problem. The *Fine* mesh size option is selected to mesh the entire geometry using the *Physics-Controlled* setting. The total number of elements is 637,366; the average element quality is about 0.7. On a 32 GB RAM, Intel® Core™ i7-10,700K 3.8 GHz Windows 10 computer, it required about 89 min to solve this stationary model, using COMSOL Multiphysics 5.6. The maximum physical memory required by the solution was 3.1 GB, which can be found under the *Log* tab.

Statistics

Mesh
　Build All

　　Geometric Entity Selection

Geometric entity level:　Entire geometry

　　Element Quality

Quality measure:　Skewness ▼

　　Statistics

Complete mesh
Mesh vertices:　131788

Element type:　All elements

Tetrahedra:　572854
Prisms:　64512
Triangles:　46970
Quads:　296
Edge elements:　854
Vertex elements:　8
　　Domain element statistics
Number of elements:　　637366
Minimum element quality:　0.1503
Average element quality:　0.6986
Element volume ratio:　0.008731
Mesh volume:　1.28E-4 m^3

　　Element Quality Histogram

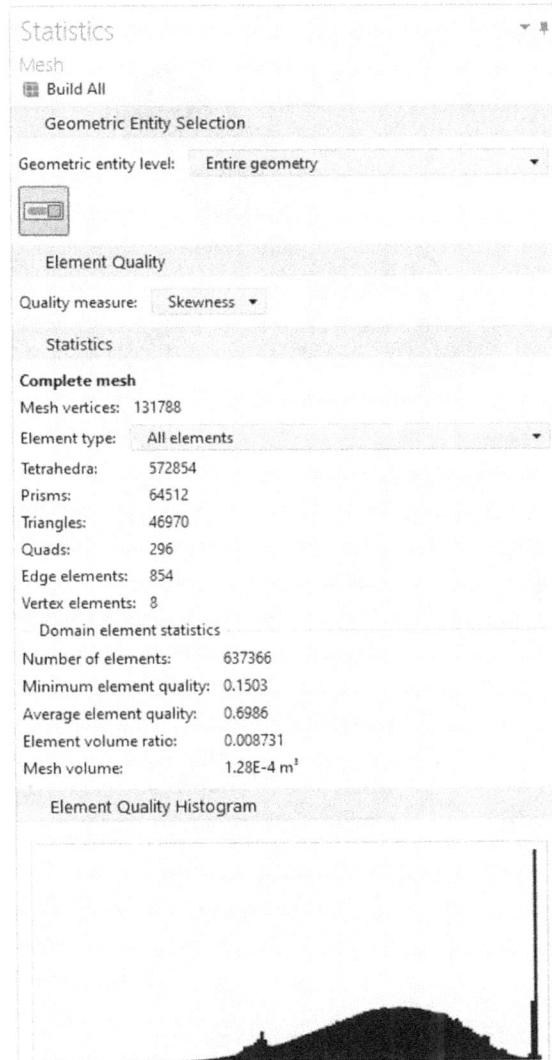

FIGURE 8.4. Mesh statistical data for the not-finned pipe.

Figure 8.5 presents the volume temperature contours. As a reminder, water enters the pipe at the inlet centered at $(0,0,-0.2,)$m and exits at the outlet at $(0,0,0)$. Thus, the hottest area on the plot (80 °C) is on the lower left, where the inlet is located. The pipe exterior can be seen to progressively cool to about 79.2 °C, from the inlet to outlet, due to the convective heat transfer to the environment.

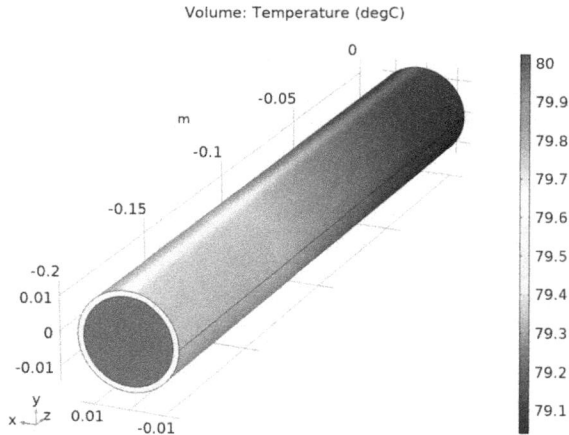

FIGURE 8.5. Volume temperature contours.

The next plots investigate the distribution of the temperature in the pipe's interior. This can be done by looking at the surface contours or at the variation along any line in space. First, the plane of interest is created by the *Cut Plane* command under the *Results / Datasets* node. Then, a *2D Plot Group* is created under the *Results* node, the previously defined surface is selected as the plot's dataset, and a *Surface* plot type is added. The figures shown also display the fluid velocity streamlines. A line dataset selection is created similarly under the *Results, Datasets* node by the *Cut Line 3D* command. A *1D Plot Group* is created and a *Line Graph* is added to it showing the variation of the temperature along this line.

Figure 8.6 shows the temperature contours and flow lines over the horizontal zx-midplane. The plot shows pipe walls being cooler than the fluid; the walls cool from left to right; a cooler region develops near the interior wall surface as the fluid flows from left to right. Details of the temperature variation over the transverse pipe section can be seen in the surface plots for the vertical transverse xy-plane located at the pipe's midpoint (Figure 8.7). Observing the fluid flow streamlines, one can see circulatory flow, with the fluid descending along the left and right walls as it is cooled there and ascending through the pipe's middle, forming two eddies. This shows the effect of adding to the model the gravity force along the negative y-coordinate.

Figure 8.8 shows the temperature contours and flow lines over the vertical yz-plane. The plot is zoomed to the outlet side of the pipe. Of interest is the observation that the flow streamlines are curving upwards towards the exit. One can relate this behavior to the circulatory flow observed in the transverse section. As the cooled water flows along the side walls, it collects towards the bottom of the pipe, forcing upward the flow through the middle. Thus, the warmer central fluid passing through the midplane is effectively rising, as indicated by the flow lines.

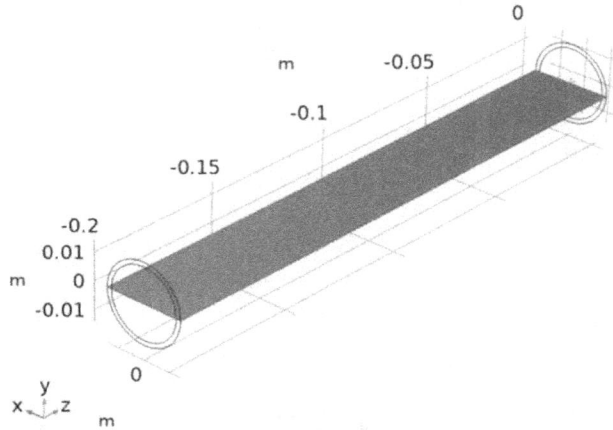

(a)

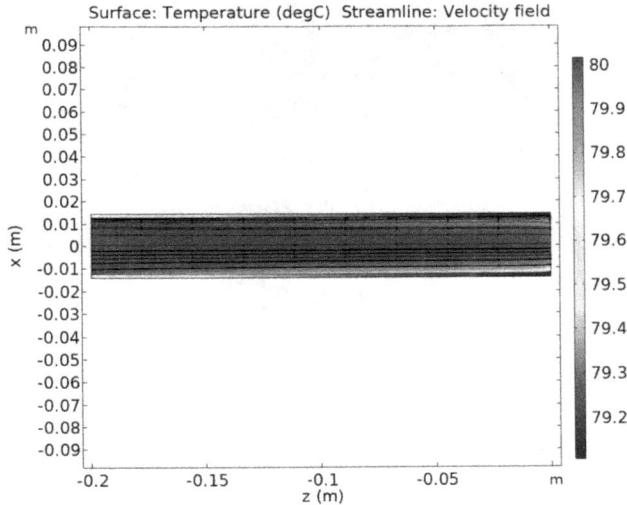

(b)

FIGURE 8.6. The *zx*-plane: (a) Selected cross-section for thermal data, (b) Temperature contours at the selected cross section, including the streamline velocity field.

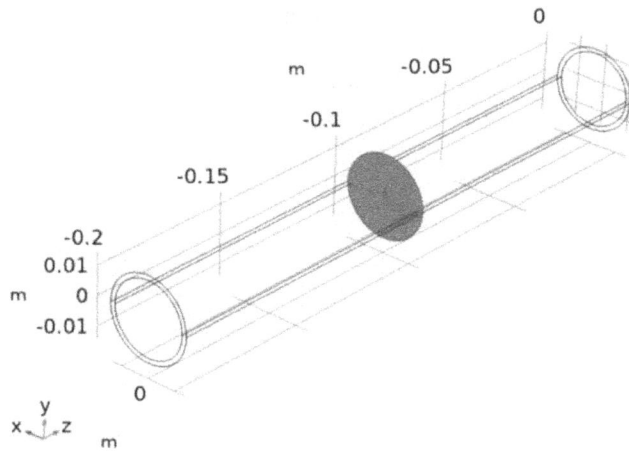

(a)

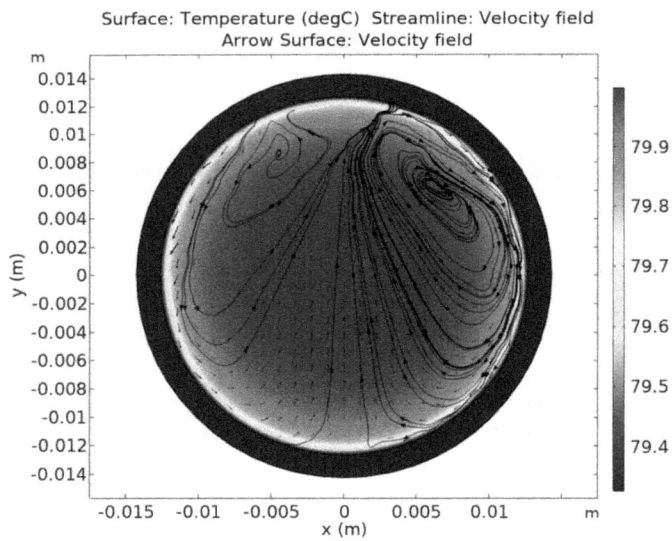

(b)

FIGURE 8.7. The *xy*-plane: (a) Selected cross-section for thermal data,
(b) Temperature contours at the selected cross section, including the streamline velocity field.

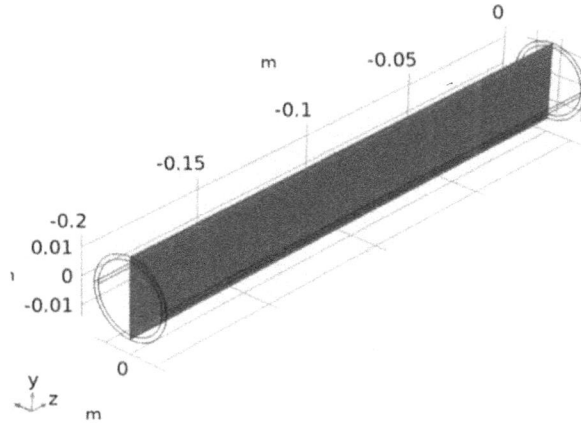

(a)

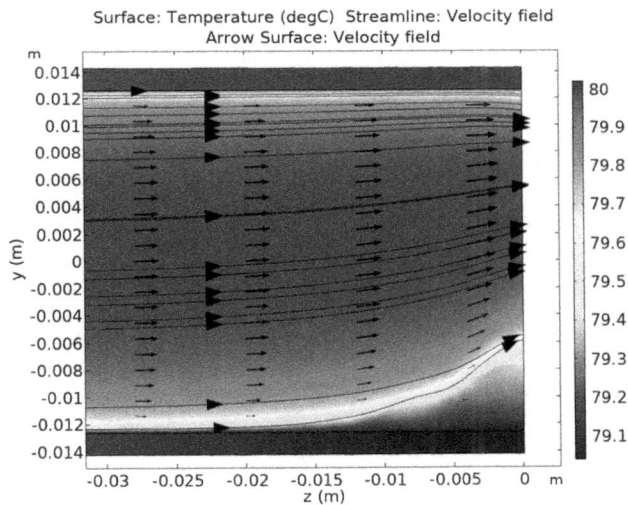

(b)

FIGURE 8.8. The yz-plane: (a) Selected cross-section for thermal data, (b) Temperature contours at the selected cross section, including the streamline velocity field (outlet at $z = 0$).

More precise numerical data about the temperature variation can be obtained from the line plots. Figure 8.9 shows temperatures along three vertical (y-coordinate) lines placed along the diameter at the inlet, middle, and outlet of the pipe. The inlet profile shows a rectangular shape with the horizontal middle segment at 80 °C, specified by the inlet temperature boundary conditions. The middle and outlet profiles show a progressive reduction in the temperature near the walls, with a parabolic profile being developed. Near the exit, the temperature in the lower pipe wall is 79.05 °C.

Similar effects can be seen for the temperature profiles along the horizontal (*y*-coordinate) lines in Figure 8.10.

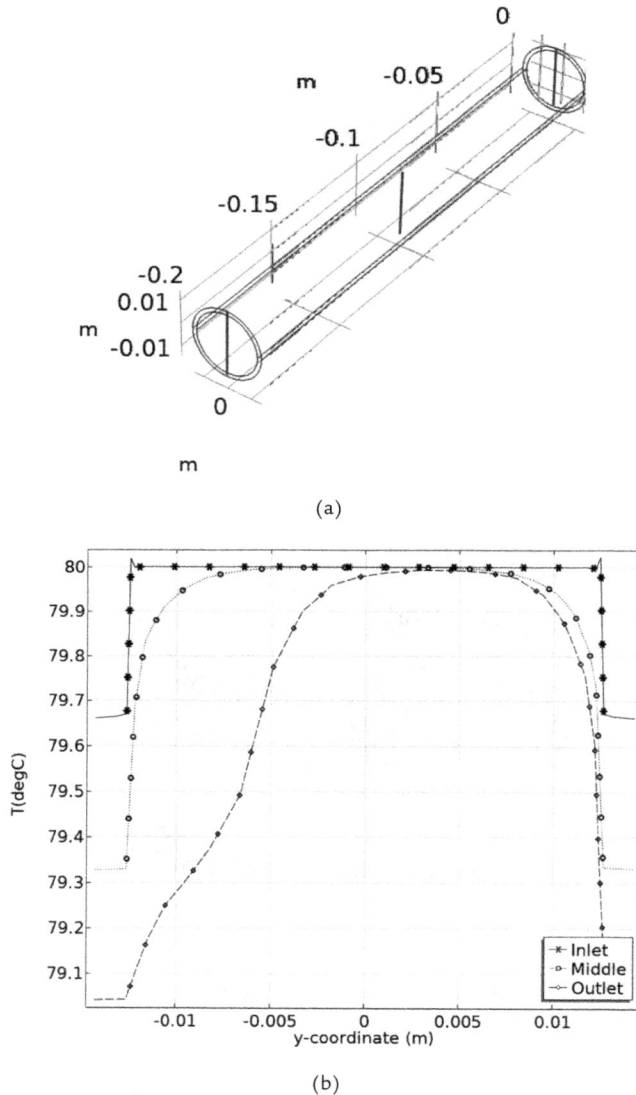

(a)

(b)

FIGURE 8.9. The *y*-coordinate: (a) Selected lines along the pipe diameter, (b) Temperature profiles along the pipe diameter at the selected lines.

(a)

(b)

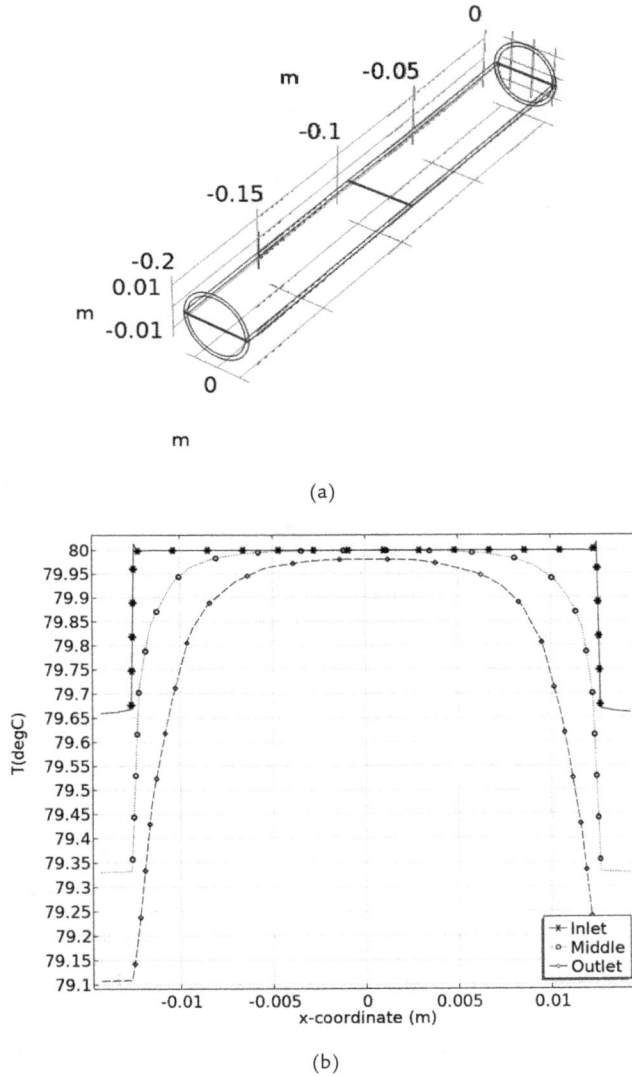

FIGURE 8.10. The x-coordinate: (a) Selected lines along the pipe diameter, (b) Temperature profiles along the pipe diameter at the selected lines.

To calculate the heat loss from the pipe to the environment, the heat flux across the exterior surfaces of the pipe needs to be integrated. To accomplish this, the *Surface Integration* node is added under the *Results/Derived Values*. Within the *Surface Integration* node, one selects the surfaces of interest and the variable to be integrated over these surfaces. In this case, there are two exterior surfaces, one facing downwards (*surface 1*) and the other one upwards (*surface 2*), as shown in Figure 8.11. The variable to be integrated is the normal total heat flux (W), *ht.ntflux*. To select it, place cursor in the cell of the table found in the *Expressions* window, under the *Expression* column (Figure 8.11, inset) and click on the *Insert Expression* icon circled in the inset figure. Search for the variable name within the dialog box that appears. Then, click on the *Evaluate* button (indicated by the red arrow) in Figure 8.11. The result appears in the table below the *Graphics* window (indicated by the red square).

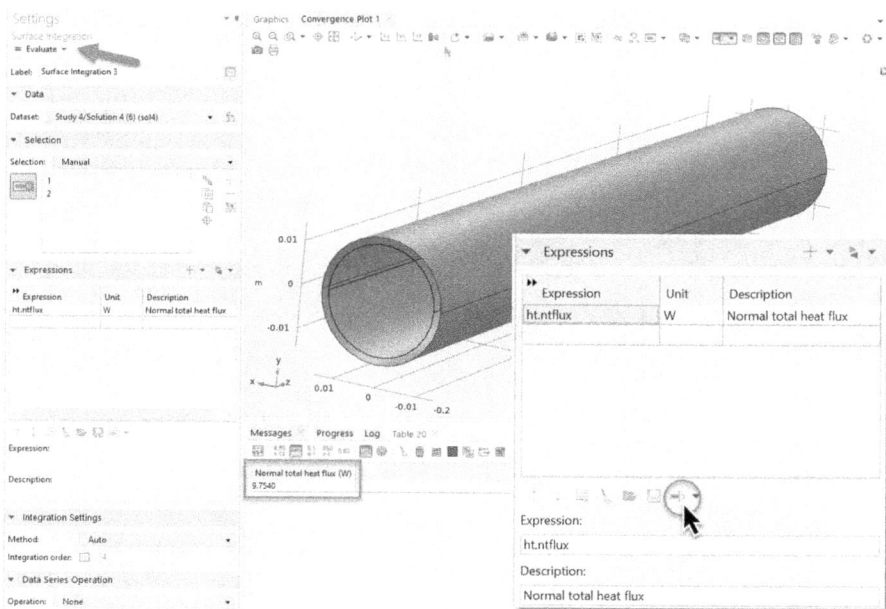

FIGURE 8.11. Normal total heat flux over *surface* 1 and *surface* 2.

The result of the above heat flux integration over the exterior surfaces is 9.75 W. The heat rate per unit length of the pipe is obtained by dividing this value by the pipe length of 0.2 m to obtain 48.76 W/m. Comparing this value with that obtained in the MATLAB Case Study 1 in Section 8.1 above (49.4 W/m) for the same conditions, the COMSOL result is 1.3% smaller than that of the MATLAB model. One possible explanation for this small difference is that the MATLAB model assumed a fixed temperature BC on the interior pipe surface while in this COMSOL model water cools as it moves along the pipe length. This slightly reduces temperature difference between environment and the pipe and surface and should lead to a lower heat loss rate.

8.2.2   Validation—Comparison with 2D Pipe Model

In this section, the 3D pipe model used in this case study is replaced with an equivalent 2D axisymmetric geometry. Solving the same problem with two different geometries and comparing the results provides validation for both approaches. As a reminder, the model input parameters are listed in Table 8.1. The solution takes about 1 min, which is considerably faster than that of the 3D model.

The 2D axisymmetric model is shown in Figure 8.12a. The inlet is located at $z = 0$ and the outlet at $z = 0.2$. The gravity force is not accounted for in this model because it assumed axisymmetric setup. The 2D free triangular mesh is generated with the *Physics Controlled* method. The *Finer* mesh size is used for the solid domain, which results in the maximum and minimum element sizes of 0.0074 m and 2.5E-5 m, respectively), and the *Fine* mesh size is used for the fluid domain (maximum and minimum element sizes of 5.0E-4 m and 1.43E-5 m, respectively). The boundaries between the fluid and solid domains (fluid-solid interface) are meshed using the *Extra Fine* elements (giving the maximum and minimum element sizes of 1.86E-4 m and 2.14E-6 m, respectively). The corners are refined using the element size scaling factor of 0.25. There are two boundary layers and maximum element growth rate is 1.25 (Figure 8.12b). The mesh consists of the total number of 57,217 elements, with the average element quality of about 0.9 (a high number).

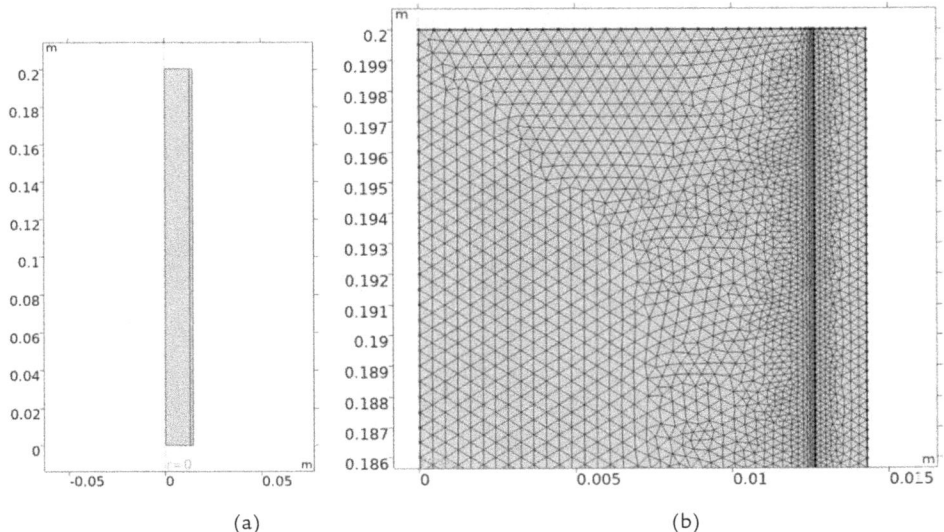

(a)   (b)

FIGURE 8.12. The 2D axisymmetric geometry for the pipe presented in Figure 8.1: (a) Geometry, (b) Mesh.

The model is solved in the stationary mode. The 2D revolved surface tempeature profiles are presented in Figure 8.13. The results of this 2D geometry are plotted using the revolution feature that rotates the solution through 360 degrees about the axis of symmetry (z-coordinate) to create a 3D representation that helps with visualization. These results show that the temperature variation for the entire pipe is between 78.7 °C and 80 °C.

A comparison can be now made between the 2D and 3D model results. Figure 8.14 shows the temperature profiles along the pipe diameter at its middle (along the y-coordinate for the 3D model). The temperature distributions are similar but the 2D model predicts slightly lower pipe wall temperature of 79 °C versus the 79.3 °C for the 3D model. The heat rate per unit length of the pipe in this case is 48.52 W/m, which is only 0.5% less than that of predicted by the 3D pipe model.

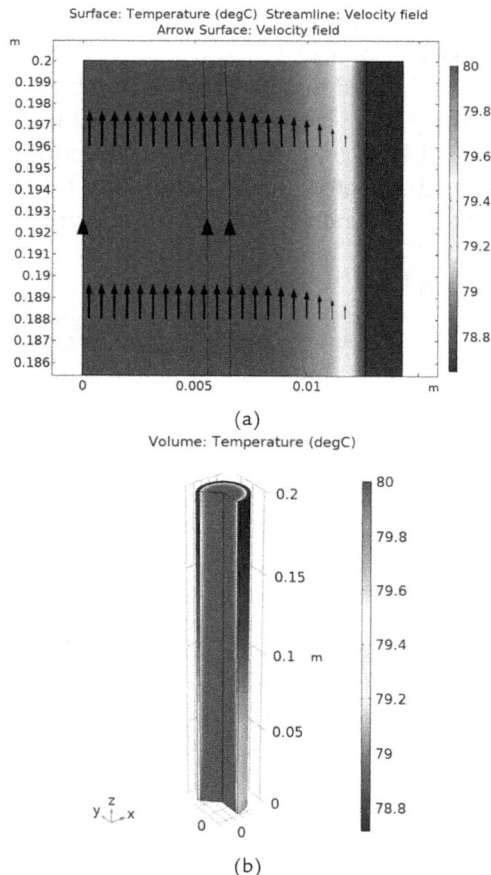

(a)

(b)

FIGURE 8.13. 2D temperature contours: (a) Included the arrow line and streamline velocity field, (b) Revolved.

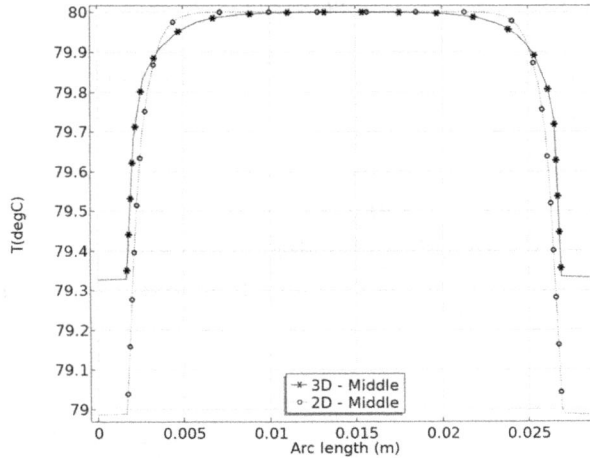

FIGURE 8.14. Comparison between the temperature profiles along the pipe diameter at the midplane for the 2D and 3D models.

8.3 Case Study 2—Internally-Finned Pipe

The geometry used in this study is presented in Figure 8.15. This case study investigates the effect of the heat dissipation by the pipe due to the addition of the extended surfaces to the pipe's interior surface. This pipe geometry adds sixteen rectangular fins, about 2-mm wide and 2.5-mm high, protruding towards the interior.

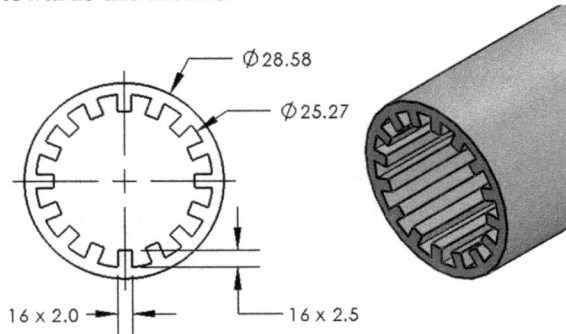

FIGURE 8.15. Geometry of the internally finned pipe (dimenions in mm).

Figure 8.16 presents the mesh for the finalized geometry. For this mesh, the maximum and minimum element sizes are 0.03 m and 0.0056 m, respectively. The fluid domain maximum and minimum element sizes are 0.00378 m and 0.00116 m, respectively, with the maximum element growth rate of 1.6. The *Coarse* mesh size option is selected to mesh the entire geometry based on the *Physics-Controlled* approach. There are a total of 1,385,085 elements, with the average element quality of 0.6. Note the increased number of elements compared to the previous simple pipe case; this effect can be attributed to the many narrow regions created by the interior fins, where finer mesh is required. On a 64 GB RAM, Intel Core i7-5,820 K 3.9 GHz Windows 10 computer, it required about 122 min to solve this stationary *Conjugate Heat Transfer* model using COMSOL Multiphysics 5.6, with a maximum of about 37 GB physical memory used.

Figure 8.17 presents the volume temperature contours. As in the previous case study, the exterior wall temperature decreases due to the convection heat transfer, from the inlet in the lower left to the outlet in the upper right.

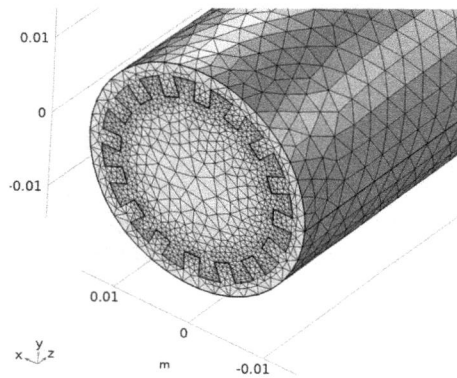

FIGURE 8.16. Mesh distribution for the internally finned pipe.

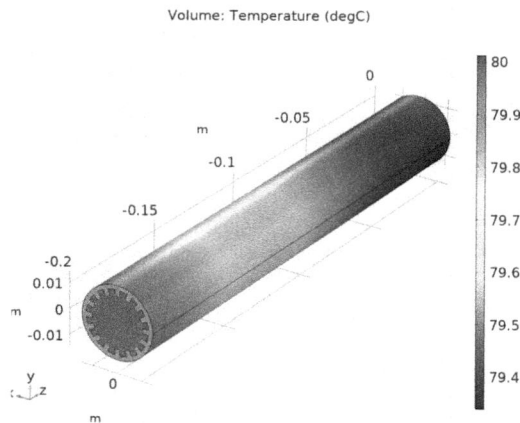

FIGURE 8.17. Volume temperature contours.

Interior variation of the temperature on the horizontal midsection zx-plane in Figure 8.18a is shown by the temperature contour plot in Figure 8.18b. The appearance is like that of the simple pipe, except the walls at the midplane are thicker due to the fins protruding towards the interior. Figure 8.19b shows the temperature contours and flowlines on the transverse xy-plane in the pipe's middle (Figure 8.19a). Here the fins protruding into the interior create cooler pockets; the flowlines indicate upward flow over most of the cross section, converging at a point near the upper end of the pipe.

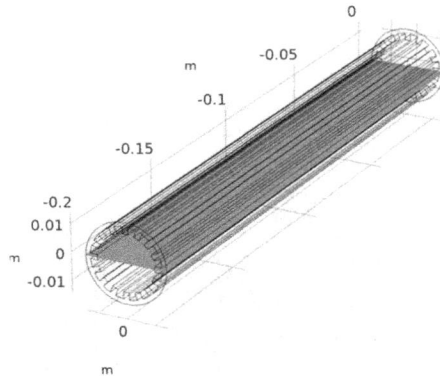

(a)

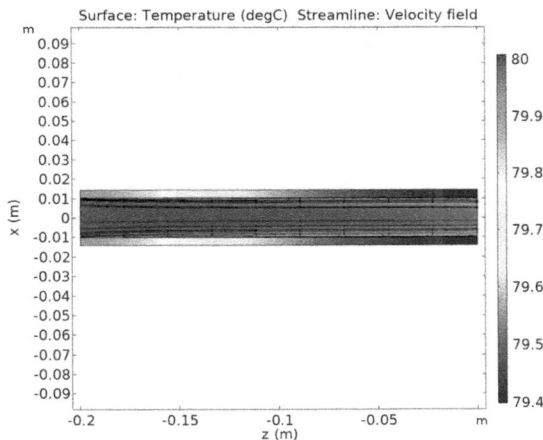

(b)

FIGURE 8.18. The zx-plane: (a) Selected cross-section for thermal data, (b) Temperature contours at the selected cross section including the streamline velocity field.

(a)

(b)

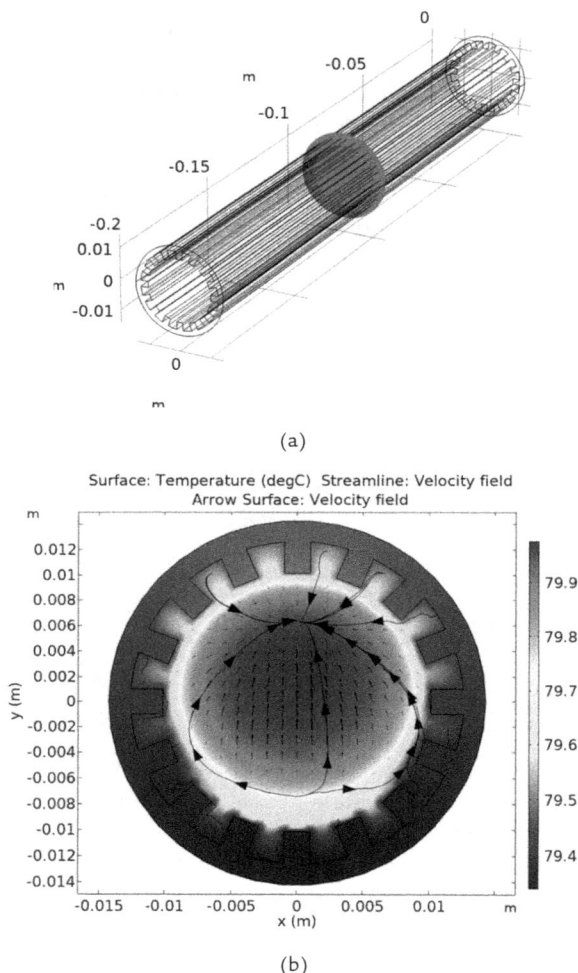

FIGURE 8.19. The *xy*-plane: (a) Selected cross-section for thermal data, (b) Temperature contours at the selected cross section, including the streamline velocity field.

Figure 8.20b zooms in on part of the vertical *yz*-plane near the pipe outlet. As in the simple pipe case, the flowlines for most of the fluid's volume are rising upwards; the warmer water around the flow channel center appears to be pushed upwards by the cooler water collecting near the bottom.

Figure 8.21b shows the temperature profiles along the three vertical diameter lines at the inlet, middle, and outlet locations. The inlet and middle curves are symmetrical about their vertical center line, but the outlet curve shows a lower temperature on the left side, corresponding with the bottom of the flow channel, where the cooler water is concentrated.

The exterior wall temperatures in general are higher for this case that of the simple pipe. Near the outlet, the wall exterior is at 79.35 °C near the bottom (versus the 79.05 °C for the simple pipe) and 79.47 °C near the top. Three horizontal line temperature profiles are displayed in Figure 8.22b. The curves are symmetrical about the vertical center line and show exterior wall temperature at the outlet equal to 79.4 °C.

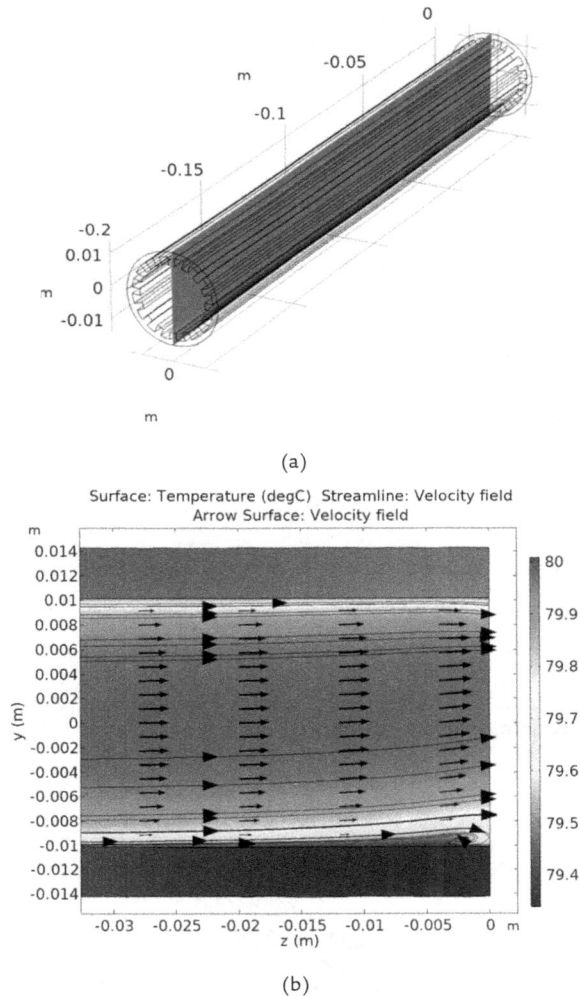

(a)

(b)

FIGURE 8.20. The yz-plane: (a) Selected cross section for the thermal data, (b) Temperature contours at the selected cross section, including the streamline velocity field.

(a)

(b)

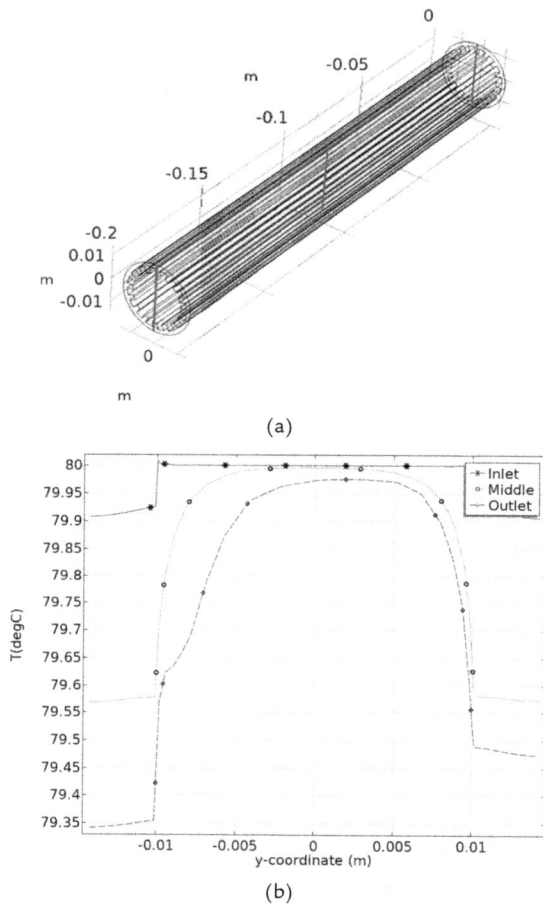

FIGURE 8.21. The y-coordinate: (a) Selected lines along the pipe diameter, (b) Temperature profiles along the pipe diameter at the selected lines.

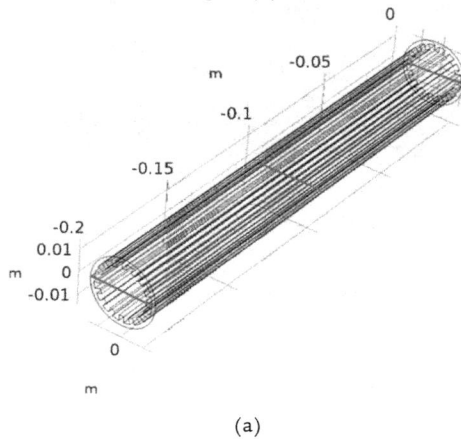

(a)

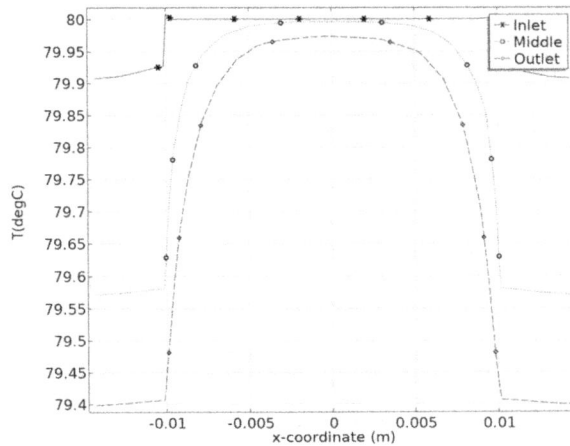

(b)

FIGURE 8.22. The x-coordinate: (a) Selected lines along the pipe diameter, (b) Temperature profiles along the pipe diameter at the selected lines.

Heat flux integration across the exterior surface gives the total heat loss to the environment equal to 9.79 W. This is equivalent to 48.95 W/m heat rate per unit length of the pipe. The result is very close to the simple not-finned pipe case (48.76 W/m), being only 0.4% higher.

8.4 Case Study 3—Externally-Finned Pipe

The geometry used in this study is presented in Figure 8.23. This case study investigates the effect of the heat dissipation due to addition of the

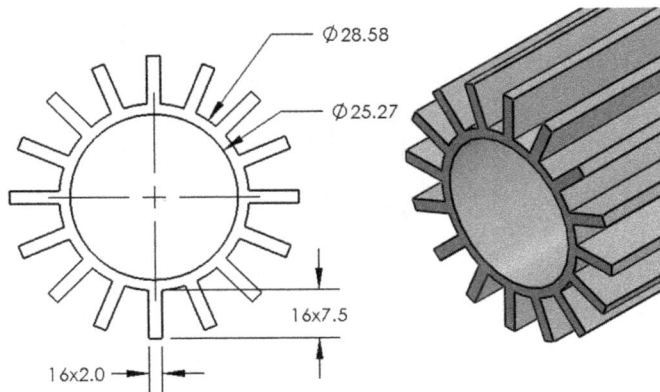

FIGURE 8.23. Geometry for the externally finned pipe (dimenions in mm).

extended surfaces to the pipe's exterior. Sixteen fins of 2-mm wide and 7.5-mm high are added. The pipe, to which the fins are joined, has the same dimensions as in Case Study 1.

Figure 8.24 presents the meshed geometry. The maximum and minimum element sizes are 0.016 m and 0.002 m, respectively. The fluid domain maximum and minimum element sizes are 0.00292 m and 8.72E-4 m, respectively, with the maximum element growth rate of 1.45. The *Normal* mesh size option was selected to mesh the entire geometry based on the *Physics-Controlled* approach. The total number of elements is 267,363, with the average element quality of 0.7. Note the much smaller number of elements in this mesh compared to the previous cases (637,366 for Case 1 and 1,385,085 for Case 2). The *Fine* mesh size was chosen for Case 1 versus the *Normal* chosen here, and there are no corners and narrow regions within the fluid domain as in Case 2. On a 32 GB RAM, Intel Core i7-10,700K 3.8 GHz Windows 10 computer, it required about 8 min to solve this stationary *Conjugate Heat Transfer* model in COMSOL Multiphysics 5.6, with a maximum of 8.4 GB physical memory used.

Figure 8.25 presents the volume temperature contours. The pipe and fins exterior are again cooling progressively from the inlet (bottom-left) to the outlet (top-right). However, there is a greater temperature difference between the water entering at the inlet (visible in red, at 80 °C) and the surrounding pipe/fins structure.

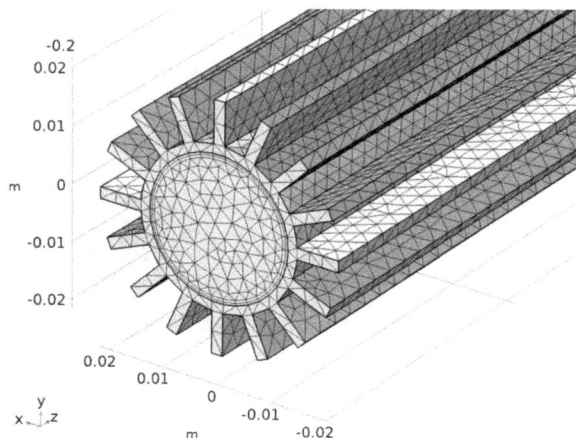

FIGURE 8.24. Mesh distribution for the externally finned pipe.

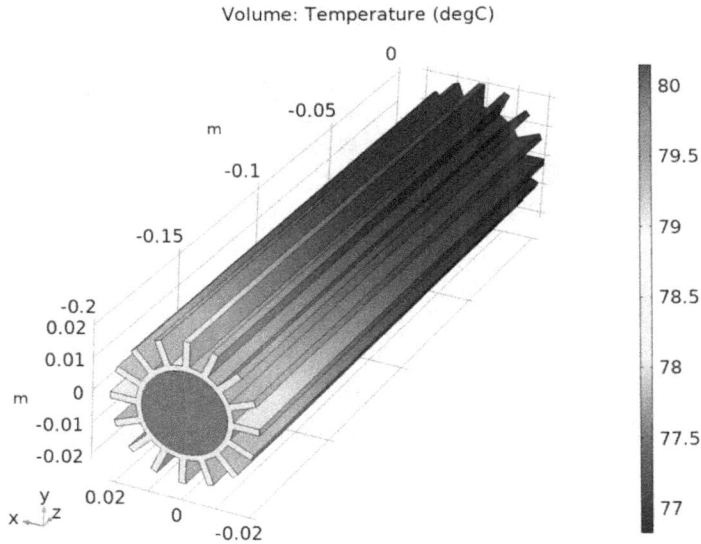

FIGURE 8.25. Volume temperature contours.

The horizontal zx-plane displays much wider cool regions within the solid domain along its length (Figure 8.26b). The horizontal plane cuts through the fins and thus their full extent is visible. The transverse xy-plane at the pipe's middle shows much stronger circulatory fluid motion within the left and right halves of the pipe (Figure 8.27b). This can be attributed to the greater temperature difference between the walls and the fluid in this case compared with the simple pipe case (at the pipe middle being 2.6 °C versus 0.7 °C). The cooler water along the walls creates a stronger convective flow. Vertical yz-plane shows upward trending flow lines and the collection of the cooler water near the bottom of the outlet (Figure 8.28b).

The temperature profiles along the vertical lines (yz-plane) in Figure 8.29b show developing asymmetry towards the outlet; at the end of the lowest fin near the outlet, the temperature is 76.85 °C; it is 77.15 °C at the end of the opposite fin, at the topmost point. Profiles along the horizontal lines (Figure 8.30b) show symmetry about vertical center line, with the fin end's temperature of 77 °C at the outlet line.

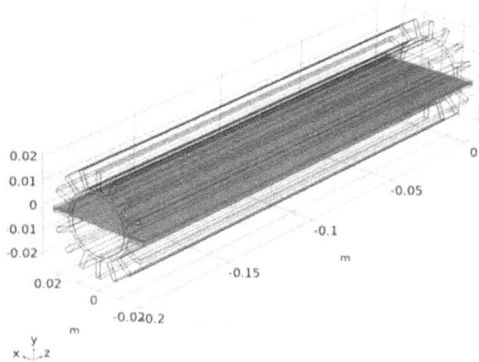

(a)

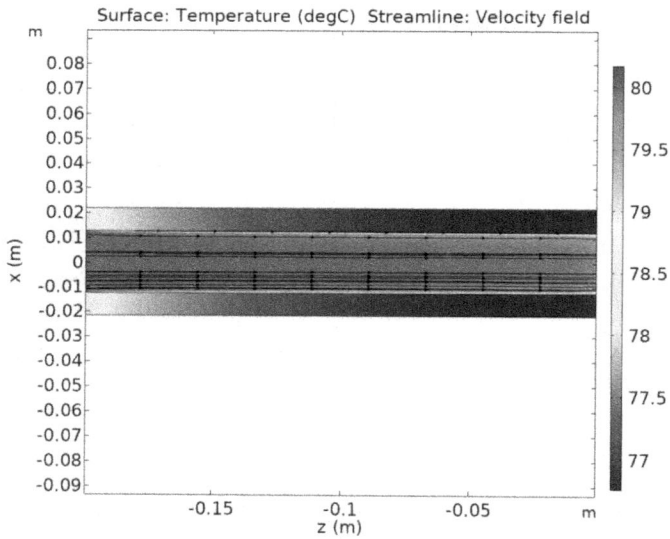

(b)

FIGURE 8.26. The zx-plane: (a) Selected cross section for the thermal data,
(b) Temperature contours at the selected cross section, including the streamline velocity field.

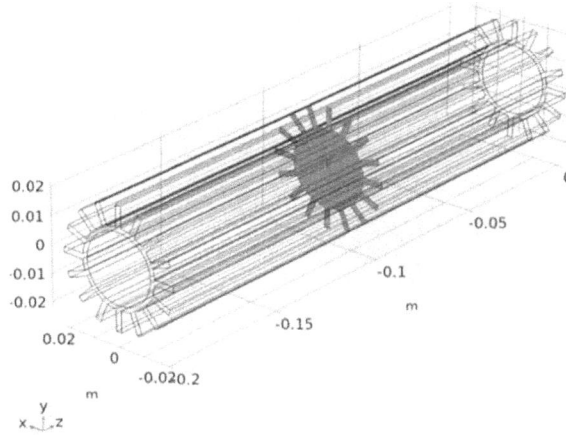

(a)

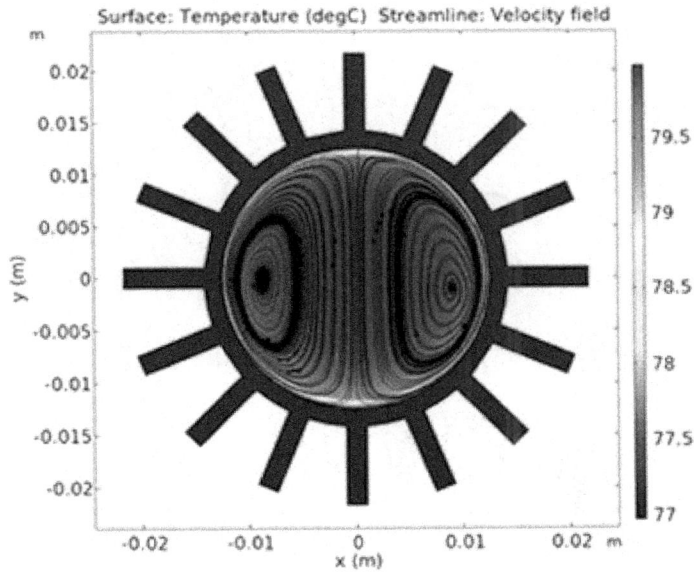

(b)

FIGURE 8.27. The *xy*-plane: (a) Selected cross section for the thermal data,
(b) Temperature contours at the selected cross section, including the streamline velocity field.

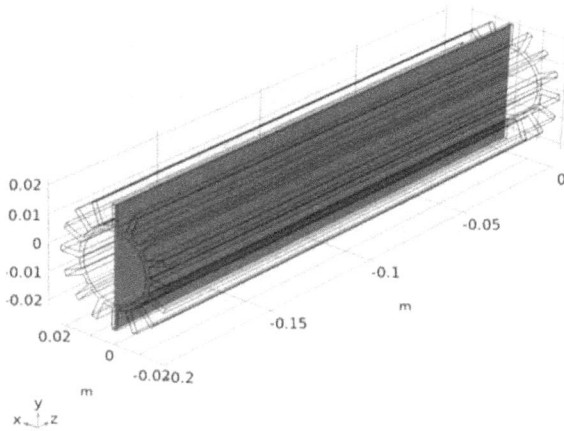

(a)

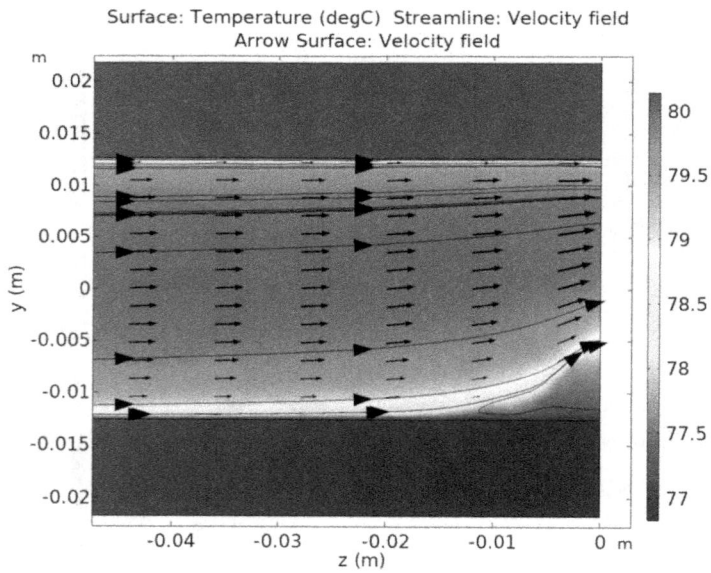

Surface: Temperature (degC) Streamline: Velocity field
Arrow Surface: Velocity field

(b)

FIGURE 8.28. The *yz*-plane: (a) Selected cross section for the thermal data,
(b) Temperature contours at the selected cross section, including the streamline velocity field.

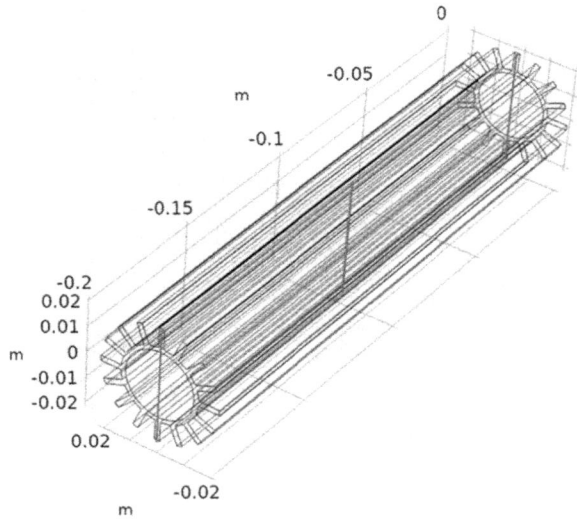

(a)

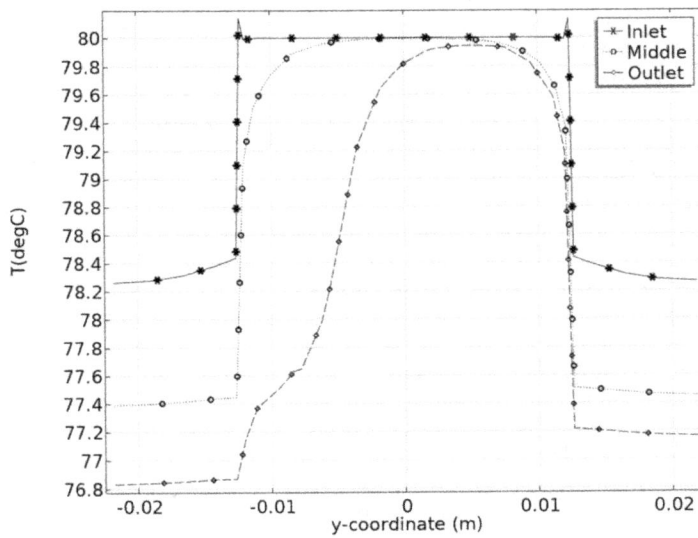

(b)

FIGURE 8.29. The y-coordinate: (a) Selected lines along the pipe diameter, (b) Temperature profiles along the pipe diameter at the selected lines.

(a)

(b)

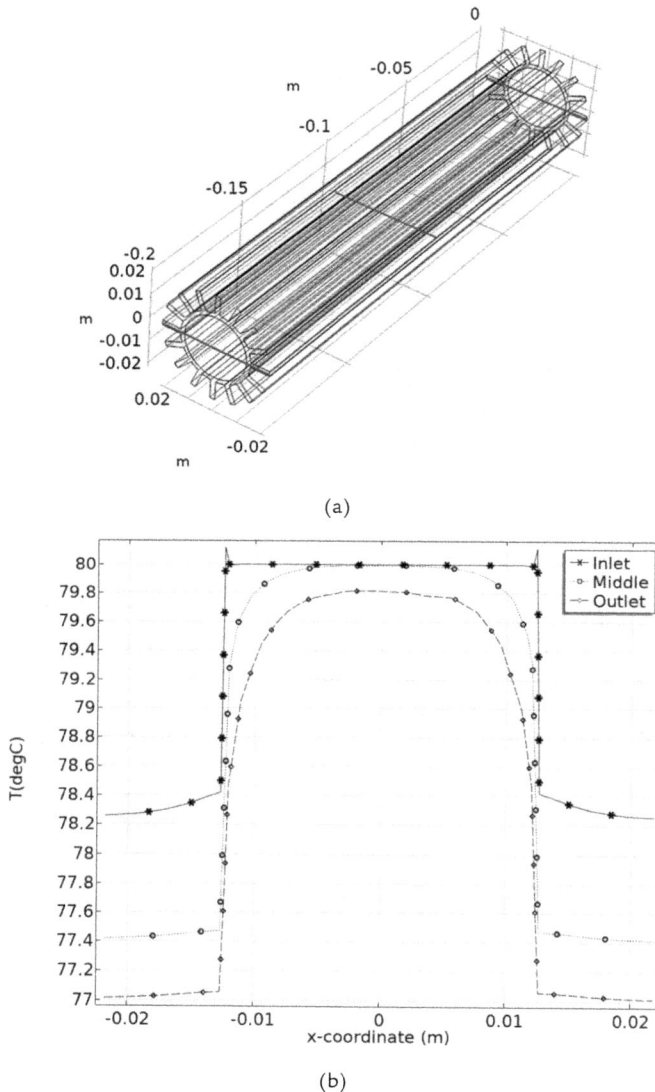

FIGURE 8.30. The x-coordinate: (a) Selected lines along the pipe diameter, (b) Temperature profiles along the pipe diameter at the selected lines.

Integration of the heat flux across all the exterior surfaces (pipe and fins) gives 34.75 W. Normalizing this value to the specific heat loss per pipe's unit length (dividing by 0.2 m) gives 173.74 W/m. The value is significantly higher (by 3.56 times) than that of the simple pipe (48.76 W/m). This difference highlights the effectiveness of the extended surfaces in increasing the heat dissipation.

8.5   Case Study 4—Internally-Externally-Finned Pipe

The geometry used in this study is presented in Figure 8.31. This geometry investigates whether having both internal and external fins increase the heat dissipation from the pipe. Sixteen interior and exterior fins structures employed in Case Studies 2 and 3 are combined here.

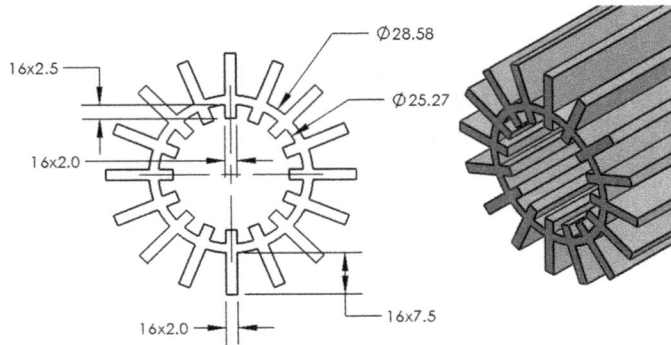

FIGURE 8.31. Geometry of the internal-external finned pipe (dimenions in mm).

Figure 8.24 presents the mesh for this geometry. It has the maximum and minimum element sizes of 0.03 m and 0.0056 m, respectively. The fluid domain and boundary maximum element sizes are 0.00566 m and 0.00174 m, respectively, with the element maximum growth rate of 1.6. The *Coarse* mesh size option is selected to mesh the entire geometry based on the *Physics-Controlled* approach, resulting in a total of 536,137 elements; the average element quality is about 0.6. Unexpectedly, the number of elements is much smaller than the 1,385,085 elements obtained for the mesh of geometry with the same interior fins (Case Study 2), which also used the *Coarse* setting. A possible explanation is that, for the current geometry, there is a greater area available for meshing within the solid part due to the addition of the external fins. This means larger elements can be used. Narrower regions require smaller element sizes. On a 32 GB RAM, Intel Core i7-10,700K 3.8 GHz Windows 10 computer, it required about 11 min to solve this stationary *Conjugate Heat Transfer* model in COMSOL Multiphysics 5.6, with a maximum of about 13.4 GB physical memory used.

Figure 8.33 presents the volume temperature contours that appear like the previous (external fins only) case. The horizontal zx-plane temperature contours follow the general trend of the previous solution; however, the fin temperature near the inlet appears to be higher than that of the external fins only case (Figure 8.34b). The transverse xy-plane displays flow circulation pattern with an upward flow around the middle and a more chaotic flow around the walls, likely caused by the internal fin structures (Figure 8.35b). The vertical yz-plane shows the rising flowlines and a cooler pocket near the bottom of the outlet (Figure 8.36b). The temperature profiles along the pipe diameter (x- and y-coordinates) are presented in Figure 8.37 and Figure 8.38, respectively. From the horizontal profile at the pipe's middle, one can observe the temperature at the end of the fins to be equal to 78.9 °C. This is about 1.5 °C higher than 77.4 °C for the same location in the previous study (with the external fins only).

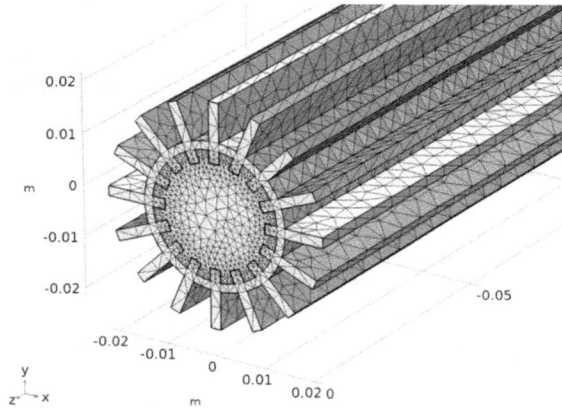

FIGURE 8.32. Mesh distribution for the internally-externally-finned pipe.

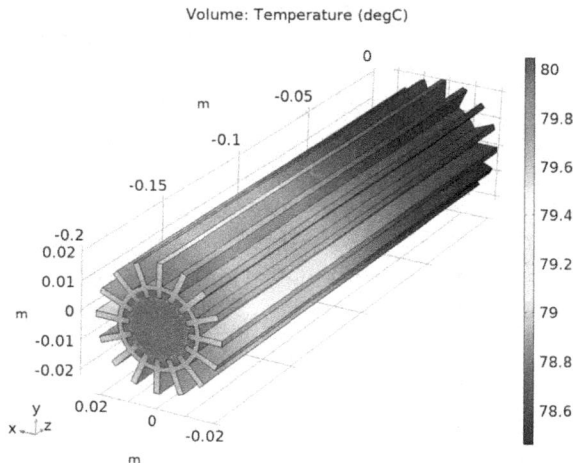

FIGURE 8.33. Volume temperature contours.

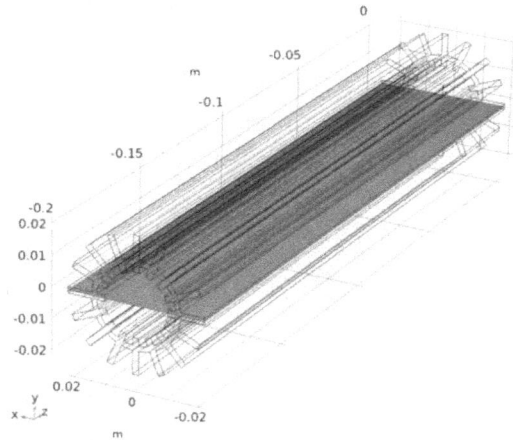

(a)

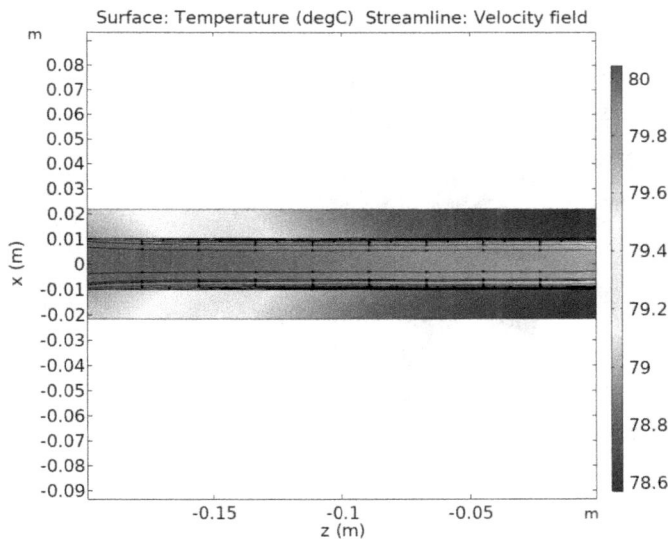

(b)

FIGURE 8.34. The zx-plane: (a) Selected cross section for the thermal data,
(b) Temperature contours at the selected cross section.

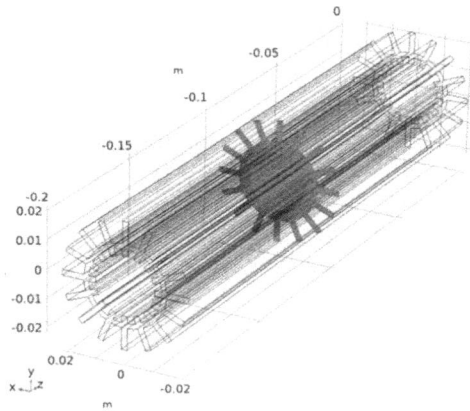

(a)

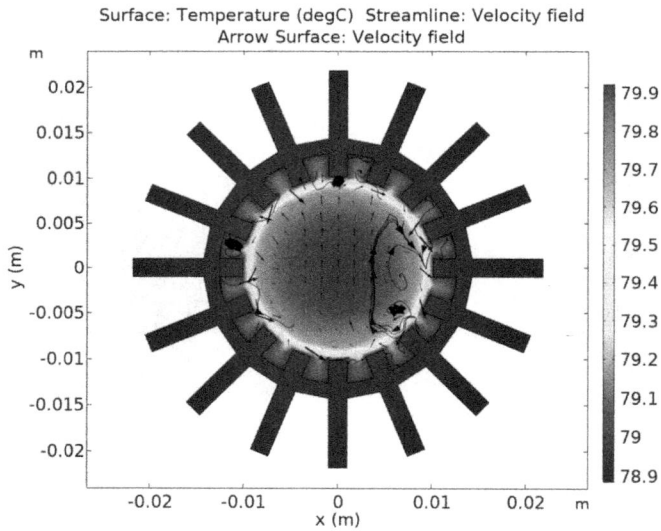

(b)

FIGURE 8.35. The *xy*-plane: (a) Selected cross section for the thermal data, (b) Temperature contours at the selected cross section.

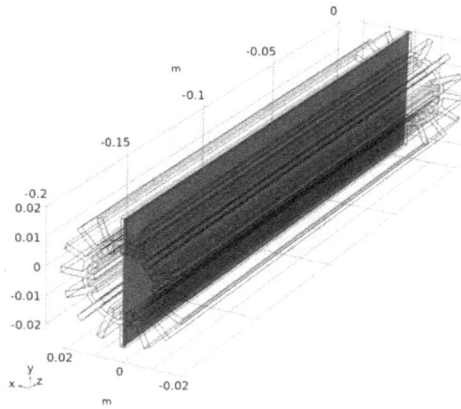

(a)

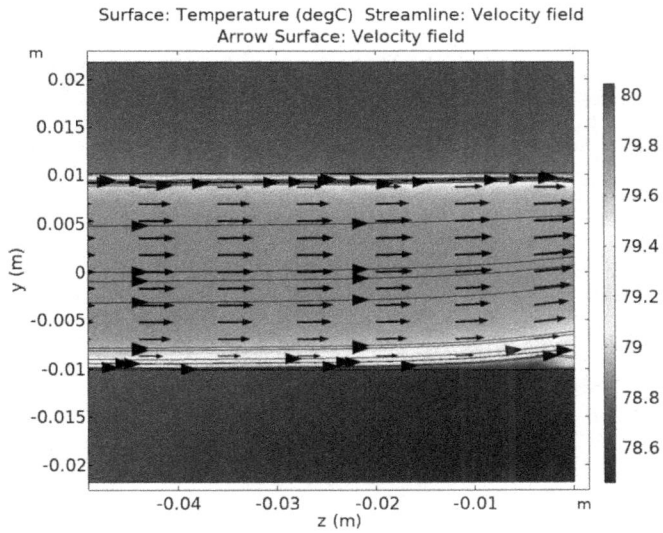

(b)

FIGURE 8.36. The *yz*-plane: (a) Selected cross section for the thermal data, (b) Temperature contours at the selected cross section.

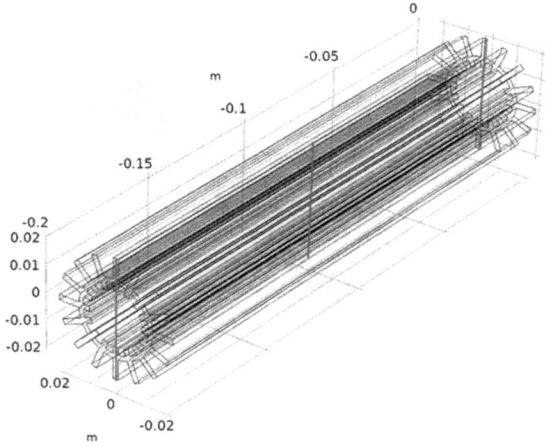

(a)

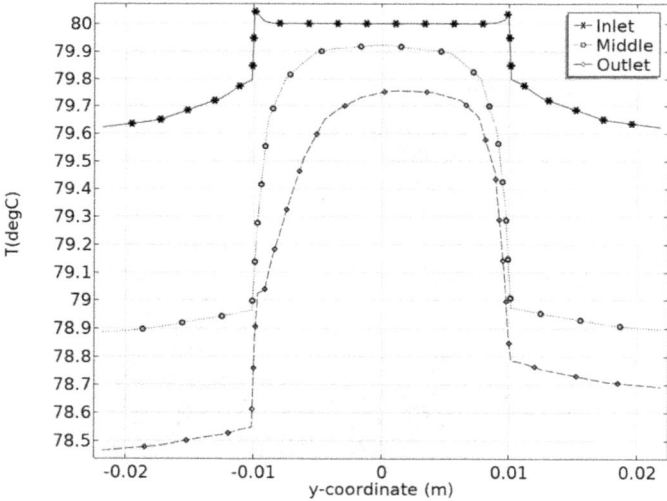

(b)

FIGURE 8.37. The *y*-coordinate: (a) Selected lines along the pipe diameter, (b) Temperature profiles along the pipe diameter at the selected lines.

(a)

(b)

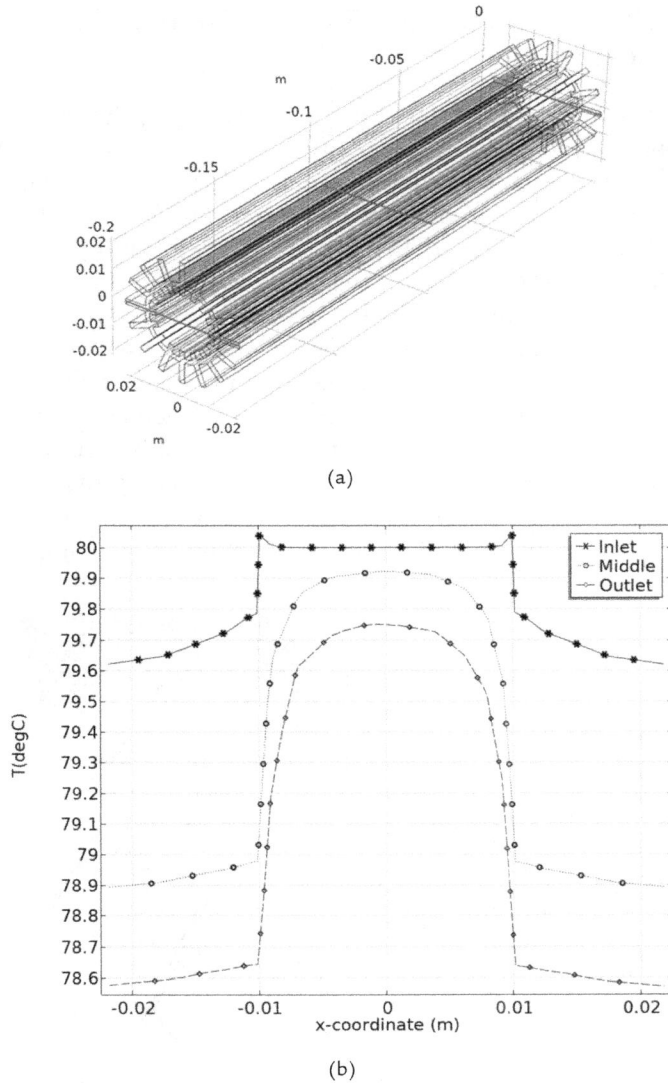

FIGURE 8.38. The x-coordinate: (a) Selected lines along the pipe diameter, (b) Temperature profiles along the pipe diameter at the selected lines.

Heat flux integration over the exterior surfaces (pipe and fins) gives 35.73 W. Dividing this by the pipe length of 0.2 m, gives the normalized heat loss rate of 178.6 W/m. This is 2.8% higher than that of the heat loss calculated for the pipe with only external fins (173.7 W/m). Thus, while there is an increase in the heat dissipation due to the addition of interior fins, the difference is very small and may not justify the additional material and manufacturing expense.

8.6 Case Study 5—Externally-Twisted-Finned (Rotini) Channelled Pipe

The geometry used in this study is presented in Figure 8.39. The fin is 0.2-m long, and its cross section can be inscribed in a 70.3 mm-diameter circle. The diameter of the internal channel is 25.3 mm. The case study presented herein is a variation of the rotini fin geometry that the author presented in her earlier work [4]. Rotini pastas are short and are corkscrew shaped. *Rotini* is an Italian term meaning *small wheels*. It is not only a shape that is geometrically interesting with its twists and turns, but it also works well as a pasta, with its large convective surface areas taking up all that sauce. If an observant reader ever made rotini , they would soon learn that these pasta shapes cool faster than other types, such as spaghetti (the long stranded thin ones) [59]. The rotini pasta piece can be considered a type of fin structure. One can conjecture that the fast cooling must be due to the very good heat dissipation properties of the rotini's large convective surface area. This study aims to determine how this fin shape compares with the straight fins explored in the previous studies.

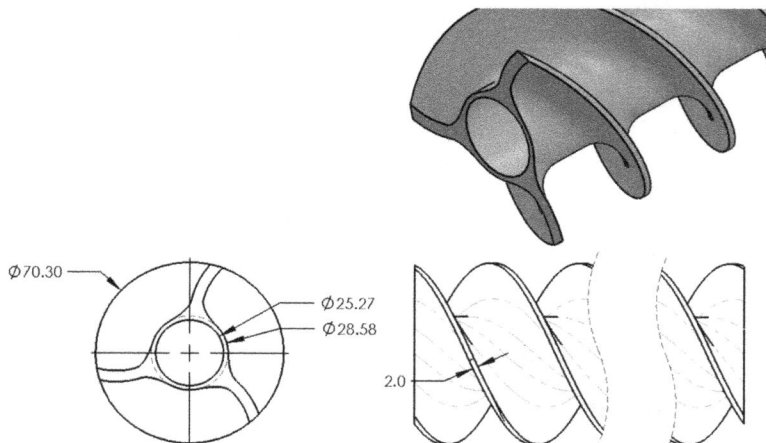

FIGURE 8.39. Geometry for the externally-twisted-finned, rotini pipe (dimensions in mm).

Figure 8.40 presents the meshed geometry. The maximum and minimum element sizes are 0.0172 m and 0.00214 m, respectively. The fluid domain maximum and minimum element sizes are 0.0048 m and 0.00143 m, respectively, with the maximum element growth rate of 1.45. The *Fine* mesh size option was selected to mesh the entire geometry based on the *Physics-Controlled* approach. The mesh contains a total of 78,504 elements, with the average element quality of about 0.7. Surprisingly, the element count here is much lower than that of the simple pipe case (637,366), which was also meshed with the same *Fine* setting. On a 32 GB RAM, Intel Core 10,700 K 3.8 GHz Windows 10 computer, it took only 1 min to solve this stationary *Conjugate Heat Transfer* model in COMSOL Multiphysics 5.6, with a maximum of 3.4 GB physical memory used. The short solution time is due to the filleted smooth surfaces on the exterior, which are much easier to mesh than the sharp corners within the geometry, and due to a smooth interior surface, which improves the flow model solution convergence by reducing the flow disturbances.

Figure 8.41 presents the volume temperature contours, with fins appearing to have similar cool temperature through large part of the pipe's length. The horizontal *zx*-plane temperature contours display cool exterior solid surfaces and parallel flow lines extending through the length of the

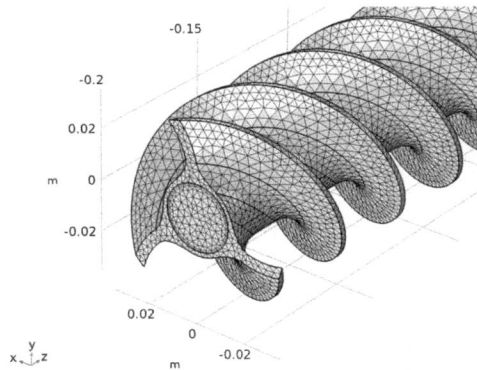

FIGURE 8.40. Mesh distribution for the channeled externally-twisted-finned, rotini pipe.

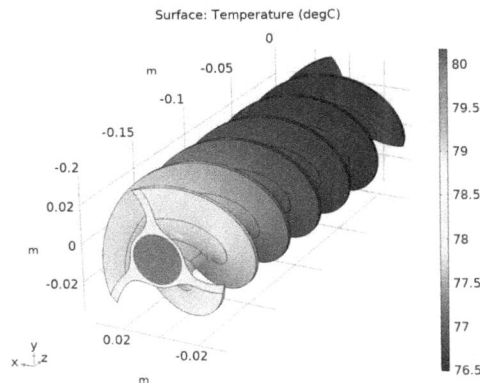

FIGURE 8.41. Volume temperature contours.

channel (Figure 8.42b). Transverse xy-plane plot in Figure 8.43b shows the low fin temperature and the well-developed circulation pattern, like the case with the external fins only. The vertical yz-plane in Figure 8.44b displays trends like those for Case Study 3 with the external fins only.

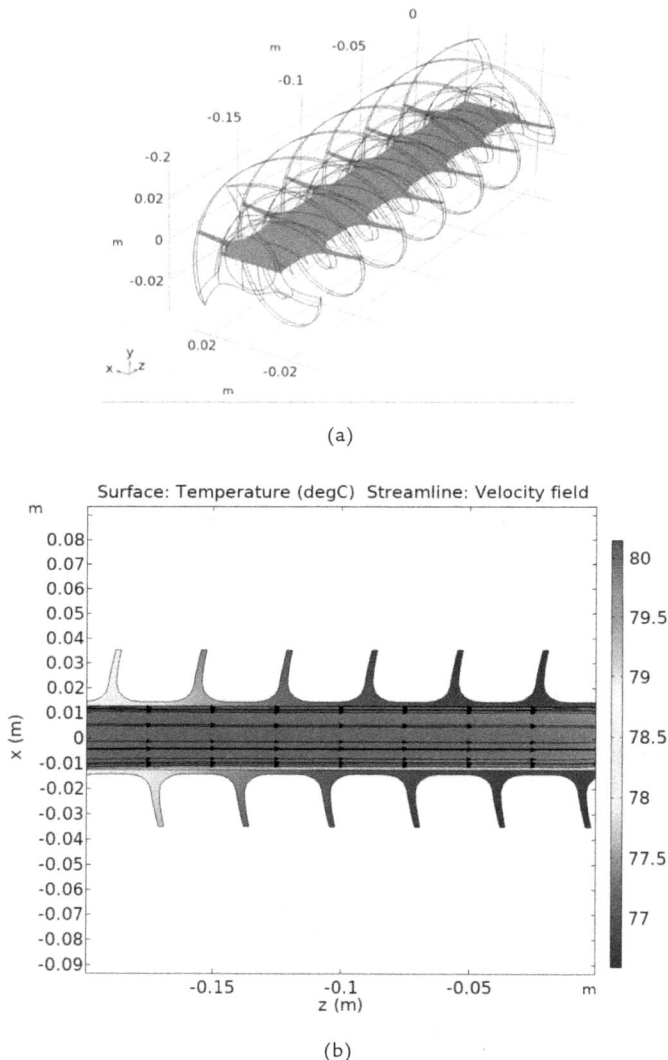

(a)

(b)

FIGURE 8.42. The zx-plane: (a) Selected cross section for the thermal data, (b) Temperature contours at the selected cross section, including the streamline velocity field.

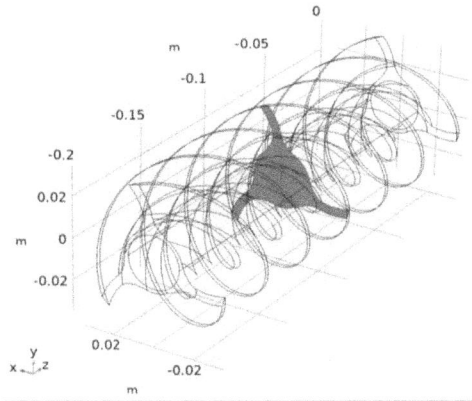

(a)

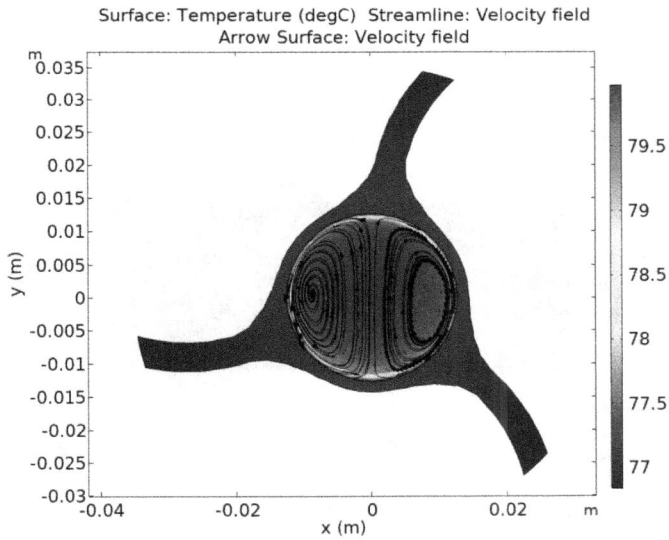

(b)

FIGURE 8.43. The *xy*-plane: (a) Selected cross section for the thermal data,
(b) Temperature contours at the selected cross section including the streamline velocity field.

(a)

(b)

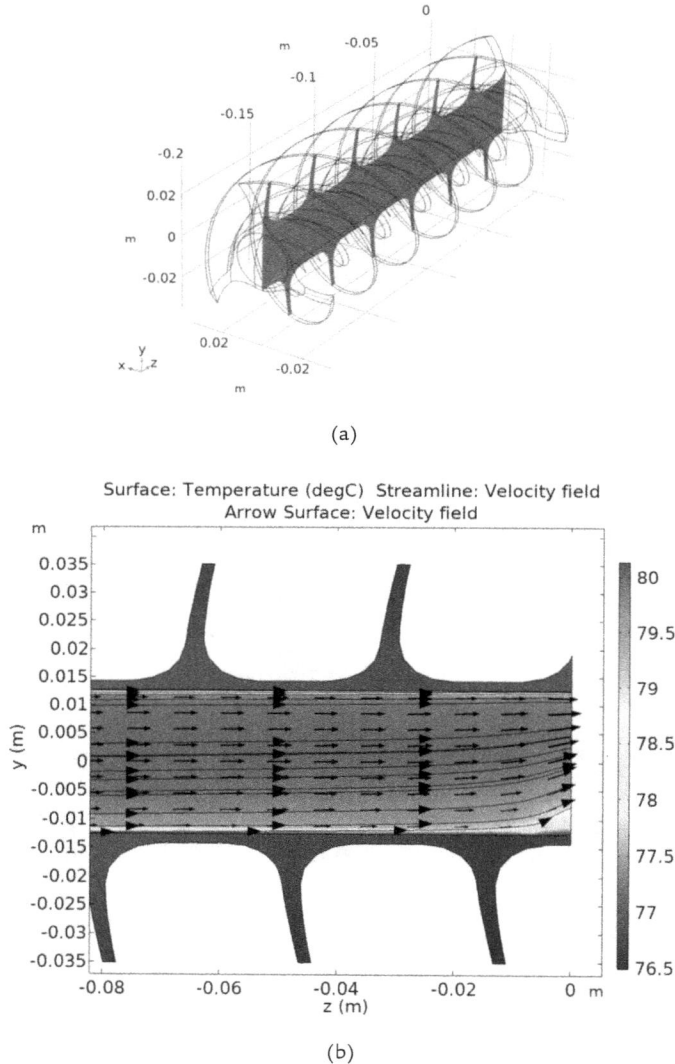

FIGURE 8.44. The *yz*-plane: (a) Selected cross section for the thermal data,
(b) Temperature contours at the selected cross section, including the streamline velocity field.

Due to the curvature of the rotini fin, the vertical and horizontal lines used to sample the temperature distribution at the inlet, middle, and outlet locations do not pass through the full extent of the fins. However, the temperature contour plots shown indicate that there is very little temperature gradient within the highly conductive fins. The vertical temperature profile at the outlet shows the asymmetry seen in the earlier cases due to the cooler fluid accumulation near the pipe's bottom (Figure 8.45b). The horizontal profiles are symmetrical (Figure 8.46b) and show the temperature in the exterior fins at pipe's middle to be 77.1—0.3 °C lower than that of the case with only exterior fins.

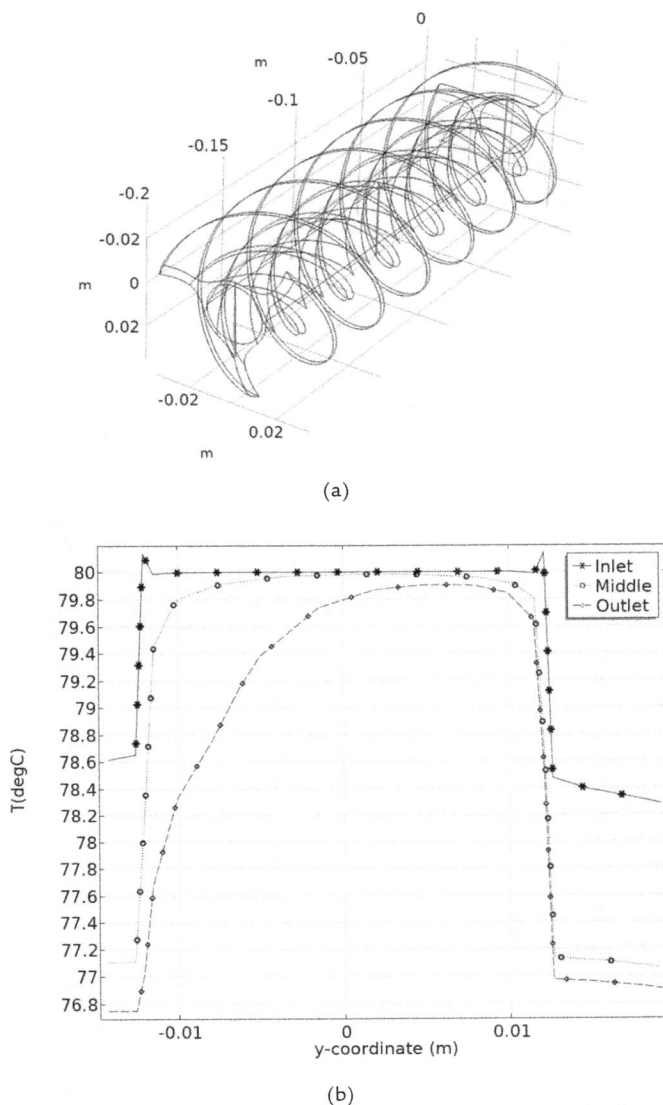

(a)

(b)

FIGURE 8.45. The y-coordinate: (a) Selected lines along the pipe diameter, (b) Temperature profiles along the pipe diameter at the selected lines.

(a)

(b)

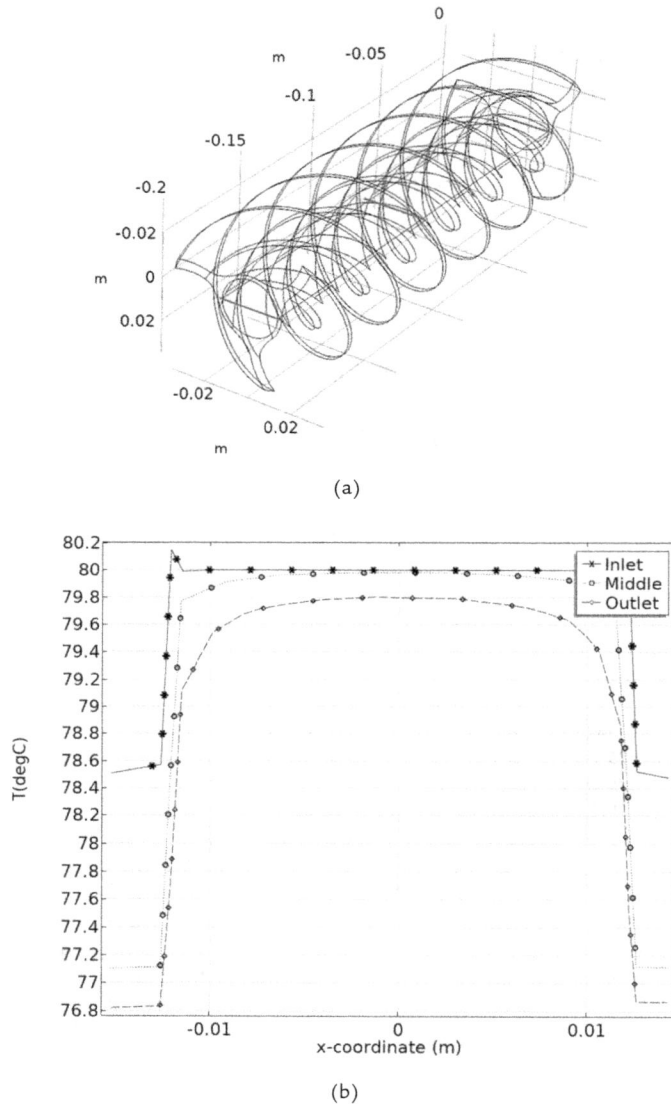

FIGURE 8.46. The x-coordinate: (a) Selected lines along the pipe diameter, (b) Temperature profiles along the pipe diameter at the selected lines.

Heat flux integration over the exterior surfaces gives 32.3 W. Divided by the 0.2-m length, the normalized heat loss per pipe's unit length is 161.4 W/m. This heat rate is only 7.1% less than of the pipe with straight exterior fins (173.7 W/m). Thus, both fin types offer similar heat dissipation ability.

8.7   Comparison between Case Studies 1 to 5

The heat loss in the various pipe geometries examined is due to the convective heat transfer to the air surrounding the pipe. This heat transfer rate is proportional to the area over which it occurs and the difference in temperature between the surface and the environment. In all cases considered here, the surface temperature is very close to the 80 °C of the incoming water flow (80 °C). Thus, it is logical to conclude that the rate of this heat loss should be primarily influenced by the exterior surface area (for the same pipe length). The external area measurements were done for all the pipe geometries examined using the COMSOL geometry measurement feature, as explained in Section 8.1.1, and the results are listed in Table 8.2.

TABLE 8.2. Pipe area and volume for the presented case studies.

Fin Type	Volume (m³)	Exterior Surface (m²)	Interior Surface (m²)	Total Surface (m²)	Exterior Surface/ Volume (1/m)	Interior Surface/ Volume (1/m)
Not-finned	2.79E-05	0.0179	0.0159	0.0341	643.09	568.73
Internal Fin	4.37E-05	0.0179	0.0316	0.0500	410.93	723.97
External Fin	7.61E-05	0.0662	0.0159	0.0828	869.32	208.51
Internal-External Fin	9.19E-05	0.0662	0.0316	0.0987	720.15	344.10
Rotini	9.67E-05	0.0620	0.0159	0.0788	641.03	164.17

Another quantity that can be of interest to calculate, and is included in the table, is the ratio of the exterior surface area to the volume of the pipe. This can give an indication of how efficiently the finned structure achieves the heat dissipation relative to the total volume of the structure. The higher the ratio, the more efficient the structure is in terms of its use of materials and consequently cost and weight. The same table also lists other geometrical characteristics extracted from the geometries.

Figure 8.49 compares the exterior surface areas of all the pipes. The not-finned and internal fin pipes both have the same areas. The external only and external-internal fin pipes have the same, but much larger, external areas. The rotini fin pipe has a slightly smaller area (by 6.3%) than that of the pipes with straight external fins. The heat rates per unit length are summarized in Table 8.3 and plotted for comparison in Figure 8.48. Similar variation in the heat rates as that in the external area can be observed. The chart shows that the heat rate per unit length is the highest for the internal-external fin pipe (178.6 W/m), followed by that of the external fin pipe (173.7 W/m). The rotini fin pipe takes the third place, with 161.4 W/m.

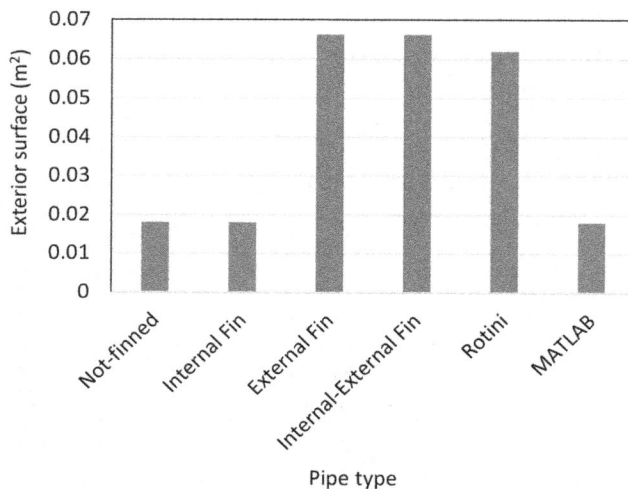

FIGURE 8.47. Comparison between the exterior surface areas for the case studies.

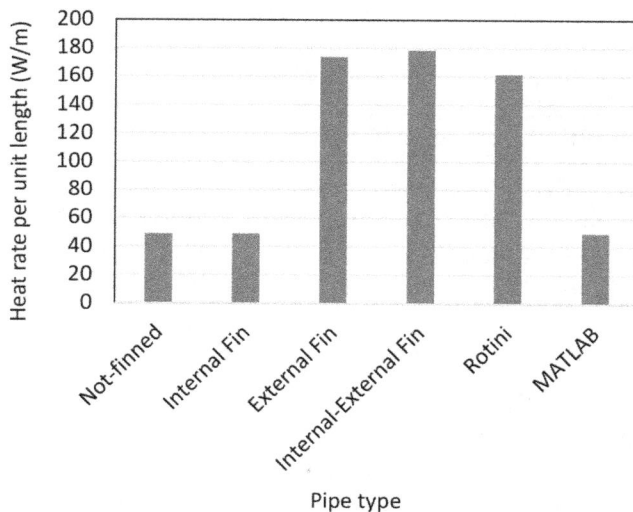

FIGURE 8.48. Comparison between the heat rates per unit length of the pipe for the case studies.

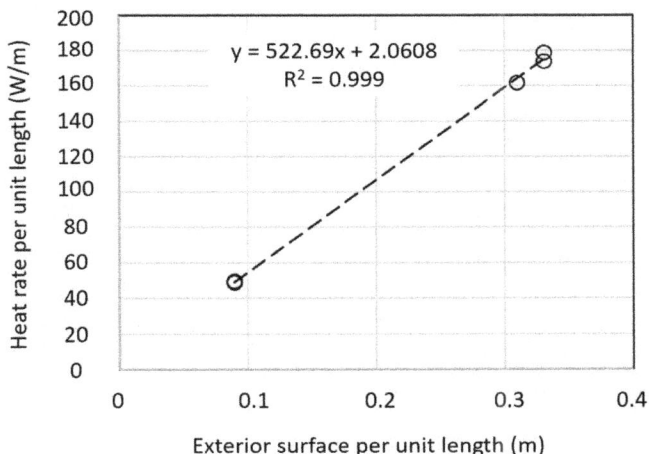

FIGURE 8.49. Heat rate per unit length of the pipe versus the exterior surface area for the case studies.

TABLE 8.3. Heat rate comparisons between the case studies.

COMSOL	Heat Rate (W)	Heat Rate perUnit Length (W/m)	Change in Heat Rate (W/m)	Exterior Surface (m²)
Not-finned	9.75	48.76	0.0%	0.0179
Internal Fin	9.79	48.95	0.4%	0.0179
External Fin	34.75	173.74	71.9%	0.0662
Internal-External Fin	35.73	178.64	72.7%	0.0662
Rotini	32.28	161.40	69.8%	0.0620
MATLAB	9.88	49.40	1.3%	0.0179

The relationship between the normalized heat rate and the exterior surface area can be better visualized by plotting one against the other (Figure 8.49). For the not-finned and the internal fin pipes, the heat rates are very similar, with the identical exterior surface areas, and so the points overlap.

While the points on this chart are clustered in two separate groups (not-finned and internal fin on the left and the rest on the right), a linear fit shows a very good correlation, confirming the hypothesis proposed above.

Furthermore, one can numerically verify the above relationship by using the equation for convective heat transfer, Equation (158), to relate the two parameters plotted in Figure 8.49—the heat rate per unit length and the exterior surface area. Let the former be Q_{sp} and the latter be A_{ext}.

Then, using Equation (158), one can relate these two quantities as follows:

$$Q_{sp} = [h_c(T_{ext} - T_{amb})][A_{ext}/L] \tag{158}$$

Evaluating the multiplier of the A_{ext}/L in the above linear relationship, one obtains 525 W/m, by assuming a 77.5 °C average external pipe surface temperature. This number is very close (within 0.5%) of the fitted line slope of 522.7 in Figure 8.49, confirming the accuracy of the hypothesis.

Figure 8.50 graphically compares the surface areas-to-volume ratios. The highest value is for the external fins only pipe. Most values are in the same range, except for the internal fin pipe, which shows a lower value due to the extra volume added by the internal fins that do not contribute to the exterior surface area.

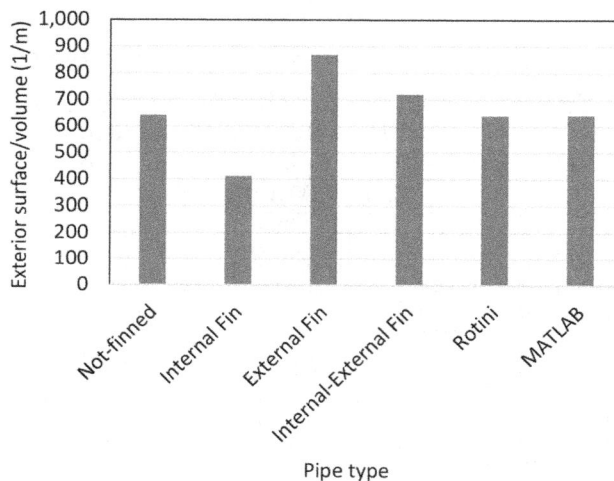

FIGURE 8.50. Comparison between the convective surface area-to-volume ratios for the case studies.

End Note

[59] https://pastafits.org/pasta-shapes/

CHAPTER 9

EXERCISES

Three sets of exercises are presented in this section. The first set contains problems related to water flow within a pipe, like the case studies presented earlier. The second set includes problems with a variety of geometries and boundary conditions, as well as problems related to the radiative energy transmission and absorption. The third set involves solving the problems introduced in the case studies using a different modeling tool than the one used earlier, and then comparing the results obtained to those that of the prior case studies.

9.1   Heat Transfer in a Pipe Exposed to the Solar Radiation

Like the case studies presented in this book, this section's exercises model water flowing through a pipe. To model the thermal response of this model accurately, a conjugate heat transfer model was developed.

Models are presented with increasing levels of complexity. Since the pipe is axisymmetric, if all the boundary conditions are also axisymmetric, a 2D axisymmetric model can be employed. However, a 3D model should be selected if non-axisymmetric boundary conditions exist. For example, if a pipe is exposed unequally to the solar radiation, a 2D axisymmetric model would not be suitable. If the flow within the pipe is modeled, the interface between the interior wall surface and fluid must be identified as a *Wall* in the flow model setup.

The cylindrical pipe is made of aluminum; it is 100-mm-long, with an inside diameter of 55 mm and outside diameter of 65 mm. The pipe exterior is exposed to the solar radiative energy specified in each exercise. The ambient (atmospheric) conditions are 25 °C and 1 atm. The pipe's initial temperature is 25 °C. The inlet water velocity is 15 mm/s, and its temperature is 35 °C. Water leaves the pipe at a fully developed regime and atmospheric pressure.

A transient solution is required for all problems. The 2D axisymmetric geometry is presented in Figure 9.1a. Figure 9.1b shows the line profile and Figure 9.1c shows the points for which the sample solutions are given in the exercises.

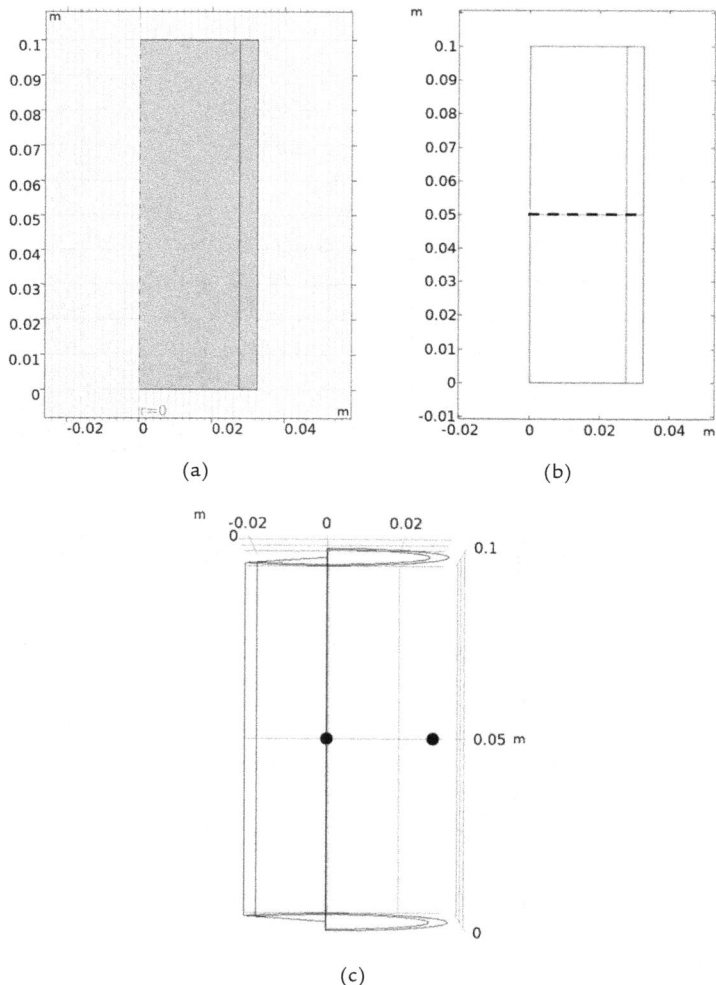

FIGURE 9.1. (a) 2D axisymmetric geometry, (b) Sample line profile, (c) Sample points.

9.1.1 Exercise 1—Constant Heat Flux and Single Surface

Obtain a transient temperature solution for the cylindrical pipe described in the introduction above. Use the 2D axisymmetric model shown in Figure 9.1a. All external cylindrical surfaces are exposed to the scattered radiation from the sun and to the ambient conditions. The exterior surfaces transfer heat by convection mechanism with the convection heat transfer coefficient of 10 W/m²K.

The pipe exterior surface emissivity is 0.8. The solar radiation intensity reaching the atmosphere at this latitude and time is equal to 1,200 W/m². Assume that this is a cloudy day, with the solar light scattered and with 35% of the incoming solar radiated energy absorbed by the moist air before reaching the modeled object.

Sample Solution Result: Presented in Figure 9.2 is a sample solution output that shows the radial variation of the temperature along the transverse midplane (Figure 9.1b) of the pipe at 1 min and 10 min.

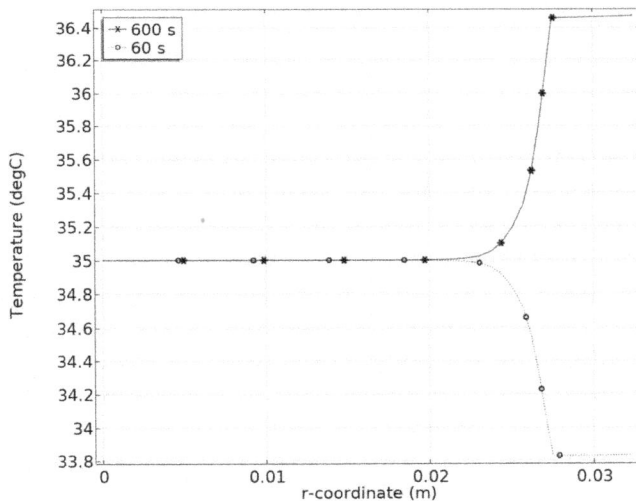

FIGURE 9.2. Radial temperature profiles in the middle of the pipe after 1 min and 10 min.

9.1.2 Exercise 2—Constant Heat Flux and Multiple Surfaces

For the problem described in Exercise 9.1.1, present a transient solution in which top, bottom, and end surfaces are exposed to the radiated energy from the sun (1,120 W/m²) absorptivity of these surfaces is 0.9. Compare the total heat absorption by the pipe in this case study with that of the Exercise 1.

Sample Solution Result: Figure 9.3 shows the sample solution output for the radial temperature variation along the transverse midplane (Figure 9.1b) of the pipe at 1 min and 10 min.

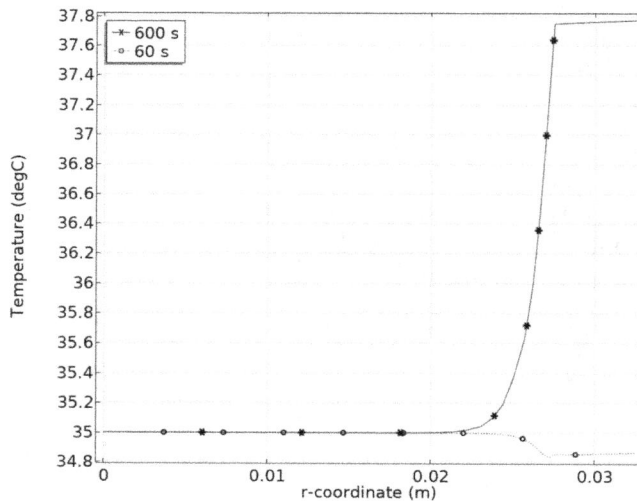

FIGURE 9.3. Radial temperature profiles in the middle of the pipe after 1 min and 10 min.

9.1.3 Exercise 3—Spatially Variable Radiative Heat Flux

For the problem described in Exercise 1, if the top, bottom, and end surfaces are exposed to the radiated energy from the sun (1,120 W/m²), present a transient solution, and predict the temperature profiles along the pipe radius and length after 120 s. Assume that 25% of the sun's radiated energy is absorbed equally by all surfaces. Choose an appropriate model type for this case. Can one use the 2D axisymmetric model or is the full 3D model required? Is it possible to take advantage of any symmetry to reduce the model size or dimension?

Sample Solution Result: Figure 9.4 shows the sample solution output for the radial temperature along the transverse midplane (Figure 9.1b) of the pipe at 1 min and 10 min.

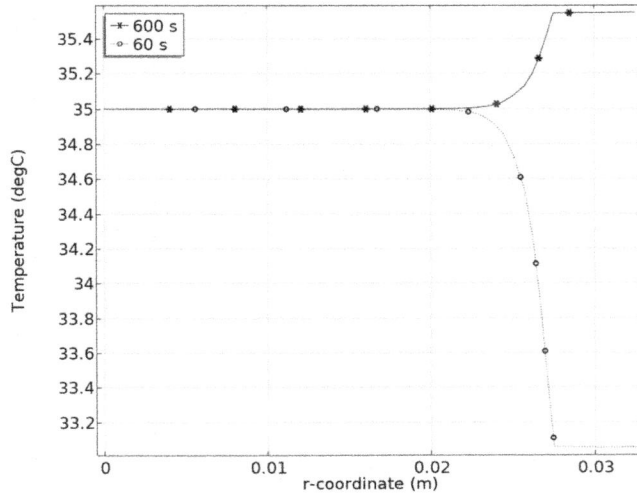

FIGURE 9.4. Radial temperature profiles in the middle of the pipe after 1 min and 10 min.

9.1.4 Exercise 4—Variable Ambient Temperature

For the problem described in Exercise 1, assume that the top, bottom, and end surfaces are exposed to the radiated energy from the sun $(1,120 \text{ W/m}^2)$, and 25% of the radiated energy is absorbed equally by all surfaces. Assume that the ambient temperature increases linearly from 25 to 35 °C during 10 min, beginning from the start of the modeled time. Present a transient solution that predicts the transient temperature over an extended time interval during which the system has nearly reached the steady-state. Predict the heat flux from the upward and downward interior surfaces. Plot the radial temperature at the time the steady condition is reached.

Sample Solution Result: Presented in Figure 9.5 is a sample result showing how the temperature varies over time at two points located at the fluid and pipe wall centers located in the middle of the pipe's length (Figure 9.1c). As a reminder, note that the pipe thickness is 5 mm, and the external diameter is 65 mm.

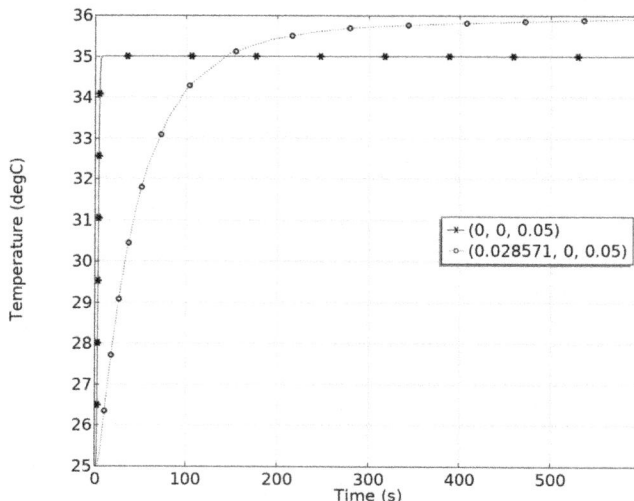

FIGURE 9.5. Transient temperature profiles in the middle of the fluid and pipe at the fluid and pipe centers.

9.1.5 Exercise 5—Variable Heat Convection Coefficient and Ambient Temperature

For the problem described in Exercise 1, assume that the top, bottom, and end surfaces are exposed to the radiated energy from the sun (1,120 W/m^2), and 25% of the radiated energy is absorbed by all surfaces, except one of the pipe's ends is partially shaded and thus only receives 10% of the radiated energy. The top and bottom surfaces as well as the other end of the pipe receive equal amount of energy.

Assume that the ambient temperature increases linearly from 25 to 35 °C in 10 min. The heat transfer convection coefficient varies linearly during the same time as well, increasing from 5 to 15 W/m^{2}K. Present a transient solution, after reaching the steady conditions to predict the heat flux from the upward and downward facing surfaces. Identify the time at which the temperature stabilizes, within 2% of the steady value. Calculate the heat flux at the bottom of the pipe when the solution becomes steady.

Sample Solution Result: The sample solution output is presented in Figure 9.6 along the transverse midplane (Figure 9.1b) of the pipe at 1 min and 10 min.

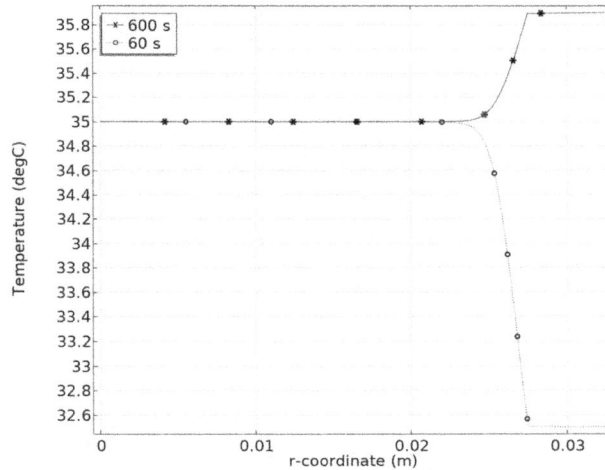

FIGURE 9.6. Radial temperature profiles in the middle of the pipe after 1 min and 10 min.

9.1.6 Exercise 6—Temperature-Dependent Thermophysical and Ambient Properties

For the problem described in Exercise 1, the ends of the pipe are partially shaded and only receive 10% of the radiated energy. The top and bottom surface emissivities are 0.75 and receive only 75% of the sun's radiated energy of 1,120 W/m². The thermal conductivity of the pipe increases linearly 0.5% for every degree Celsius above the ambient temperature of 25 °C, from 5 to 15 W/m²K and then remains constant. At the ambient temperature, the conductivity is equal to 201 W/mK.

The heat capacity of the pipe increases linearly 1% with every degree Celsius temperature increase, from 25 to 225 °C and then remains constant. The heat capacity at the ambient temperature (25 °C) is 922.5 J/kgK. Assume that the ambient temperature increases linearly from 25 to 35 °C over the first 10 min and then remains constant. The heat transfer convection coefficient varies linearly during the same time as well, increasing from 5 to 15 W/m²K.

Present a transient solution, after reaching the steady conditions to predict the heat flux from the upward and downward facing surfaces. Identify the time at which the temperature stabilizes. Calculate the heat flux at the upward and downward surfaces of the pipe when the solution becomes steady.

Sample Solution Result: The sample solution output is presented in Figure 9.7 along the transverse midplane (Figure 9.1b) of the pipe at 1 min and 10 min.

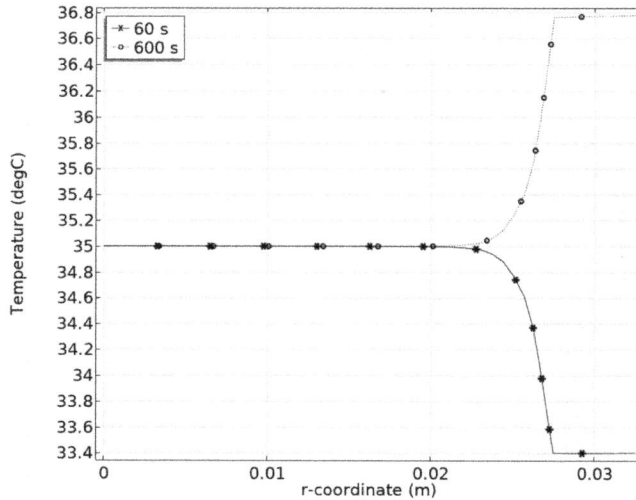

FIGURE 9.7. Radial temperature profiles in the middle of the pipe after 1 min and 10 min.

9.1.7 Exercise 7—Non-Axisymmetric Model

For the problem described in Exercise 6, the water temperature increases linearly from 35 to 55 °C over the first 10 min.

Sample Solution Result: A sample solution is presented in Figure 9.8 along the transverse midplane (Figure 9.1b) of the pipe at 1 min and 10 min.

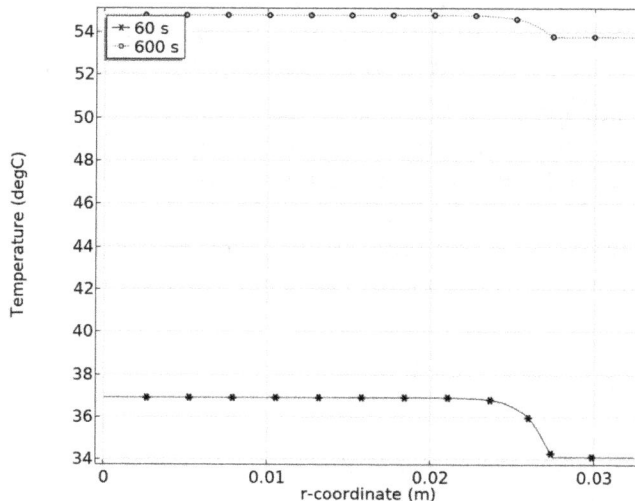

FIGURE 9.8. Radial temperature profiles in the middle of the pipe after 1 min and 10 min.

9.2   Heat Transfer in Various Geometries

9.2.1   Exercise 8—Heat Transfer from a Pipe with Extended Surfaces

In this exercise, the effect of extended surfaces on the heat transfer is investigated. The intention is to validate the hypothesis that the heat transfer rate from the pipe's exterior surface is linearly related to the extended surface area.

Create a pipe with four 1.5-mm-thick straight fins, equally distributed around the pipe exterior, each fin of geometry like that in Case Study 3 in Section 8.4. The pipe material, dimensions, and the boundary conditions are the same as those of Case Study 1 in Section 8.1 (Figure 8.1). Adjust the fin length so that the exterior surface area is twice that of the pipe in Case Study 1. Represent the water flowing through the pipe by a fixed interior pipe temperature of 80 °C. Compare the result obtained with that of predicted by the linear fit in Figure 8.49.

9.2.2   Exercise 9—Heat Transfer from a Pipe in a Heat Exchanger

A 150-mm long aluminum pipe with an external diameter of 30 mm and internal diameter of 25 mm is located inside a heat exchanger. The pipe's exterior surface is exposed to a constant temperature of 95 °C. Water at 10 °C enters the pipe at one end with a flow velocity of 5 mm/s. Water leaves the pipe at the atmospheric conditions. The initial temperature for water and pipe is 10 °C. Perform a transient analysis and predict the water temperature versus the time at three points located at the distance ratios of 0.5, 0.75, and 0.95 of the pipe's length from the inlet, along the pipe's center axis. Calculate the average water temperature at the outlet.

Sample Solution Result: A sample solution is presented in Figure 9.9b along the transverse midplane (Figure 9.9a) of the pipe at 5 min. The average water temperature at the transverse midplane is 45.2 °C.

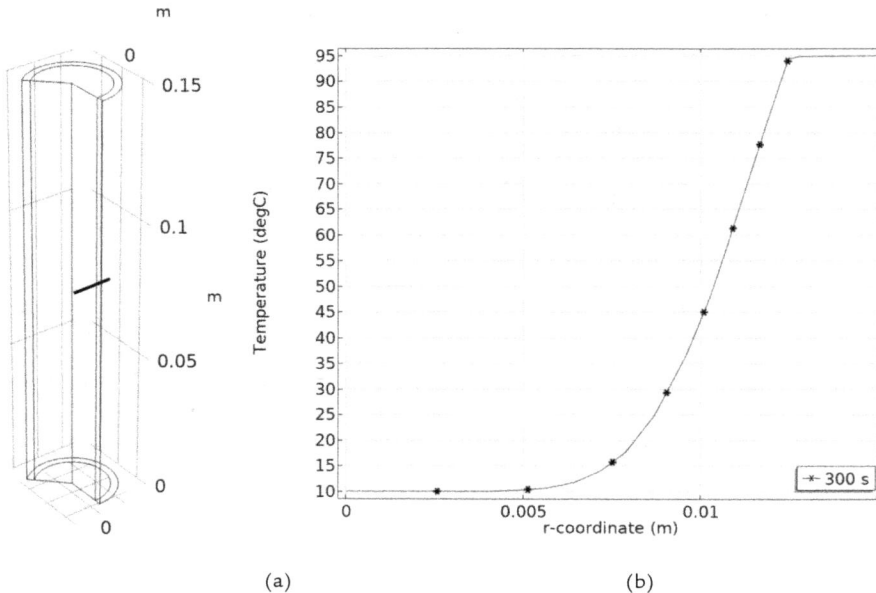

(a)   (b)

FIGURE 9.9. (a) Line profile, (b) Radial temperature profiles in the middle of the pipe after 5 min.

9.2.3 Exercise 10—Heat Transfer from a Solid Cylinder

A glass (silicate) cylinder with a diameter of 55 mm and length of 100 mm is taken out of a furnace and is at the initial temp of 700 °C. The cylinder is exposed to the ambient conditions. Model how the cylinder is cooling. (a) How long will it take for the maximum temperature in the cylinder to reach 100 °C? Incorporate the radiative heat transfer; assume the emissivity of 0.9 from all the emitting surfaces. (b) How are the results affected if the radiation is neglected? (c) Investigate the effect of introducing forced convection, through a parametric study, with the convection coefficient values of 10, 20, and 30 W/m²K.

Sample Solution Result: A sample solution is presented in Figure 9.10 along the z-coordinate at the midplane (Figure 9.9a) of the cylinder at 2 min and 10 min.

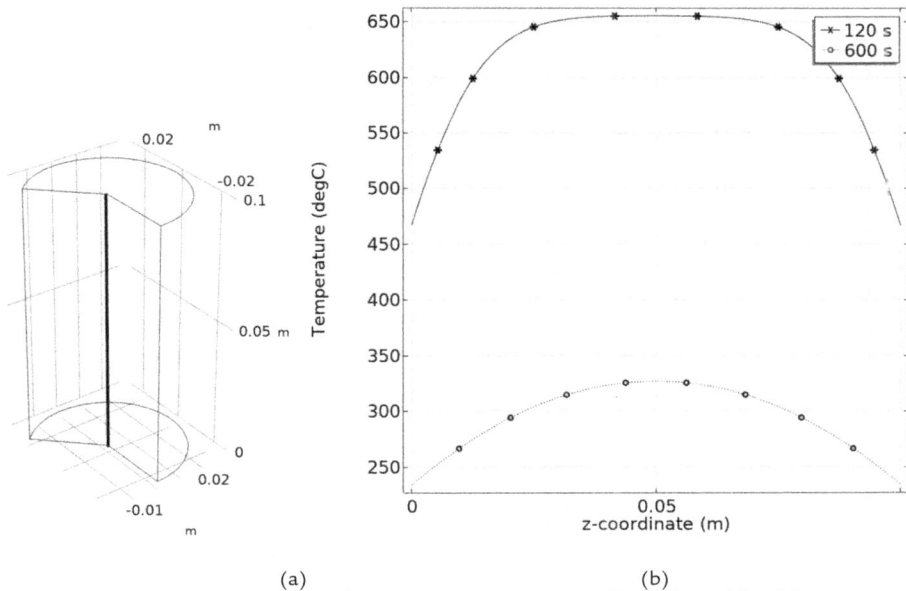

(a) (b)

FIGURE 9.10. (a) Line profile, (b) Axial temperature profiles in the middle of the cylinder after 2 min and 10 min.

9.2.4 Exercise 11—Energy Absorbed in a Cavity

The absorption coefficient of the interior surface of a black-body cavity located in an environment at the standard conditions is 0.25. Only 3% of the energy hitting the interior surface is transmitted to the environment (Figure 2.9). Assume that radiant energy beam is directed through a 1-mm diameter hole into the sphere, remaining inside it, until its energy is fully dissipated. Calculate: (a) The number of times the beam hits the interior surfaces until 99% of the initial energy is absorbed by the cavity; (b) The number of times the beam hits the interior surface until its energy is at 5% of the initial value; and (c) The total energy leaked to the environment outside the sphere.

Sample Solution Result: A sample solution is presented in Figure 9.11.

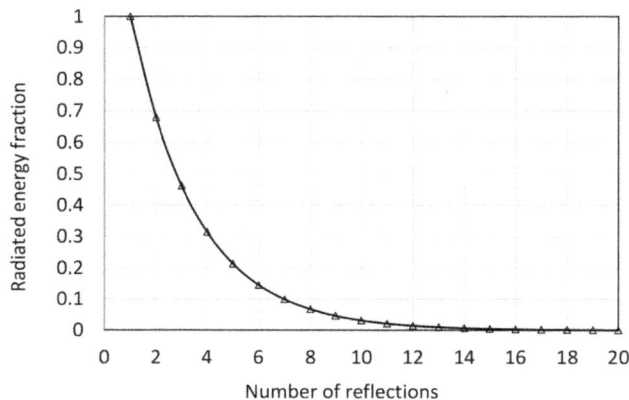

FIGURE 9.11. Solution sample for the radiated energy versus the number of number of reflections.

9.3   Modeling Approach Comparisons

9.3.1   Exercise 12—The MATLAB Heat Transfer Problems Solved with COMSOL

For the MATLAB case studies presented in Chapter 6, create models in COMSOL that represent the transverse cross-sections of the pipes. Assume the same materials and exterior boundary conditions as were used in the studies. Assume a constant temperature of 80 °C for the pipe's interior surfaces. List the simplifying assumptions if any were required to complete the analyses. Estimate the heat loss per unit length of the pipe for these models and compare them to that of the MATLAB.

9.3.2   Exercise 13—The COMSOL Heat Transfer Problems Solved with MATLAB

For the COMSOL case studies presented in Chapter 8, create models in MATLAB that represent the transverse cross-sections of the pipes. Assume the same materials and exterior boundary conditions as were used in the studies. Assume a constant temperature of 80 °C for the pipe's interior surfaces. List the simplifying assumptions if any were made to complete the analyses. (a) Estimate the heat loss per unit length of the pipe for these models and compare to that of the COMSOL. (b) Vary the interior pipe temperature from 75 to 85 °C and plot the heat loss rate versus the interior temperature. Is there a point where the predicted heat loss rate matches that of the COMSOL?

9.3.3 Exercise 14—The MATLAB and COMSOL Heat Transfer Problems Solved Analytically

Using an approach like that employed in Section 6.2.3, develop solutions to the MATLAB and COMSOL Multiphysics models presented in the Section 9.1 exercises above. You can also try to apply any of the analytical approaches including the ones described in Appendix A. State the simplifying assumptions when attempting the problems analytically. Compare the analytical results to those of the MATLAB and COMSOL solutions.

10

LEAN SIX SIGMA IMPLEMENTATION

P rojects or product lifecycles start with the idea conception and end when they are fully operational, with the last stage encompassing product disposal. The main objective for such a cycle should be adding value to the company's bottom line, creator, consumer, and environment (i.e., all involved). For the products or processes to improve, the baseline characteristics should be identified, or measured; otherwise, the deviations from the initial state will not be revealed. If these historical data are collected and organized in a meaningful fashion, past, current, and future trends can be identified. This way, the producer can evaluate the tipping points of the trends and make educated decisions regarding product management. You may have noticed that some products face a quicker decline than others after they become mature. The maturity has different levels for different products—even products within the same category do not follow the same trends. Developing effective methodologies to make this assessment possible are important steps as part of product improvement cycle.

10.1   Introduction to the Concepts

The Lean Six Sigma concepts provide tools to achieve such optimization by making it possible to measure process progress, quantify the deviations from the baseline, and predict its effects on the process trend. In these studies, the three characteristics to consider are *quality*, *time*, and *cost*. Internal and external surveys are conducted to determine how a service or product has performed, how it has progressed in terms of usefulness, and how successful it has been to attract and to maintain the market support. Having said that, the majority of the products that are both *wanted* and *needed* may not make it in the long run for the lack of support from their environment. Supply and demand are inversely related; therefore, in a *healthy* organization, *meritocracy* should be the priority when introducing a product. This is directly related to the *culture* of the place. If the culture is not conducive to such vision, neither good products nor good people can *thrive* or *survive*.

Empirical statistical techniques are used to analyze the collected *qualitative* and *quantitative* data. These data are then to be analyzed to identify the *critical-to-quality* characteristics or the variables that serve as input to the process model. This approach also helps to isolate those variables that are *trivial* and should be eliminated or emphasized less in the decision-making process. There are occasionally qualifying factors that affect the decision, such as ethical considerations, which will be discussed later in the concluding remarks.

The concept of a quantifiable study based on critical variables can be explored further by the example relevant to the topic that currently occupies most of the world: COVID. Most of the world is dreaming of returning to their normal lives, and vaccination is perhaps the way to get there. Consider then the vaccination process. Vaccination sites are set up, information about them is posted on-line, and so people can register for an appointment to have the vaccine product administered to them. Suppose that the authorities in charge of a particular region plan to have 1,000,000 people vaccinated within 90 days of the product's release. Assuming a constant vaccination rate, each day, an equal number of doses would be administered by the health professionals.

Administrators record the people who received the vaccine, keeping track of the daily delivered dose numbers. For the first couple of weeks, the vaccination process is going as planned. Imagine then that after two weeks, something goes wrong: The infection rate that was under control seems to experience a reverse trend, with the rate of infections increasing. This change can only be observed because this process is being monitored. Further investigation shows that new virus variants, with increased transmission rates, have been discovered circulating within the population. The officials now understand what has happened, including the human factor of some people losing their *trust* in the vaccine product's effectiveness—they believe it does not work as expected. To address the issue, either the current vaccine is to be modified, a new one is to be introduced to the market, or one must accept the associated risks and carry on with the current vaccine.

From the consumer point of view, they can choose to follow or not to follow the suggested hygiene regimens, stay away from the populated places, and comply with the posted public health protocols. Assume the infection rates continue rising, and people know about it because of the accurate recordkeeping system. If the initial reasoning has not changed, the officials may start looking into how people are dealing with the situation. If people keep organizing large parties, which are known to spread the virus, then they are the main source of the problem. In other words, for whatever reason, the disease control trend has deviated from the set path, and the variation is significant. It is then concluded that something is not working with the current approach and a correction should be made as soon as possible to be able to experience the *normalcy* again soon. One can then propose measures, such as only allowing remote interactions that use on-line technologies.

The same approach can be brought to engineering applications. Engineers decide upon a target value and strive to achieve it by: (a) Knowing that the improvement is necessary and feasible; (b) Identifying the areas in which the improvements are needed; (c) Deciding if the improvements can be made; (d) Determining if the enhancements should be made; (e) Measuring or predicting the maximum rate of return in the identified improvement areas; (f) Analyzing how the changes affect the overall bottom line; (g) Refining the methodologies to introduce revisions by making educated decisions; (h) Controlling the output by monitoring the process; (i) Standardizing the processes and establishing new best practices; and (j) Integrating the methodologies throughout the process(es) or operation(s) by allocating appropriate resources, such as expertise, time, and funding.

The successful product of this effort is an improved relationship between the *cost*, *quality*, and *time* achieved by eliminating the unnecessary steps (*wastage* or *redundancy*)—the visible or hidden steps that add no value to the experience [60]. The main potential sources of waste are *Transportation*, *Inventory*, *Material*, *Waiting*, *Overproduction*, *Overprocessing*, *Defects*, and *Skills* (TIMWOODS). Being responsible citizens, engineers strive to reduce waste to the extent possible to respect the (a) nature, (b) people, (c) surrounding world, including the resources they indirectly interact with, and (d) the immediate environment. The value entitlement defines this interaction in the form of services, products, or experiences, processes, and the responsibilities of individuals to respect the said steps.

Lean Six Sigma defines quality as the state of the realization of the full value of the entitlement in all aspects of the relationships. *Entitlement* is the right value of the expectation, which takes the form of *utility* (form, fit, and function), *access* (volume, time, and location), and *worth* (economical, emotional, and intellectual). Entitlement is what one should obtain given the available resources. It is the rightful level of expectation of every aspect of a relationship.

One way to implement Lean Six Sigma is to design *smart* experiments. Saying *smart* here not only means a synonym for *clever*, but it is also a memory aid, standing for *Specific, Measurable, Attainable, Relevant,* and *Timely* (SMART):

(1) *Specific*: One knows exactly what they will be doing—the scope is defined clearly.

(2) *Measurable*: One would be able to collect good quality data.

(3) *Attainable*: Everything one plans to do is within their capabilities and they are aware that there are things out of their control that may interfere with the process.

(4) *Relevant*: The project addresses some of the needed deficiencies and its usefulness is confirmed.

(5) *Timely*: It can be completed within an acceptable or set timeframe.

For example, one can try to apply these considerations to the modeling work reported in this book, where thermal models were developed using three methodologies.

(1) *S*: The model geometry, boundary conditions, and desired results are clearly defined: temperature at specific locations and times needs to be predicted.

(2) *M*: The MATLAB PDE, COMSOL Multiphysics FEA, or analytical tools can be employed to calculate spatial-temporal data within a reasonable accuracy.

(3) *A*: The infrastructure for the three possible tools is in place (i.e., the required software is available, the user has the needed skills, after reading this book, and the computer has adequate memory); however, not all the infrastructure is available, neither is the expertise to use it professionally.

(4) *R*: It is known why the temperature information is needed and how it will benefit the project.

(5) *T:* Some of the tools are available on-site, but there are uncertainties (e.g., the solution may take too long, there is not enough computer memory available, the processor is not sufficiently powerful, or if they do not have the skills to solve the problem in time). The reservations are related to the tool's costs, human resources, and equipment. This step should be carefully investigated, and the pros and cons identified.

Let us assume that the company decides to use one of the three analysis tools; they find that the results cannot be interpreted due to the lack of expertise. The root cause analysis should be performed so that the source of the error can be identified. Brainstorming may be used to start this process. The ideas generated by brainstorming that examine the process and methodology can be organized by means of the 5S methodology (*Sort*, *Set in order*, *Shine*, *Standardize*, and *Sustain*).

Imagine you organized your graduation party. You brought all the supplies: the teacups, plates, cutlery, napkins, bowl, cake, drinks, and party hats. After the party, you were left with cleaning up the premises. It looked like an *intimidating* proposition, but you remembered the 5S methodology you learned during your Lean Six Sigma training and decided to apply the knowledge to this scenario as follows:

(1) *Sort*: You sort items into appropriate categories and identify which ones you need to deal with immediately and which ones you can take care of at a future time.

(2) *Set in order*: You take those items that require immediate attention and separate them into categories, such as to *donate*, *recycle*, *throw away*, and *stored away*.

(3) *Shine*: You clear the premises to create an area where you can move and work safely.

(4) *Standardize*: You make a note of your procedure so that you can repeat it later when similar circumstances arise.

(5) *Sustain* (safety): You verify that the developed procedures follow the regulations (premises, organization, city, and country), can be easily followed, and are therefore sustainable. Furthermore, the new standards are safe to adhere to and respect the well-being of the parties in-

volved. For this reason, design standards were developed that include systems of measurements and acceptable tolerances.

10.2   Good Practices

The term *best practices* is a well-known expression in a variety of engineering disciplines in which the product CDIO (*Conceive*, *Design*, *Implement*, and *Operate*) lifecycle concept is used. When working on a model, analysis, or process of any kind, a variety of techniques may be employed, revised, and expanded upon. Implementing good practices is a systematic approach for planning, executing, and reporting design-related tasks to comply with certification requirements. The term *best practices* is the one most used. However, the author believes there are never best practices. There are only good practices that may be obsolete tomorrow and these will be replaced by more effective ones at a forward-thinking company or be still held (knowing that they are not effective, for a *backward* thinking company).

After recognizing the parameters affecting the process or product outcome, tests may be conducted and their results may be selected for further review. Good practices are more likely to lead to a useful outcome. In general, sensitivity analyses characterize the rate at which the dependent variables (*outputs*) change as the function of the significant critical variables (*inputs*). Select the most important contributing items, the ones that make the most impact—the few *critical-to-quality* variables—and eliminate the rest that are *trivial*. Improving processes and designs is an ongoing challenge. This process ensures performance improvements and the elimination of waste, while focusing on *critical-to-quality* characteristics. The concept of waste was introduced as part of the Toyota Production System that created management strategies where every employee was empowered to reduce waste or *muda*, a Japanese term for futility, uselessness, and wastefulness. The concept is also part of the Lean Manufacturing concepts.

When designing experiments, create a table encompassing the critical variables and decide on the tests and the number of repeats. The rows (m) of the table are associated with the experiments and the columns (n) with the critical process parameters. You may decide to run experiments for the complete sets of variables along with their combinations ($m \times n$); for example, for two and three sets of process parameters you can set up six sets of experiments. The effect of each critical variable on the dependent variable can then be analyzed using a regression analysis—a mathematical relationship that identifies the *goodness of fit* to the data by statistical tools. The next step is to report the key performance indices.

There is no preferred approach to design geometries; the process usually fits into three categories: (a) approach, (b) order, and (c) interface. The approach tells the story of the origin of the assembly or part, where and how it is created, and the environment in which it is grown to its full maturation. The order informs the successive steps that have been taken for the geometry to be generated—if it is *ordered* (each step is the steppingstone for the next steps) or *unordered* (steps are independent of one another). The *interface* tells the interconnectivity between the assembled parts and their relation to the new environment. The host environment in these scenarios may be the FEM specialized tools, while the originator can be either the CAD tool, FEA tool, or a combination of both.

As science progresses, the approach, order, and interface improve through the introduction of new commercial software packages in the analysis and geometry-generation fields. With this knowledge progression, the concept of standardization becomes even more important, since the cost associated with converting the geometries generated in the prior revisions of the specialized tools (FEM or CAD) becomes prohibitive. Projects are delayed when the geometries created with an older-version CAD tool cannot be easily translated to the ones compatible with the new CAD tool— the only acceptable version to a newly developed FEA tool. This concept may be extended to other types of models, where physics of any kind are investigated (e.g., Computational Fluid Dynamics, CFD). Although the community of the fields' specialists may propose workarounds—and the vendors attempt to introduce compatible products and added modules— costly challenges remain both in terms of human effort and project delivery timelines.

There are multiple steps to be taken on the way to an accurate heat transfer model. The geometry creation is among the first steps, and so it will affect all the subsequent ones. Thus, one must devote appropriate care to this stage of the model development. The geometry must be carefully reviewed before and after import into the analysis tool. One should be particularly careful if there is any change in the units used, such as a change from meters to millimeters.

Confusion with units has been known to cause trouble in the past. This has been the case in one well-known airplane accident in 1983. An Air Canada jet ran out of fuel at 41,000 feet but the plane's pilots managed to make a safe landing by gliding into an airport in Gimli, Manitoba. This became known as the *Gimli Glider case* [61]. The investigation found the cause was due to an error with fuel quantity calculations, which confused pounds and kilograms; this led to the accident because the plane's fuel gauges also malfunctioned. While this example is not related to the thermal analysis, it still shows how one small error can have enormous consequences.

Employing dimensionless analysis or variables (or parameters) when setting up the models is always a good practice. This approach facilitates the interface between multiple platforms, allowing for synchronization between the tools. Following this practice facilitates carrying out sensitivity analysis studies. When selecting parameter names, take care to choose meaningful ones that will allow you to correctly recognize each variable.

Another good practice to follow is to watch out for devoting excessive resources in the pursuit of negligeable issues. One needs to keep in mind the overall sense of the model uncertainty and avoid working on the areas, which are likely to have minute effects on the model predictions. Thus, if one can only estimate heat transfer coefficient to within 10 percent of the actual value, there is little benefit to measuring density to eight decimal places. Resources devoted to pursuit of the issues with little impact on the outcome could be better spent in other areas, producing the greater Return on Investment (ROI).

The last note is that the designers should always try to think ahead while they are in the middle of the creation process. They need to remember to occasionally step away from the day-to-day details they are focusing on and take a broader outlook. They should be asking themselves the following questions:

(1) What is happening next?

(2) What kind of accessibility features do I need to include in my design?

(3) Will I need to define additional boundary conditions or reinforce the structure?

(4) Do I need to incorporate redundancy systems for safety purposes, such as the ones seen in the Boeing 747 design?

(5) Do I need to check the historical data for lessons learned and compatibility of my design with the environment, such as the incident that occurred with the Challenger space shuttle's O-ring?

11

CONCLUSION

One common question that arises when a specialist decides to analyze systems for their thermal or mechanical responses is which FEA tool to select. As noted earlier, the more complex the geometries are, the more accurate analysis methods are needed. Finite element is an approach that can handle most of the shapes with challenging geometries. There are tools that take advantage of this technique. These tools can take on either an independent or hybrid approach. Most commercial tools (e.g., ANSYS and COMSOL Multiphysics) follow this strategy. There are other tools, such as programing languages, that can be employed to solve the problems by working directly with the physics relations (e.g., C++). Other tools are designed to integrate special capabilities by assisting with analyzing the problems incorporating the coding approach with specialized built-in functions. Thus, they take a semi-hybrid approach (e.g., the MATLAB *PDE Toolbox*). There are circumstances in which these tools can be combined with the commercial tools and further improve the modeling capabilities (e.g., COMSOL Multiphysics in combination with MATLAB).

11.1   Choice of FEA Tools

The question is determining which tool is a *better* one: it simply depends on the available resources and on the problem and its applications. For example, for educational purposes, it is important for students to become familiar with the fundamental physics before using any tools as a *black box*. If the analysts already gained a deep understanding within the field, such as senior or graduate students, they are to be guided differently and will likely approach the problems differently. In these cases, the first consideration is almost always the tool availability, which is often determined by the budget. Some tools, such as MATLAB, are relatively more education-friendly than the commercial FEA tools. Given the widespread usage of such tools, universities often have special agreements and perpetual licenses for them. This means that professors or students do not pay for them from their research grants, and if they had to, the prices would be affordable. In addition to the lower price, they do include discounts or trial versions for the students and for educational purposes.

Even though commercial FEA tools do have educational pricing, they are usually purchased as specialized tools as part of a research grant. On occasion, several academics can share the license and associated costs given that the floating license cost, which makes it accessible from multiple sites, is relatively more expensive than one tied to a single machine. In many cases, there are only a few seats available, meaning that a very limited number of students can simultaneously run the programs. When solving large models, High Performance Computer Virtual Laboratories (HPCVL) may be required. The problem files are sent in batches to these computers and are placed in a queue so that the solution may be attempted, and possibly achieved within hours or even days for more complex cases.

Using these virtual labs requires special memberships even within the university community and fees are to be paid per seat to maintain the special administration costs. An annual fee is normally paid to maintain the licensing rights and/or support for a commercial FEA tool; this is either in addition to the original perpetual fee (e.g., COMSOL Multiphysics, which requires annual fees for support and upgrades) or as an ongoing cost, where an annually updated license is required to operate the software.

Performing any consulting work requires special permission from the service provider (FEA company); this means that the license must be a commercial type versus the educational type. Some of these commercial tools let the user have multiple installations, with the possibility to simultaneously run several FEA models (e.g., four installations and two simultaneous runs for the commercial COMSOL Multiphysics users). Typically, FEA tools include free trials of up to 30 days.

The next consideration, and perhaps the first one in educational settings, is the technical support. Professors and students require support both in installation and use of the tools. Universities or academics provide the support the students need in most of the cases; however, the educational institutions and often professors need technical support from the software provider or the company that owns the product. In most of the cases, the technical support provided by the education-friendly tools is better than their commercial counterparts. It is also possible that the technical support provided by the FEA software providers is available through sharing the inhouse expertise, technical publications, and conferences, where users share their expertise. Ease of administration and flexibility to work remotely may also affect the decision as to which tool to use.

It is possible to suggest an improvement to the tool by submitting an enhancement ticket to the tool's technical support. These enhancements can be related to the documentation, presentation method, usage, formula, or analysis method and presentation. Almost all these tools are regularly updated, considering the submitted enhancement requests and based on the product roadmap. Commercial tools may follow a faster pace, depending on their product vision.

Some of the tools are equipped with geometry-creation facilities (e.g., COMSOL Multiphysics), making it possible to import, manipulate, and create the geometry within the tool. Geometry revision is a very important feature, especially when dealing with complex models. Occasionally, specialized CAD tools (e.g., SOLIDWORKS) are employed as separate tools in parallel with the commercial FEA tools to facilitate the geometry creation process. Dedicated connectivity can be provided between these two tools (e.g., *Livelink* for SOLIDWORKS or one of the specialized COMSOL Multiphysics *Modules*), establishing a real-time linkage between the CAD and FEA environments. Editing and rendering capabilities are other considerations when selecting the FEA tool. The possibility to write scripts or input files for the FEA tool to avoid repetitive tasks to minimize error and to easily modify the models as required, are also to be considered. Examples include the APDL input files that can be written and fed as inputs to the FEA models created in the ANSYS classic environment. Many novice users may not find this method user-friendly; therefore, the graphical user interface (GUI) becomes a user distinct feature (e.g., the ANSYS Workbench).

The user can (to some extent) modify the problems' physics by revising the underlying basic relationships. MATLAB is such a tool, where the user can interact through a GUI (e.g., the *PDE Modeler* application) and also can revise the underlying formulae as needed in the form of scripts. Programs such as MATLAB excel at carrying out any computations, particularly matrix manipulations.

Plotting options and quality of the plots that can be embedded in reports or be exported as images or PDFs, are to be considered when selecting the FEA tool. These diagrams can be either part of a script, such as the ones written for MATLAB, where the diagram type, labels (e.g., title, axis information and limits) are identified or the results can be exported from the FEA tool and imported to other tools to be processed (e.g., Tecplot, MATLAB, and statistical tools such as Minitab). The processing steps may include plotting data in different forms, curve-fitting, and performing regression and statistical analyses.

Both MATLAB and COMSOL provide facilities for setting up applications that provide a graphical user interface to simplify user interaction with the software functionalities. The MATLAB *PDE Modeler* application, which is supplied as part of the MATLAB *PDE Toolbox*, is an example of this; however, users can create their own applications that have buttons, dialog boxes, and mouse interactivity. Normally, these applications would be executed from the MATLAB command line; with the MATLAB *Compiler Module*, one can create stand-alone applications that do not require MATLAB to be installed.

COMSOL introduced its own application building capability with version 5.0 release in 2015. As with MATLAB, this capability is part of the base software package. Tools are provided to simplify the creation of these apps. The apps allow users who are not skilled in use of FEA software to carry out pre-defined analyses after varying inputs via the app interface. Thus, engineering designers may explore a variety of scenarios on their own, without having to interact with an analyst. For example, the pipe wall thickness can be defined as a parameter and the designer may experiment with different values to see what effect they may have on the pipe wall temperature.

As with MATLAB, these apps can run on any machine where COMSOL is installed; however, the intended audience for them are users who are likely not to have a full software installation. Thus, two options are provided (as extra modules): the COMSOL *Compiler* to create stand-alone applications or the COMSOL *Server* to allow running the apps remotely by connecting with the server via internal or external network.

In terms of carrying out FEA approach, one can summarize the comparison between MATLAB and COMSOL as follows. There is no doubt that a dedicated FEA tool, such as COMSOL, is better at its task than a more general tool, such as MATLAB. If both are available, the choice is clear. However, it is much more likely that a user with an FEA problem to solve has access to MATLAB and is wondering if they need to obtain a dedicated FEA tool (like COMSOL) to solve it, at some additional, substantial cost.

The answer depends on the complexity of the model. This includes the complexity of the geometry, boundary conditions, and physics interactions. For example, as was demonstrated in this book, even complex 3D geometries can be exported from a CAD tool in format and then imported easily into MATLAB. However, this is not the end of the task. One now must identify all the faces and cells (3D regions) of the model to assign the applicable boundary conditions and material properties. If one is importing something shaped like a simple pipe, that is not a problem, but if it is a component with several dozen faces, the challenge quickly multiplies, making errors much more likely.

Other aspects of carrying out FEA in MATLAB, such as the meshing process control and types of elements as well as the solution control, also present substantially fewer options, as expected. Finally, one must not forget that post-processing is also significantly facilitated in COMSOL. Identifying points or lines for which plots are to be generated is easy, with interactive graphical interface provided. In MATLAB, one must manually identify the exact coordinates of interest and then execute commands to extract data points and plot them, requiring knowledge of the right commands to use.

However, the good news is that if your problem is not exceedingly complex and you already have access to MATLAB, you can use this excellent general-purpose software to carry out FEA and obtain plenty of valuable results, without incurring any additional costs. Presumably, having read this book, you are also well-equipped with the knowledge of the techniques required to carry out this task.

11.2   Sustainable Designs

When creating thermal designs, creativity is as important as adhering to known and tested methods. If designers, engineers, doctors, and educators were to just follow *old-fashioned* knowledge and manufacturing techniques, humans would still live in caves. Although contemporary cave homes offer modern amenities within a primeval setting [62,63,64]. Thinking divergently is the reason for the exceptional creations.

Independent thinking in an unrestricted environment is an indispensable part of this process. The fuel of resources and experiences available to a creative mind ignited by its imagination drives the development of new ideas. Interconnections among diverse fields of study, such as art, engineering, design, and health, have brought us the innovative products to enrich our lives. Think about the lifestyle changes brought about by the introduction of smart phones and tablet computers such as the iPad.

Creative and independent thinking require *valor*, as there are often *pressures to conform* to the accepted practices. Historically, the brave and curious scientists and innovators made sacrifices to bring new ideas and a better life to humanity. In the seventeenth century, Galileo Galilei realized that the old concepts of planetary motion did not make sense. He considered the ideas proposed by Copernicus, as well as what he saw himself with the telescope he built. However, Galileo lived in a time when the Church wanted to protect the *status quo*, and so he was punished for his ideas. However, the ideas could not be suppressed and flourished despite all the reactionary efforts, for the *light of wisdom cannot be turned off*.

Innovative and responsible designs are not only rewarding for the designers who create them, but also beneficial to humanity and the environment. Looking around us, we observe numerous examples in which this brilliance of the human mind is seen. These projects show how our natural resources can be used responsibly. Here are some examples of ethical leadership in the thermal management field:

(1) Leadership in Energy and Environmental Design (LEED) certified designs that improve efficiency and health to achieve a sustainable environment—These initiatives have transformed a tornado-hit American city, Greensburg, Kansas, into a model of a green village [65,66]. They have transformed a fading clay pit in Cornwall, England, into a thriving green community through the construction of an eco-friendly park, museum, and indoor rainforest that educate people about the responsible use of the natural resources, such as composting waste, water treatment, and the use of geothermal and wind energy [67].

(2) Harvesting the energy of the Sun in the most remote and under-privileged villages, in places that are exposed to sunshine most of the year—Sichanloo is a remote location in arid rural Iran, with simple clay houses that have been decorated with high-tech rooftops made from photovoltaic cells provided by a government-subsidized project. These are part of the growing efforts to provide steady power for a fossil-fuel country that relies on oil and natural gas sources for 40% and 37%, respectively, of its energy usage. Sichanloo and similar communities are recovering from the noise and pollution that gas-fueled power generators have imposed on their lives for decades [68].

(3) Vertical gardens as a platform for planting, working, and shading environments—An example is Supertrees Grove at Gardens in central Singapore, which improves the quality of life by introducing greenery into this densely populated city. These trees not only provide homes with exotic plants and birds, but also exist in harmony with their surroundings, imitating a living tree by harvesting the solar energy with the photovoltaic cells and collecting rainwater for irrigation and fountain displays [69].

(4) Harvesting the energy from the waves using *PowerBuoy*, which can be connected to an electrical grid using power transmission cables in a deep-water environment—The *PowerBuoys* installed in Cromarty Firth, Scotland, can generate 3 MW of power. They convert the rising and falling of the waves into electricity. The *PowerBuoys* are aesthetically pleasing due to their low surface profile and small horizontal footprint, and can operate in severe conditions [70,71].

(5) Efficient residential apartments that people wish to live in, even though they lack basic amenities (such as a parking space or air conditioning, or they are located next to a train track)—An apartment building in Melbourne, Australia, was designed to keep warm in the winter and cool in the summer with the ultra-thick exterior walls shielding it from train noise. The building has a rooftop garden to provide additional insulation and a green environment [72].

(6) Passive house designs being incorporated into new building architecture or as a retrofit for the existing ones—These designs heat and cool the structure so as to minimize its ecological footprint. Examples are the Vauban residences in Freiburg, Germany, and Cornell's green 26-story high-rise campus on Roosevelt Island in New York City [73,74,75].

(7) Sustainable cities that are both aesthetically pleasing and functional—An example is the Dubai Smart Sustainable City project. It looks like a flower in the middle of a desert, with shiny roofs covered by the solar panels that generate about 200 MW of electricity [76].

11.3   Ethical Designs

As the concluding remarks of this work were being written, Virgin Galactic and Blue Origin had their first flights. Soon anyone (with sufficient funds) can become an astronaut [77,78]. While at this time, it appears that a lot of precious non-renewable resources are expended on just a few minutes of fun, one can expect that in the long term, the technology developed will help the humanity with space exploration for more practical purposes. For those among us who have always dreamed of exploring outer space, one can hope that the price of these adventures will someday become reasonable enough so that people without millions of dollars can go into space.

Over the past couple of years, it feels like we have transitioned into a new era, which we perhaps can designate as *AC* (After COVID-19). It is humbling to think that a microscopic, technically non-living entity wields such transformative power. As with any major crisis, it has speeded up technological development to a degree that did not seem possible before. Communication companies developed new ways for us to work and interact remotely. Delivery companies ramped up their capacity and created new approaches to shipping (for example, putting perishable items in insulated containers that are affordable and recyclable). Human ingenuity has been harnessed to its full capacity to fight this crisis, with unprecedented resources poured into vaccine development, bringing about positive results within record time. As the fraction of fully vaccinated is hovering around 60% to 70%, we are all waiting to see what happens next. What will the new AC era be like? Will we ever be able to stop wearing masks?

While the professionals may be rushed and pressured to deliver new designs and processes, they must remember their responsibilities to the public. Consider, for example, the case of asbestos. It is a naturally formed silicate material and was widely used in the thermal industry. This material is a very good electrical insulator and can resist extreme heat. Therefore, it had been used extensively in the construction of buildings until 1970s, even though its use had been under scrutiny as early as 1924, when English textile workers' poor health conditions were reported in the medical literature.

Due to its proven adverse health effects on those who handled it, leading to a condition known as asbestosis, it has been prohibited in many countries, even though its use goes back to the Stone Age, when it was employed to strengthen ceramic pots. It was reported that every year, 100,000 people lose their lives globally due to exposure to this material. It is the number one cause of work-related fatalities in the world; however, the grievous effects take years to develop, making it difficult sometimes to link them to the original cause [79].

The production of asbestos in the United States stopped in 2002. In Quebec, Canada, there is a small mining town which until 2020 was named Asbestos in honor of its primary product. It is now known as Val-des-

Sources. Its asbestos mines reportedly closed in 2012, with Canada banning the mineral's use in 2018. It appears that the city *rebranded* their new product, which is a specific type of the mineral, to chrysotile, also known as *white asbestos* (used in brake pads, asbestos cement roof sheets, and industrial tanks) to disassociate it from the carcinogenic mineral, insisting that it is less *dangerous* than the original asbestos. Nevertheless, there are still concerns about safety of this product [80,81].

Based on the World Health Organization (WHO) report, this material was widely used in cement building materials (90%) and friction materials (7%), mainly in the developing countries by the end of 2014 [82]. Some developing countries, such as India, still encourage the use of this material in the construction of their buildings and other countries, such as Russia, continue to mine it.

The asbestos example shows the challenging decisions faced by professionals. When making these decisions, the efficiency of the process should not have the *veto* power over all other considerations. As everything else in history, humanity's ethics have been evolving. While people may have felt that some practices were not ethical, they may have been permitted to carry on as part of the majority-based ethics, despite their detrimental effect on some minority groups (think of slavery and cotton, of women and voting rights, of indigenous peoples and colonization, of intimidation practices and employment). With the evolution of ethics, there is some hope, for it appears that new virtue-based ethics have been taking hold in recent years.

Professionals who are to deliver the work ethically ought to acquire all the information needed to develop safe processes that use safe materials. Those who produce materials must make accurate information about them available to all. Those involved must be given the necessary training for safe handling of these materials, for their own good and that of the product. Based on the virtue-based approach, professionals must resist peer-pressure and, within regulations, do their best to avoid use of unsafe materials.

Any of the ethical-based decisions in the thermal management field, which is a part of heat transfer science, requires an in-depth understanding of the thermal process (heat-material interaction) for there will be occasions in which the material or the process must be redesigned to safeguard the environment and those who depend on it. In this work, they can rely on the rock-solid foundation of the thermal sciences.

The complexity of the world around us is unimaginable. Even a vacuum, which is empty space, may not be as empty as we thought. Physicists are now hypothesizing it is full of energy. They are discovering patterns that can connect the microscopic world of quanta with the large-scale phenomena that we can experience with our senses. The most radical ideas are often brought to light by *misfits*, those who are blessed with the power of curiosity, who can think critically about their surroundings and are not preoccupied with fitting in. They fight the darkness with their perseverance, patience, and prudence, with their inner light; for, *the inner light cannot be distinguished*.

End Notes

[62] https://www.bing.com/images/search?q=cave+homes&qpvt=cave+homes&FORM=IGRE

[63] https://www.britannica.com/topic/The-Flintstones

[64] https://inhabitat.com/amazing-transformation-of-a-decrepit-cave-into-a-beautiful-modern-home/

[65] https://new.usgbc.org/leed

[66] https://www.usgbc.org/articles/rebuilding-and-resiliency-leed-greensburg-kansas

[67] https://en.wikipedia.org/wiki/Eden_Project#Environmental_aspects

[68] https://www.thenational.ae/world/iran-looks-to-solar-alternative-for-energy-1.456219

[69] https://www.gardensbythebay.com.sg/

[70] https://www.oceanpowertechnologies.com/

[71] https://www.waveenergyscotland.co.uk/

[72] https://www.nbcwashington.com/news/national-international/Five-Innovative-Green-Projects-from-Around-the-World-375893561.html

[73] https://www.theatlantic.com/magazine/archive/2016/01/the-height-of-efficiency/419124/

[74] https://tech.cornell.edu/campus

[75] https://www.greenbuildermedia.com/buildingscience/passive-house-is-it-worth-the-upfront-cost

[76] http://www.middleeastgreenbuildings.com/11864/top-10-green-building-projects/

[77] https://www.virgingalactic.com/mission/

[78] https://www.blueorigin.com/

[79] https://www.asbestos.com/asbestos/statistics-facts/.

[80] https://www.nytimes.com/2020/10/21/world/americas/asbestos-quebec-canadian-town.html.

[81] https://www.washingtonpost.com/world/the_americas/asbestos-quebec-val-des-sources-canada/2020/10/19/29e32e5e-1244-11eb-bc10-40b25382f1be_story.html.

[82] www.who.int/ipcs/assessment/public_health/chrysotile_asbestos_summary.pdf.

BIBLIOGRAPHY

[1] Mark Ahlers, *Aircraft Thermal Management: Integrated Energy Systems Analysis*, SAE International, 2016.

[2] Mark Ahlers, *Aircraft Thermal Management: Systems Architectures*, SAE International, 2016.

[3] Vedat S. Arpaci, *Conduction Heat Transfer*, Addison-Wesley, 1966.

[4] Andrew Aziz, T. Y. Na, *Perturbation Methods in Heat Transfer*, Springer, 1984.

[5] Thomas Bertels, *Rath and Strong's Six Sigma Leadership Handbook*, John Wiley & Sons, 2003.

[6] Marshall Brain, *The Engineering Book: From the Catapult to the Curiosity Rover, 250 Milestones in the History of Engineering*, Sterling, 2015.

[7] Yunus A. Cengel, M. A. Boles, *Thermodynamics: An Engineering Approach*, Eighth Edition, McGraw-Hill Education, 2014.

[8] Yunus Cengel, Afshin Ghajar, *Heat and Mass Transfer: Fundamentals and Applications*, Fifth Edition, McGraw-Hill Education, 2014.

[9] Clifford A. Pickover, *The Physics Book: From the Big Bang to Quantum Resurrection, 250 Milestones in the History of Physics*, Sterling, 2011.

[10] Iain G. Currie, *Fundamental Mechanics of Fluids*, Fourth Edition, CRC Press, 2012.

[11] Mark Denny, Alan McFadzean, *Engineering Animals: How Life Works*, Belknap Press of Harvard University Press, 2011.

[12] Abraham S. Dorfman, *Applications of Mathematical Heat Transfer and Fluid Flow Models in Engineering and Medicine*, First Edition, Wiley-ASME Press Series, 2017.

[13] Russell C. Eberhart, Avraham Shitzer, *Heat Transfer in Medicine and Biology: Analysis and Applications*, Volume 2, Springer, 2008.

[14] Federal Aviation Administration (FAA)/Aviation Supplies & Academics (ASA), *Pilot's Handbook of Aeronautical Knowledge: FAA-H-8083-25B*, Aviation Supplies and Academics, Trotter Publishing, 2016.

[15] Richard Haberman, *Elementary Applied Partial Differential Equations with Fourier Series and Boundary Value Problems*, Third Edition, Prentice Hall, 1997.

[16] William L. Haberman, James E. A. John, *Introduction to Fluid Mechanics*, Third Edition, Prentice Hall, 1998.

[17] Mikel Harry, Richard Schroeder, *Six Sigma: The Breakthrough Management Strategy Revolutionizing the World's Top Corporations*, Crown Business, 2006.

[18] Russell C. Hibbeler, Kai Beng Yap, *Mechanics for Engineers, Dynamics*, Thirteenth Edition, Pearson, 2013.

[19] Jack P. Holman, *Heat Transfer*, McGraw Hill India, 2011.

[20] John R. Howell, M. Pinar Menguc, Robert Siegel, *Thermal Radiation Heat Transfer*, Sixth Edition, CRC Press, Taylor and Francis Group, 2015.

[21] Frank P. Incropera, David P. DeWitt, *Introduction to Heat Transfer*, Fifth Edition, John Wiley & Sons, 2000.

[22] Frank P. Incropera, *Fundamentals of Heat and Mass Transfer*, Sixth Edition, John Wiley & Sons, 2006.

[23] Walter Isaacson, *Leonardo da Vinci*, Simon & Schuster, 2017.

[24] Walter Isaacson, *Einstein: His Life and Universe*, Media Tie-In Edition, Simon & Schuster, 2017.

[25] Erwin Kreyszig, *Advanced Engineering Mathematics*, Tenth Edition, John Wiley & Sons, 2011.

[26] P. E. Liley, *2000 Solved Problems in Mechanical Engineering Thermodynamics*, First Edition, McGraw Hill, 1989.

[27] Seymour Lipschutz, Murray R. Spiegel, J. Liu, *Schaum's Outline of Mathematical Handbook of Formulas and Tables*, Fifth Edition, McGraw- Hill Education, 2017.

[28] Norman G. McCrum, Craig P. Buckley, *Principles of Polymer Engineering*, Second Edition, Oxford University Press, 1997.

[29] Bruce R. Munson, Alric P. Rothmayer, Theodore H. Okiishi, Wade W. Huebsch, *Fundamentals of Fluid Mechanics*, Seventh Edition, John Wiley & Sons, 2012.

[30] John Murphy, *Numerical Analysis, Algorithms, and Computation*, Ellis Horwood Ltd, Publisher, 1988.

[31] Glen E. Myers, *Analytical Methods in Conduction Heat Transfer*, Second Edition, Amch, 1656.

[32] National Aeronautics and Space Administration (NASA), *Inside the International Space Station (ISS): NASA Thermal Control System (TCS) and Simplified Aid for EVA Rescue (SAFER) Astronaut Training Manuals*, World Spaceflight News, Progressive Management, 2011.

[33] Necati Ozisik, *Heat Transfer: A Basic Approach*, ISE Edition, McGraw-Hill Education, 1985.

[34] Suhas Patankar, *Numerical Heat Transfer and Fluid Flow*, First Edition, Taylor and Francis, 1980.

[35] Junuthula N. Reddy, *An Introduction to the Finite Element Method*, Fourth Edition, 2018.

[36] Shames, *Mechanics of Fluids*, Third Edition, McGraw Hill Exclusive, 2014.

[37] Frederick S. Sherman, *Viscous Flow*, ISE Edition, McGraw-Hill Education, 1990.

[38] Tien-Mo Shih, *Numerical Heat Transfer*, First Edition, CRC Press, Hemisphere Publishing Corporation, 1984.

[39] Richard E. Sonntag, Gordon J. Van Wylen, *Fundamentals of Statistical Thermodynamics*, Series in Thermal and Transport Sciences, Ninety-Ninth Edition, John Wiley & Sons, 1966.

[40] Richard E. Sonntag, Claus Borgnakke, Gordon J. Van Wylen, *Fundamentals of Thermodynamics*, Sixth Edition, John Wiley & Sons, 2002.

[41] George B. Thomas, *Elements of Calculus and Analytical Geometry*, Fourth Edition, Addison-Wesley Educational Publishers, 1981.

[42] Kenneth Wark, *Advanced Thermodynamics for Engineers*, ISE Edition, McGraw-Hill Education, 1995.

[43] Joel Hass, Christopher Heil, Maurice Weir, *Thomas' Calculus*, Fourteenth Edition, Pearson, 2017.

[44] Frank M. White, *Heat and Mass Transfer*, First Edition, 1988.

[45] Mark W. Zemansky, *Basic Engineering Thermodynamics*, Second Edition, McGraw-Hill, 1975.

A

MATHEMATICAL METHODS TO SOLVE HEAT AND WAVE PROBLEMS

There are several analytical techniques that a skilled analyst can employ to solve physics equations instead of applying numerical methods. Some 2D and 3D problem types may be solved by means of specialized analytical practices that simplify the problem and represent the physics with acceptable accuracy. This section summarizes some of the more commonly employed methods.

A.1   Analytical Approaches to Solve Heat Equations

The first step when starting to work on a solution using the analytical and numerical approaches is to simplify the problem to the extent possible without compromising its integrity. For example, this simplification may comprise ignoring the second and third dimensions. Most of the examples presented in this chapter assume that heat is transferred along the length of the geometry (e.g., x-coordinate), and therefore the problem is a 1D case. An additional step is to perform a dimensionless analysis. This helps with the problem dimensions, redefining the problem in terms of the variable ratios and meaningful and occasionally dimensionless parameters such as the *Fourier Number*. Parametrizing the model by means of dimensionless analysis allows for the effect of the important process parameters, such as dimensions or material thermophysical properties, to be studied.

The above steps are related to sensitivity analysis like the ones that may be carried out when modeling partial differential equations using the FEM commercial tools such as COMSOL Multiphysics. For example, when modeling heat transfer in a pipe, to which fins may be attached, it is possible to define its thermal performance including a dimensionless number (m^2), which is the ratio of the convective to conductive forces weighed by the ratio of the perimeter to the area of the fin ($\sqrt{hp/kA}$). Additionally, there are several assumptions that can be made to further simplify the analytical approach. For example, if the width of a conduit with a rectangular cross section is considerably larger than its thickness, the latter (thickness) can be ignored when calculating its perimeter. As a result, the area to perimeter ratio discussed earlier $wt/2(w + t)$ is simplified to $t/2$, where t and w are the conduit's thickness (m) and width (m). An interpretation for a conduit with an insulated tip is to include its corrected length in calculating the surface areas. *Corrected length* is the initial length plus a characteristic length, which is the ratio of the area of the fin to its perimeter. Using this analogy, the characteristic length can be as simple as 50 percent of the conduit's thickness ($t/2$), while for a fin with circular cross section, this value is 25 percent of its diameter ($D/4$). These assumptions facilitate heat transfer calculations based on the convective surface areas. If the fin width is small compared to its length, the area of the fin tip may be ignored when calculating its surface area. These assumptions are particularly useful when modeling semi-infinite conduits with arbitrary cross sections.

A.2   General Analytical Approaches

This section presents analytical techniques that can be employed to solve heat transfer equations. The most general form is a wave equation, which includes second-order linear partial differential equations, describing the heat waves with respect to time and space (temporal and spatial). The application of wave equations extends from heat waves to sound waves, light waves, and water waves, and it is important in fields such as acoustics, electromagnetics, and fluid dynamics.

A.2.1   Separation of Variables

This method involves separating the variables. By doing so, two or more state variables (e.g., time and distance) used to define a dependent variable (e.g., temperature in heat transfer problems) are separated, so that they can independently represent the dependent variable. In other words, their combined effect has been discretized to show their individual impact. For 3D equations, this involves defining an energy equation, Equation (159), where each dependent variable is a function of a single space (x, y, z) or time (t) variable, Equation (160).

$$\frac{d}{dx}\left(k_x\frac{dT}{dx}\right) + \frac{d}{dy}\left(k_y\frac{dT}{dy}\right) + \frac{d}{dz}\left(k_z\frac{dT}{dz}\right) + q_{\text{gen}} = \rho C_p\left(\frac{dT}{dt} + v_i\frac{dT}{dx}\right) \quad (159)$$

$$T(x, y, z, t) = X(x)Y(y)Z(z)t(t) \quad (160)$$

There are m number of linearly independent boundary conditions matching the number of the highest number of derivatives times the number of independent variables in a differential equation. For instance, for the second order three-dimensional steady-state heat transfer problem, six boundary conditions are required, representing the conditions for each side of the brick. To facilitate solving these problems, a change of the variables resulting in homogeneous differential equations or boundary conditions is recommended. In most cases, this results in dimensionless equations.

For example, assuming that a boundary is kept at the surrounding temperature (e.g., $T_{x=L} = T_\infty$), the difference between the main dependent variable (T) and ambient temperature (T_∞) may be defined as a new variable ($\theta = T - T_\infty$), which can be substituted for its counterpart in the heat transfer equation. Note that in this case, the derivatives are to be revised for the new variable to implement this change e.g., $\left(\dfrac{dT}{dx} = \dfrac{dT}{d\theta}\dfrac{d\theta}{dx} = \dfrac{d\theta}{dx}\right)$.

A number of these mathematical relations can be solved using the *Fourier Transform*, which represents a complex function for the real dependent variable. If the newly-defined variable (θ), is then divided by its equivalent at the initial condition (temperature, $\theta_0 = T_0 - T_\infty$, where T_0 is the initial temperature), a dimensionless temperature is obtained (θ/θ_0).

As an exercise, you can attempt the following 2D problem—Equation (161)—that simulates the conditions presented by the boundary conditions given by Equation set (162). The solution is provided by Equation (163). The first step is to define a new dependent variable for the temperature (θ), where $\theta = T - T_\infty$. You may apply the dimensionless approach presented above by using θ/θ_0.

$$u_{xx} + u_{yy} = 0 \quad (161)$$

$$\begin{cases} u(0,y)_{x=0} = 0 \\ u(1,y)_{x=1} = 0 \end{cases} \quad \begin{cases} u_y(x,0)_{y=0} = 0 \\ u(x,1)_{y=1} = 1 \end{cases} \quad (162)$$

$$u(x,y) = \sum_{n=1}^{\infty}\frac{2}{n\pi\cosh(n\pi)}(1-(-1)^n)\sinh(n\pi y)\sin(n\pi x) \quad (163)$$

A.2.2 Variation of Parameters

The concept for partial solutions and variation of parameters is like the *separation of variables* method. These are the steps to be taken to solve such problems: (a) A problem that represents the homogeneous case for (x, θ) is set up, where x and θ are dimension and time, respectively; (b) The eigenfunctions are determined; (c) A solution using the function $u(x, \theta) = \sum_m A_m(\theta) \varphi_m(x)$ is constructed; (d) $A_m(\theta)$ is evaluated by the orthogonality of $\varphi_m(x)$; (e) An ordinary differential equation is set up; (f) $A_m(\theta)$ is solved; and (g) The solution is completed. This method is particularly useful for transient analysis.

Attempt Equation (164), given the boundary conditions presented by Equation set (165). Note that the second-order equation with respect to the space variable (x) and the first order with respect to time (θ) require three boundary conditions. The solution is provided by Equation (166).

$$u_{xx} = u_\theta \tag{164}$$

$$\begin{cases} u(0,\theta)_{x=0} = 1 \\ u(1,\theta)_{x=1} = 0 \\ u(x,0)_{\theta=0} = 0 \end{cases} \tag{165}$$

$$u(x,\theta) = \frac{2}{\pi} \sum_{n=1}^{\infty} \left[\frac{\sin(n\pi x)}{n\pi} - \frac{\sin(n\pi x)e^{-(n\pi)^2 \theta}}{n} \right] \tag{166}$$

A.2.3 Duhamel's Theorem

The problems in this category are essentially similar to the previous scenarios except that $u(x, \theta)$ is the response to a boundary condition that is initially zero and then progresses to a constant value, or the problem is non-homogeneous in general terms, as shown in Equation (167).

Attempt the problem presented by Equation (168) with the boundary conditions presented by Equation set (169). The solution is given by Equation (170).

$$u(x,\theta) = \int_{\tau=0}^{\theta} u(x,\theta-\tau)F'(\tau)d\tau + \sum_{i=1}^{N} u(x,\theta-\tau_i)\Delta F_i \tag{167}$$

$$u_{xx} = u_\theta \tag{168}$$

$$\begin{cases} u(0,\theta)_{x=0} = \cos(w\theta) \\ u(1,\theta)_{x=1} = 0 \\ u(x,0)_{\theta=0} = 0 \end{cases} \quad (169)$$

$$u(x,\theta) = 2\pi \sum_{n=1}^{\infty} \left[\frac{n\sin(n\pi x)}{(n\pi)^4 + w^2} \left[(n\pi)^2 \cos(w\theta) + w\sin(w\theta) \right] \right] -$$

$$2\pi^3 \sum_{n=1}^{\infty} \left[\frac{n^3 \sin(n\pi x)}{(n\pi)^4 + w^2} e^{-(n\pi)^2 \theta} \right] \quad (170)$$

A.2.4 Complex Combinations

The following steps may be adopted when solving differential equations: (a) A new variable that is 90° out of phase with that of the main dependent variable is defined (v)—this variable is the imaginary component of the ultimate solution; (b) A new variable is presented, which is the conjugate of the real and imaginary parts $w = (u + iv)$—this variable is the ultimate solution; (c) The ultimate solution is defined as $w = X(x)e^{\pm iw\theta}$ using the Euler's formulae, where $e^{ix} = (\cos x + i\sin x)$; (d) The problem is solved $X(x)$; and (e) The final complex variable (w) is obtained.

Attempt Equation (171), along with the boundary conditions given by Equation set (172), is a relatively complex problem that may be solved using this technique. Equation 173) is the solution expressed as a function of the complex variable w

$$u_{xx} + \cos(w\theta) = u_{\theta} \quad (171)$$

$$\begin{cases} u(0,\theta)_{x=0} = 0 \\ u(\infty,\theta)_{x=\infty} = 0 \end{cases} \quad (172)$$

$$u(x,\theta) = \frac{1}{w}\left[e^{\left(-x\sqrt{\frac{w}{2}}\right)} \sin\left(x\sqrt{\frac{w}{2}}\right) \cos(w\theta) \right] + \left[1 - e^{\left(-x\sqrt{\frac{w}{2}}\right)} \cos\left(x\sqrt{\frac{w}{2}}\right) \sin(w\theta) \right] \quad (173)$$

A.2.5 Superposition

There are scenarios where you may superimpose multiple solutions you have attempted using different techniques. The boundary conditions may be either homogeneous, constant, or periodic.

Attempt Equation (174) given the boundary conditions presented by Equation set (175). Note that you may convert Equation (174) to four components, consisting of cases where a single non-homogeneity is taken into consideration at a time —Equation (176) and Table A.1.

$$u_{xx} + F(x,\theta) = u_\theta \tag{174}$$

$$\begin{cases} u(0,\theta)_{x=0} = g(\theta) \\ u(1,\theta)_{x=1} = h(\theta) \\ u(x,0)_{\theta=0} = f(x) \end{cases} \tag{175}$$

$$u(x,\theta) = v(x,\theta) + w(x,\theta) + p(x,\theta) + q(x,\theta) \tag{176}$$

TABLE A.1. Equation sets to be solved independently and then superimposed.

Partial Solution	$v_{xx} + F(x,\theta) = v_\theta$	$w_{xx} = w_\theta$	$p_{xx} = p_\theta$	$q_{xx} = q_\theta$
Partial Boundary Conditions	$\begin{cases} v(0,\theta)_{x=0} = 0 \\ v(1,\theta)_{x=1} = 0 \\ v(x,\theta)_{\theta=0} = 0 \end{cases}$	$\begin{cases} w(0,\theta)_{x=0} = g(\theta) \\ w(1,\theta)_{x=1} = 0 \\ w(x,\theta)_{\theta=0} = 0 \end{cases}$	$\begin{cases} p(0,\theta)_{x=0} = 0 \\ p(1,\theta)_{x=1} = h(0) \\ p(x,\theta)_{\theta=0} = 0 \end{cases}$	$\begin{cases} q(0,\theta)_{x=0} = 0 \\ q(1,\theta)_{x=1} = 0 \\ q(x,\theta)_{\theta=0} = f(x) \end{cases}$

A.2.6 Laplace Transform

This transformation is very similar to the *Fourier Transform*; however, it is more comprehensive in the sense that both function and variable (i.e., frequency) are complex. The inverse transformation is also possible, where a complex variable (such as the frequency) is transformed to a real variable (i.e., time).

Attempt Equation (177) with the boundary conditions presented by Equation set (178). The result is the solution given by Equation (179).

$$\frac{dT}{d\theta} + \frac{hA}{\rho CV}(T - T_\infty) = 0 \tag{177}$$

$$\begin{cases} T_\infty = T_1 + T_0 \cos(w\theta) \\ T(0) = T_i \end{cases} \tag{178}$$

$$\tilde{\tilde{T}}(\theta) = \left(\tilde{\tilde{T}}(0) - \frac{1}{1+w^2}\right)e^{-\bar{\theta}} + \frac{1}{1+\bar{w}^2}\left[\cos(\bar{w}\bar{\theta}) + \bar{w}\sin(\bar{w}\bar{\theta})\right] \tag{179}$$

A.2.7 Integral Method

This method is an approximate solution to relatively complicated problems and may be attempted by taking the following steps: (a) A temperature profile as a function of the dependent variables is estimated, where one variable is incorporated as a multiplier and the other one as the variable in a polynomial relationship—$T(x,t) = a(t) + b(t)x + c(t)x^2 + d(t)x^3$; (b) A penetration depth is defined as a function of the non-polynomial dependent variable, which satisfies the initial condition $x = \rho(t)$; (c)

The multipliers are calculated using the variable defined in step *(b)*, considering the boundary and initial conditions; and (d) The final solution is obtained by integrating from the main equation, considering the variable in step *(b)* as the boundary limits—$\theta(t) = \int_0^{\rho(t)} T(x,t)dx$.

Attempt Equation (180) with the boundary conditions presented by Equation set (181). The solution is presented by Equation (182).

$$\frac{dT(x,t)}{dt} = k\frac{d^2T(x,t)}{dx^2} \quad 0 < x < \infty \tag{180}$$

$$\begin{cases} \dfrac{dT(\infty,t)}{dx}\bigg|_{x=\infty} \to 0 & x \to \infty \\ T(0,t)_{x=0} = T_0 & t > 0 \\ T(x,0)_{t=0} = 0 & 0 \leq x < \infty \end{cases} \tag{181}$$

$$T(x,t) = T_0 erfc\left(\frac{x}{2\sqrt{kt}}\right) \tag{182}$$

A.2.8 Perturbation Method

This technique assumes an approximate solution that is perturbed by introducing an infinitesimal variation to the main dependent variable to the exact solution of a simplified solution, which is similar to the original problem, as given in Equation (183). The solution is then solved for the solvable component as well as the perturbed component. The final solution is achieved when the perturbed term approaches zero.

Attempt Equation (184), which is the dimensionless form of the derived one for the heat capacitance method, where the initial temperature is given—Equation (185)—to obtain the solution presented by Equation (186).

$$\theta = \theta_0 + \varepsilon\theta_1 + \varepsilon^2\theta_2 + \varepsilon^3\theta_3 + \dots \tag{183}$$

$$(1+\varepsilon\theta)\frac{d\theta}{d\tau} + \theta = 0 \tag{184}$$

$$\tau = 0 \to \theta = 1 \tag{185}$$

$$\theta = e^{-\tau} + \left[e^{-\tau} - e^{-2\tau}\right] + e^2\left[e^{-\tau} - 2e^{-2\tau} + \frac{3}{2}e^{-3\tau}\right] \tag{186}$$

GOVERNING EQUATIONS SUMMARY

Content	Governing Equations
Dimensional Analysis	Input energy → (- Heat generation, - Energy storage) → Output energy $$\dot{E}_{in} - \dot{E}_{out} + \dot{E}_{gen} = \dot{E}_{storage}$$ $$\dot{E}_{in} = -kA_x \frac{dT}{dx}$$ $$\dot{E}_{out} = -kA_x \frac{dT}{dx} + \frac{d}{dx}\left(-kA_x \frac{dT}{dx}\right)dx$$ $$\dot{E}_{gen} = q\,dx\,dy\,dz$$ $$\dot{E}_{st} = \rho\,c_p \frac{dT}{dt}\,dx\,dy\,dz$$ $$\frac{d}{dx}\left(-k_x \frac{dT}{dx}\right) + \frac{d}{dy}\left(-k_y \frac{dT}{dy}\right) + \frac{d}{dz}\left(-k_z \frac{dT}{dz}\right) = \rho C_p\left(\frac{dT}{dt} + v_i \frac{dT}{dx}\right) + \dot{q}_{gen}$$ homogeneous material $\rightarrow \dfrac{d^2T}{dx^2} + \dfrac{d^2T}{dy^2} + \dfrac{d^2T}{dz^2} = \dfrac{\rho C_p}{k}\left(\dfrac{dT}{dt} + v_i \dfrac{dT}{dx}\right) + \dfrac{1}{k}\dot{q}_{gen}$

Content	Governing Equations
Dimensionless Analysis	**Case 1—Semi-infinite Solid:** $$T_{\text{tip}(x=L)} = T_\infty$$ $$T_\infty = T_{\text{ambient}} = T_{\text{surroundings}} \text{ (assumption)}$$ **Case 2—Insulated Tip:** $$Q = -k_x A \frac{dT}{dx}\bigg|_{ti(x=L)} = 0$$ **Case 3—Convective Surfaces:** $$Q = -k_x A_1 \frac{dT}{dx}\bigg|_{\text{tip}(x=L)} = hA_2(T_s - T_\infty) + A_2 h_r \sigma \varepsilon (T_s - T_\infty)$$ $$h_r = \left(T_s^2 + T_\infty^2\right)\left(T_s + T_\infty\right)$$ $$m^2 = \frac{hP}{kA}$$ $$\theta = T - T_\infty$$ $$\theta_{x=0} = \theta_0 = T_0 - T_\infty$$ $$\theta_L = T_L - T_\infty$$ $$\theta_{t=0} = \theta_i = T_i - T_\infty$$ $$\frac{d^2\theta}{dx^2} - m^2\theta = 0$$ $$\theta = C_1 e^{mx} + C_2 e^{-mx}$$ $$\eta_f = \frac{q_0}{q_{\text{fin}(\theta=\theta_0)}}$$ $$\eta = \frac{\eta_f A_f h\theta_0}{A_b h\theta_0}$$ $$\frac{dq}{dx}\bigg|_{x=L\_\text{optimum}} = 0$$

Content	Governing Equations
General Curves	$\theta = T - T_\infty$ $\theta_{x=0} = \theta_0 = T_0 - T_\infty$ $\theta_{tip(x=L)} = L = T_L - T_\infty$ $y = f_2(x)$ width function $A = f_1(x)$ area function: cross-sectional area $A = 2y \rightarrow 2Lf_2(x) = f_1(x)$ $L = 1 \rightarrow 2f_2(x)\dfrac{d^2\theta}{dx^2} + 2\dfrac{df_2(x)}{dx}\dfrac{d\theta}{dx} - \dfrac{2h}{k}\theta = 0$ $Ap = 2\displaystyle\int_0^b f_2(x)dx$
Cylinder	$P = \pi D$ $A = \pi D^2/4$ **One-dimensional:** $\dfrac{d^2T}{dx^2} - \dfrac{hP}{kA}(T - T_\infty) = 0$ $\theta_{x=0} = \theta_0 = T_0 - T_\infty$ $\theta_{tip(x=L)} = L = T_L - T_\infty$ $m^2 = \dfrac{hP}{kA} = \dfrac{4h}{Dk}$ $\theta = C_1 e^{mx} + C_2 e^{-mx}$ **Case 1—Semi-infinite Solid:** $\theta_{tip(x=L)} = \theta_\infty = T_\infty - T_\infty = 0$ $\dfrac{\theta}{\theta_0} = e^{-mx}$ $q = \sqrt{hPkA}\,\theta_0 e^{-mx}$ $q_0 = \sqrt{hPkA}\,\theta_0$

Content	Governing Equations
Cylinder	**Case 2—Insulated Tip:** $\theta_0 = c_1 + c_2$ $\dfrac{\theta}{\theta_0} = \dfrac{\cosh[m(L - x)]}{\cosh mL}$ $q = \sqrt{hPkA}\,\theta_0 tgh[m(L - x)]$ $q_{fin} = q_0 = \sqrt{hPkA}\,\theta_0 tgh(mL)$ $\eta_f = \dfrac{\sqrt{hPkA}\,\theta_0 tgh(mL)}{hpL_0} = \dfrac{tgh(mL)}{mL}$ $A_b = A$ $A_f = PL$ $\eta = \dfrac{tgh(mL)}{\sqrt{\dfrac{Ah}{Pk}}}$ $N = \dfrac{Ph}{kA} = \sqrt{\dfrac{h}{k}}$ $\sqrt{Nu} = \dfrac{h}{mk} = \sqrt{\dfrac{h}{k}}$ Assume $L_c = L + \delta$: $\rightarrow$ For $0.25 \le Nu = \sqrt{\dfrac{h}{k}} \le 0.5$ $\rightarrow q_0 = -kA\dfrac{d}{dx}_{x=0} = 2\delta k N_0 tgh(NL + \sqrt{Nu})$ **Pipe's Optimum Length:** $\lambda = 0.5\dfrac{A_p}{\sqrt{\delta^3}}\sqrt{\dfrac{h}{k}}$ $q_{fin} = q_0 = -kA\dfrac{d}{dx}_{x=0} = 2_0\sqrt{kh}\left\{\delta^{\frac{1}{2}}tgh\lambda\right\}$ $\dfrac{dq_0}{d}_{=\text{optimum}} = 0$ $\rightarrow \lambda = \dfrac{1}{6}\sinh 2\lambda \rightarrow \lambda = 1.4192$

Content	Governing Equations
	Assume $L_c = \dfrac{A_p}{2\delta}$

Assume $L_c = \dfrac{A_p}{2\delta}$

$\rightarrow \delta_{opt} = \left(\dfrac{A_p{}^2 h}{4\lambda^2 k}\right)^{1/3}$

$\rightarrow L_{opt} = \dfrac{A_p}{2\delta_{opt}}$

$\rightarrow q_{0\text{-}opt} = 4_0 (A_p h^2 k)^{1/3}$

$\rightarrow Ap = \dfrac{0.5}{h^2 k}\left(\dfrac{q_0}{\theta_0}\right)^3$

Case 3—Convective Surfaces:

$q = -k\dfrac{d\theta}{dx}\Big|_{tip(x=L)} = h\theta_L$

$\dfrac{\theta}{\theta_0} = \dfrac{\cosh\left[m(L-x)\right] + \left(\dfrac{h}{mk}\right)\sinh[m(L-x)]}{\cosh(mL) + \left(\dfrac{h}{mk}\right)\sin(mL)}$

Fin's Optimum Length:

$\dfrac{dq_0}{dL}\Big|_{L=L_{optimum}} = 0$

$\rightarrow Nu = \dfrac{h\delta}{k} = 1$

LIST OF FIGURES

Content	Governing Equations

Assume $L_c = \dfrac{A_p}{2\delta}$

$\rightarrow \delta_{opt} = \left(\dfrac{A_p{}^2 h}{4\lambda^2 k}\right)^{1/3}$

$\rightarrow L_{opt} = \dfrac{A_p}{2\delta_{opt}}$

$\rightarrow q_{0\text{-}opt} = 4_0 (A_p h^2 k)^{1/3}$

$\rightarrow A_p = \dfrac{0.5}{h^2 k}\left(\dfrac{q_0}{\theta_0}\right)^3$

Case 3—Convective Surfaces:

$q = -k\dfrac{d\theta}{dx}_{tip(x=L)} = h\theta_L$

$\dfrac{\theta}{\theta_0} = \dfrac{\cosh[m(L-x)] + \left(\dfrac{h}{mk}\right)\sinh[m(L-x)]}{\cosh(mL) + \left(\dfrac{h}{mk}\right)\sin(mL)}$

Fin's Optimum Length:

$\dfrac{dq_0}{dL}_{L=L_{optimum}} = 0$

$\rightarrow Nu = \dfrac{h\delta}{k} = 1$

LIST OF TABLES